ILLUSTRATED

Annual Register

OF

RURAL AFFAIRS,

FOR

1870-1-2.

WITH 440 ENGRAVINGS.

VOL. VI.

ALBANY:

LUTHER TUCKER & SON.

1872.

RURAL AFFAIRS:

A PRACTICAL AND COPIOUSLY

ILLUSTRATED REGISTER

OF

RURAL ECONOMY AND RURAL TASTE,

INCLUDING

COUNTRY DWELLINGS, IMPROVING AND PLANTING GROUNDS,
FRUITS AND FLOWERS, DOMESTIC ANIMALS,

AND ALL

FARM AND GARDEN PROCESSES.

By J. J. THOMAS,

AUTHOR OF THE "AMERICAN FRUIT CULTURIST," AND "FARM IMPLEMENTS,"
ASSOCIATE EDITOR OF THE "CULTIVATOR & COUNTRY GENTLEMAN."

VOL. VI.

FOUR HUNDRED AND FORTY ENGRAVINGS.

ALBANY, N. Y.
LUTHER TUCKER & SON, 395 BROADWAY.
1872.

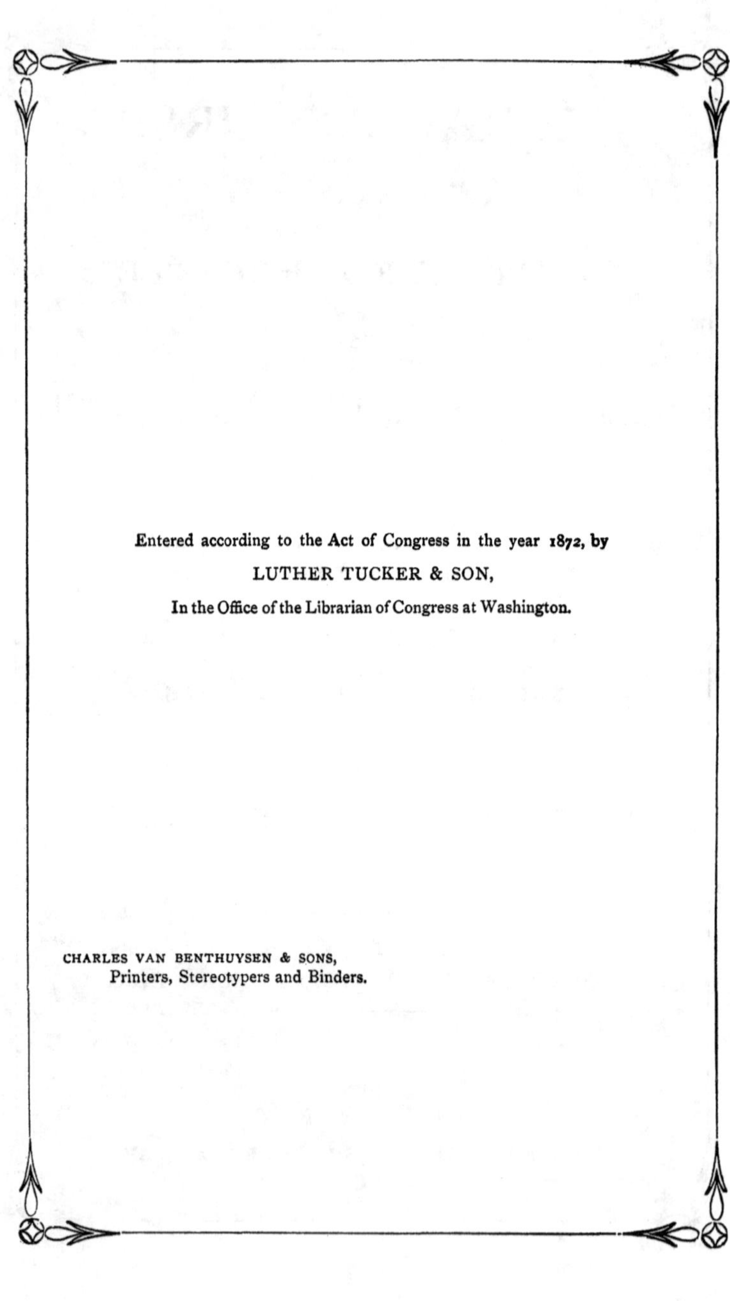

CHARLES VAN BENTHUYSEN & SONS,
Printers, Stereotypers and Binders.

PREFACE.

VOLUME SIXTH of "RURAL AFFAIRS," now presented to the public, continues the Triennial Series composed of matter published from year to year in *The Annual Register of Rural Affairs and Cultivator Almanac.* The design is to include as much practical information as can be condensed within their compass, on various topics of Agricultural, Horticultural and Domestic interest; and from copiousness of illustration, to render them the most beautiful and attractive, as well as among the most comprehensive and useful books on the subjects to which they are devoted.

The Six Volumes now ready, thus contain upwards of *twenty-five hundred engravings,* most of them from original drawings by the editor, and *about two thousand pages* of reading matter, embracing Farming, Orchard Culture, Rural Architecture, Landscape Gardening, Flower and Kitchen Gardening, Raising and Marketing the Small Fruits, Entomology, the Management of Poultry and Bees, Farm Buildings of every kind, Domestic Economy, etc. The different volumes are sold separately, there being no necessary connection in their respective contents, although it is intended that each shall be made in some degree a link between the others. While brief extracts from current Agricultural and Horticultural periodicals are not excluded, far the larger portion of their pages was written expressly for this purpose, and can be procured in no other form.

As the publication of *The Annual Register* will be continued, shortly before the opening of each year, farther volumes of RURAL AFFAIRS may be anticipated in the future at the same intervals as heretofore. They will be published in similar style and binding to those already issued; and the series will eventually constitute almost a complete library in itself, if, indeed, such is not already the case.

Sets or separate Volumes of RURAL AFFAIRS are especially commended to the attention of Agricultural and Horticultural Societies, as attractive additions to their Premium Lists—being equally appropriate as an award in all the different departments in which prizes are usually given.

Either Volume of RURAL AFFAIRS may be had by mail at any time, postage prepaid, by enclosing its price, $1.50, (or $9 for the six volumes,) to the address of

LUTHER TUCKER & SON,
Publishers of the CULTIVATOR & COUNTRY GENTLEMAN.

Albany, N. Y., April 1, 1872.

VOLUME TWO.

Four Hundred and Fifty Illustrations.

A COMPLETE COUNTRY RESIDENCE.—The Dwelling, Ornamental Grounds, Orchard, Gardens, Out-Houses, described and illustrated—concluding with an article on the APIARY, embracing the management of Bees, by M. QUINBY.

COUNTRY HOUSES.—*Twenty-Seven Designs*, including some of great merit for Workingmen's Cottages, and an illustrated Chapter on Ventilation.

FRUITS AND FRUIT CULTURE. — Farther Notes and Lists—a full Article on Pear Culture—Hardy Fruits at the West—Apples and Apple Orchards—Grafting and Grafting Knives, with upwards of Fifty Illustrations.

FLOWER AND KITCHEN GARDEN.—Annual Flowers — Vegetable Management — the Vinery and Green-House—the Verbena—also a full Article on Hedging and *Hedges*, with Directions for their Cultivation.

FARM BUILDINGS—*Eight Designs* of Barns and Stables; Stalls for Horses and Cattle—Cattle and Sheep Racks—also a full Chapter on *Iron* for Furniture and Rural Structures.

FARM MANAGEMENT.—Mr. THOMAS' Prize Essay, with new illustrations—also a Chapter on *Underdraining*, pronounced by all, the most concise and complete of its kind that has yet appeared.

FARM FENCES AND GATES—*Cheap Fences*—a full Article on Wire Fences—Modes of Construction—Hurdles—useful Hints about *Gates*, with Fifteen Engravings on the latter subject alone.

DOMESTIC ANIMALS.—Feeding—Steaming Food—Veterinary Receipts — Wintering and Stabling—Wool Table, &c., &c.

NURSERY LISTS.—A Descriptive and Illustrated List of the Principal Nurseries in the United States—Supplement to the above—Principal Nurseries in Europe.

ORNAMENTAL PLANTING. — Beautifying Country Houses—Modes of Grouping—Lawns, Walks and Rustic Objects—with *Nine Plans* of Grounds and nearly Forty Engravings.

IMPLEMENTS OF TILLAGE.—Tillage—the Gang Plow — Improvements in Plows and Harrows—Plowing and Subsoiling—Ditching Plows—Implements for Surface Tillage.

OTHER NEW IMPLEMENTS, &c. — Farm Workshops—A Horse-Power—Hay Fork—Mill—Stalk Cutter—Potato Digger—Painting Tools—with numerous hints.

RURAL AND DOMESTIC ECONOMY.—Root Crops—Good and Bad Management—Dairy Economy—Rules for Business—Early Melons—Cleaning Seed Wheat—Packing Trees for Transportation, &c.

VOLUME THREE.

Four Hundred and Forty Illustrations.

WORKINGMEN'S COTTAGES.—*Six Designs* and Seventeen Engravings—the Cottages costing from $250 up to $800.

GROUNDS AND GREEN HOUSES.—The Arrangement of small and large Gardens—Structures for Green-House Plants, including the Cold Pit, Ward Cases, &c.

FARM BUILDINGS.—General Considerations involved in their construction—*Four Designs* for Barns—Thirty Engravings.

ARCHITECTURE.—Complete Directions for One, Two or Three Story Buildings on the *Balloon Frame* System—24 Engravings—Directions as to Carpenter and Mason's Specifications, and a Glossary of Architectural Terms—48 Engravings.

FARM HUSBANDRY.—How to render Farming Profitable, is treated in one or more Chapters, and a very great variety of Hints and Suggestions are given in Practical Matters and General Rural Economy.

WEEDS AND GRASSES.—The chief varieties of Annual and Perennial Weeds, and of Useful Grasses, are described very fully, the former accompanied with 21 Engravings, and the latter with 13.

PRACTICAL ENTOMOLOGY.—Dr. FITCH'S Chapter on Insects Injurious to Fruit Trees, Grain Crops and Gardens, with 34 Engravings, and full Definitions and Descriptions.

FRUITS AND FRUIT CULTURE—The Newer Plums — Strawberries — Dwarf Pears—Management of the Grape — Summer Pears—Training Pyramids—Dwarf and other Apples—Cherries and Gooseberries—A Cheap Grapery, &c., &c.—more than 50 Engravings.

FLOWERS.—Pruning and Training Roses—Notes on New and Desirable Flowering Plants—20 Engravings.

VEGETABLE PHYSIOLOGY—Tracing Growth of the Plant from the Embryo throughout—the Principles of Grafting and Budding, &c.—61 Engravings

DOMESTIC ANIMALS.—A large variety of Hints as to Breeds and Management—*The Apiary;* different Hives and the Mode of Caring Properly for Bees.

THE DAIRY.—A full Chapter on Butter and Cheese Making and Management of Cows, with numerous Hints.

THE POULTRY YARD.—A Complete Chapter, by C. N. BEMENT, with 33 Engravings of Fowls, Poultry Houses, &c.

ALSO—*Filters* and *Filtering Cisterns*, 5 Engravings—*Lightning Rods*, 13—*Useful Tables* of Weights and Measures, &c. *Maple Sugar* Making. To these and many other subjects more or less space is devoted.

VOLUME FOUR.

Three Hundred and Eighty Illustrations.

FARM WORK.—A Calendar of Suggestions for each month in the Year, with *Fifty-Six* Engravings—including Ice-Houses and storing ice—making Stone Wall and many other incidental points often omitted—a very valuable article.

ORCHARD AND NURSERY.—Calendar for the Year, with many useful hints and *Twenty-Two* Engravings.

KITCHEN & FLOWER GARDEN and Green-House.—The labors of each successive month reviewed, with notes on varieties of different Vegetables, &c., &c., and *Fifty* Engravings.

ROAD MAKING.—With numerous Illustrations and complete directions.

CHEESE DAIRYING.—A description of the Cheese Factories and System of Manufacture—also Design for private Dairy-House, and Miscellaneous Hints for Dairy Farmers.

ENTOMOLOGY.—A full Chapter on Collecting and Preserving Insects, particularly interesting to beginners in this important science.

COUNTRY HOMES.—An article with *Eight Designs*, accompanied by Ground Plans, &c., &c.

PRUNING. — The principles and practice fully described, with over *Thirty* illustrations.

POULTRY.—Treatise on the Turkey—Poultry Houses and their arrangement, with Designs.

FRUITS AND FLOWERS.—Training Grapes The leading new Pears—New and Desirable Flowers—with a very large number of condensed hints, and select lists according to the latest authorities—fully illustrated.

DOMESTIC ECONOMY.—Full Directions for Canning Fruits and Vegetables—a large number of Useful Recipes, &c., &c.

DOMESTIC ANIMALS.—A full article on Mutton Sheep — The Management of Swine—also Hints for the Bee-Keeper, &c., &c.

IMPLEMENTS AND INVENTIONS.—Mechanical Contrivances for various purposes—the Implements of Horticulture—New Machines—largely illustrated.

WOODLANDS.—Planting Timber for Screens—the Care and Culture of the Timber Crop.

VOLUME FIVE.

Four Hundred Illustrations.

GRAPE CULTURE.—Varieties, Propagation, Grafting, Training, Transplanting, Trellises; Soil for Vineyards; Marketing, &c.—Very Complete and Practical—*Thirty-Nine* Engravings.

MILK FARMING, by the Author of "My Farm of Edgewood." Winter and Summer Feeding, Soiling, &c. With plans of Milk Barn—*Six* Engravings.

THE DUCK—its Management and Varieties, by C. N. Bement—*Fifteen* Engravings.

TURNIPS AND THEIR CULTURE.—An admirable article on the Ruta Baga, with Practical Directions — *Fifteen* Engravings.

GARDEN INSECTS, by Dr. Asa Fitch—two papers, with about *Forty* Engravings.

REAPERS AND MOWERS—the leading Machines at the Auburn Trial—*Nine* Engravings.

ROTATION OF CROPS—principles involved and rotations suggested—illustrated.

SMALL FRUITS—their Culture on the Hudson, by Prof. Burgess—*Thirty* Engravings.

SHRUBS—a Practical and Descriptive Article on Shrubberies and the Selection of the Leading Varieties—about *Thirty* Engravings.

LABOR SAVING CONTRIVANCES.—Simple and Handy Things about the Farm and House—about *Thirty* Engravings.

VERMIN about the House, and How to Drive them Away—illustrated.

WHEAT—an Essay on the Crop and its Culture, quite complete and practical—*Fourteen* Engravings.

HEDGES and their Management, Causes of Failure, &c.—*Ten* Engravings.

POTATOES—Culture, Varieties, &c., with *Twelve* Engravings.

RURAL IMPROVEMENTS, by Robert Morris Copeland — with Plans and Modes of Planting—illustrated.

FRUITS.—Practical Hints in Fruit Culture, with numerous Short Articles, and over *Thirty* Engravings.

STRAWBERRIES.—Marketing the Crop in New Jersey, by the Author of "Ten Acres Enough"—illustrated.

FLOWERING PLANTS.—Select Varieties, with Descriptions and *Twenty-Two* Engravings.

And among Numerous Shorter Articles:

Hints in Rural Economy, by S. E. Todd—*Nine* Engravings.
South-Down Sheep—illustrated.
Items in Domestic Economy.
Hay Barracks and Corn-House—illustrated.
Rain-Gauge—Protecting Melons, do.
Hot Air Furnaces, do.
Implements for Farm and Garden, do.
Improved Bee Culture, by M. Quinby.
Three-Story Barn, Grape Houses, illustrated.
&c., &c., &c., &c.

INDEX.

INDEX.

INDEX.

INDEX TO ILLUSTRATIONS.

INDEX.

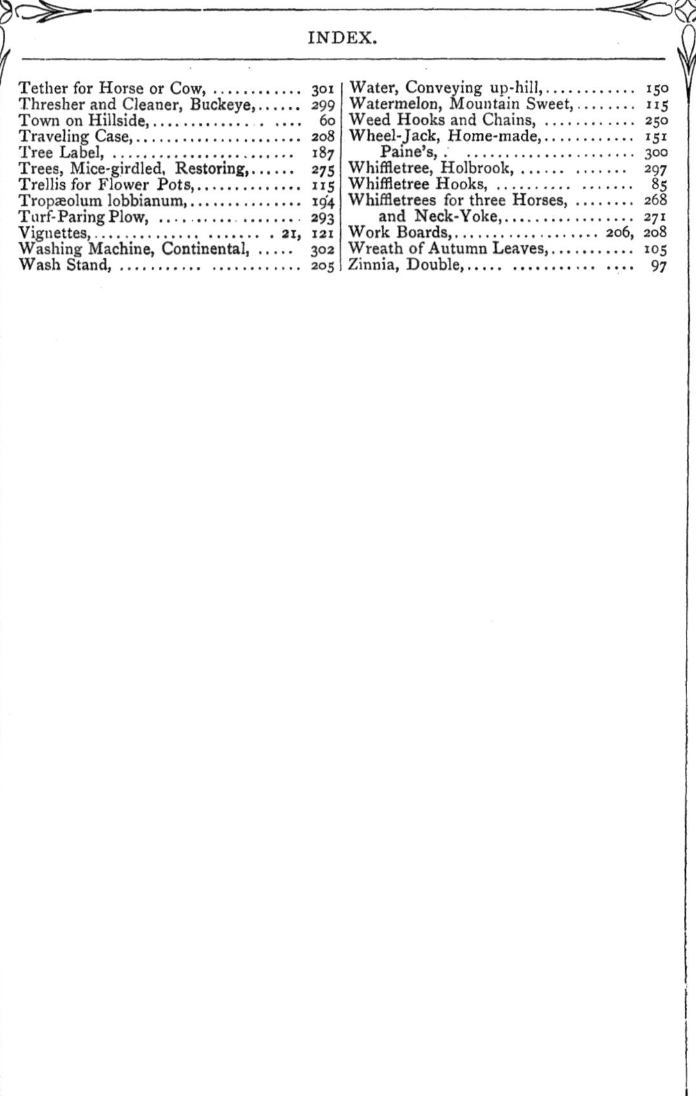

INDEX.

THE

ILLUSTRATED ANNUAL REGISTER

OF

RURAL AFFAIRS.

THE CULTURE OF INDIAN CORN.

THIS CROP, although grown in Southern Europe and some other places, assumes the greatest importance in the western world. Boussingault said that it was "the true wheat of the Americans." Nearly a thousand million bushels are turned out annually, besides twenty or thirty million tons of fodder. Under fair management, and with the variety properly adapted to the climate, it is one of the most certain and reliable of all crops, and there is perhaps less risk in obtaining a recompense for good and enriching culture. The intelligent farmer who determines to raise a big crop of corn, can be pretty sure of attaining his object.

There are however different ways of doing the thing, affecting more the cost of raising, or the net profits received, than with most other crops. A man may obtain fifty or sixty bushels per acre, with almost the certainty that summer and winter succeed each other; but he may do it in such a way that it shall cost him a dollar per bushel, by a want of proper and economical management. When Frederick Tudor of Nahant, succeeded in bringing the blowing sands of the sea-coast, by means of costly screens, into successful culture, one of his friends remonstrated with him for expending three dollars for every bushel of corn which grew, when it was worth only one dollar in market. "I get three dollars per bushel," answered Tudor, "for all that I raise; one dollar in market value, and two dollars in the satisfaction of raising it." There is not, however, a single farmer in all our wide corn territory, who would accept the last mentioned pay on a large scale. The question therefore comes up, "How can I obtain the largest return for my labor and expenditure?" The answer will embrace two chief points:

I. *To raise heavy crops.*

II. *To obviate hand-hoeing by keeping the soil clear of weeds.*

The first is self-evident to every one, it it shall require no unusual expenditure. It costs no more to plow an acre for a sixty bushel crop than for one that yields only twenty bushels—nor to harrow it, nor plant it. The only saving in the small crop is in the diminished labor of husking, and in drawing in the ears and fodder. The difference between a heavy and a light crop, the cultivation being the same, is in the richness and condition of the soil. There is very little in those influences of weather and seasons, commonly called "luck." "Does not the *richness* you speak of," asks some one, "depend on the size of the manure heap which you happen to have to apply to the land?" Not altogether. It depends much on the management of previous years, and very much on using present means to best advantage. We do not propose to tell farmers what they ought to have done years ago, but to do at the present time; and if their land is poor, to suggest to them how they can now begin to make improvement. The proper time, however, to prepare for planting corn, is in autumn, by spreading manure broadcast on sod at that time, to remain spread all winter, and plowed in shortly before planting. It is becoming well established that a load of manure thus applied is worth more than two put on the ground in the spring.

We may set down these six requisites for obtaining large crops: 1. Plant on inverted sod, turned over in spring shortly before planting. 2. Plow to a moderate depth, so that the vegetable matter of the turf may decay by exposure to air. 3. Apply the manure by spreading it on the sod in autumn. 4. Let this sod, if practicable, be a dense growth of clover, or an old pasture where clover has flourished. 5. Manure in the hill, if some reliable concentrated manure can be had that may be deposited from a planting machine. 6. Manure spread in the spring should be thoroughly harrowed before turning under.

1. The object in turning over the sod in spring before planting, is to save the expense of plowing more than once, and to prevent the growth of

weeds till the young corn has a fair start. *Plow well*—turn over flat and thoroughly if you want a clean field. Harrow immediately; a Shares harrow is best, as it cuts and pulverizes twice as deep as a common harrow, and presses down instead of tearing up the turned sod. If one of these implements cannot be had, lay all flat by means of a roller, and then harrow well. If the corn is to be planted in drills, or in rows running but one way, the planting should be done as each successive land is plowed, which will have three advantages—first, the earth will be fresh, moist and mellow; secondly, there will be no danger of the corn failing to grow in consequence of dried lumps, resulting from leaving the plowed surface many days exposed, or of delay by intervening storms; and, thirdly, the soil being freshly turned up, the corn will get ahead of the weeds, if there are any in the soil—which is nearly always the case, but never should be.

Fig. 2.—*Plowing, Harrowing and Planting with Drill and Horse.*

2. Decaying vegetable matter is an admirable manure for young corn, which is the reason that the sod should not in general be plowed more than five or six inches deep. If a layer of compact subsoil or clay is placed upon the turf, it will retard its decay so much that the crop will be slightly benefited, and at the same time will be so far down that the roots will not reach it readily. The depth must, however, vary much with the previous management. If the land has been subjected to deep, thorough and enriching culture, the grass roots will have gone down further than in a soil plowed shallow; and corn will succeed best of all on such a soil, by the deeper bed of fertility, as well as by the more equal supply of moisture from below.

3. The autumn application of manure results in its being more intimately diffused with the particles of the soil, by the autumn, winter and spring rains, than could be accomplished by the use of the most perfect harrow. The roots of the grass assist in carrying down the liquid fertilizer thus made, so that when inverted in the spring, it is just where the young corn wants it. The correctness of this reasoning is fully confirmed by experiment; and the many accurate farmers who have tried this mode, agree

that one load of autumn manure is worth at least two loads of spring manure, and some estimate a greater difference. This is one of those improved modes of treatment that are never accompanied with an exception—the superiority is invariable. Even spreading in winter on the snow is much better than to defer it till spring. The fear that the manure will be wasted or washed away by the water of thaws or rains, has been proved by trial to be nearly or quite groundless—except in the bottoms of swales or hollows, when considerable streams are made by the rains or melting snows. If spread evenly, it settles on the ground as soon as the snow disappears, at which time there is enough of the soil free from ice to absorb and retain all the dissolved portions of the manure. It is found on examination that the discolored water becomes clear after passing but a few feet over the surface; and masses of manure laid on sloping grass lands enrich and stimulate the growth of the grass only a foot or two below.

4. Inverted clover, top-dressed with manure the previous autumn or winter, never fails to give a heavy crop of corn, unless under unusual or extraordinary circumstances or very bad management. If the soil is rather heavy or clayey, the advantage is still greater, on account of the loosening effected by the roots of the clover.

5. Manure in the hill always gives the young plants a good start. It is too laborious and expensive to do by hand, except on a small scale, and should be performed by a planting machine which drops the fertilizer with the seed, but not in contact with it. Plaster thus applied is nearly always a benefit; ashes often so; superphosphate on some soils eminently so; guano is more universally beneficial, whether home-made from poultry droppings or purchased in market. A handful of fine compost dropped in each hill has a uniform and excellent effect, but we know of no machine capable of applying large enough doses.

6. When manure cannot be spread in autumn, it is quite important that it be thoroughly pulverized and well intermixed by previous harrowing. If left in lumps and masses, it can do but little good. The new Smoothing Harrow is a good implement for this purpose.

II. The second great requisite in raising corn is to obviate the necessity of hand-work. Hence the importance of a clean farm, kept so by such a rotation as shall serve to extirpate all foul stuff. Land may be made clean by hand-hoeing and hand-weeding, but it is too laborious and costly. The farmer who never allows an undergrowth of weeds among his corn, may have a good clean crop of barley to succeed it, and leave a clean field for another clover seeding after this barley, or its wheat successor. Stirring the soil is pre-eminently important for a good crop of corn; but this stirring is performed economically only by horses. It is cheaper to employ a man and two-horse team, with a good two-horse cultivator, at five dollars a day, than a man with a hand-hoe at twenty-five cents a day. Taking out weeds by hand-work is therefore only a *costly necessity*.

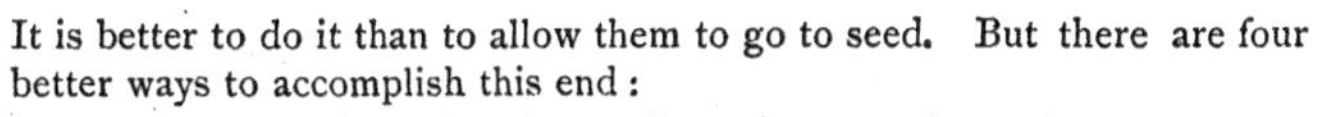

It is better to do it than to allow them to go to seed. But there are four better ways to accomplish this end:

1. To keep the farm clean by good previous rotation and management. Or, if not clean,

2. To summer fallow.

3. To plant the corn in perfectly straight rows both ways, so that a double cultivator may be run closely to the rows, and cultivation kept up weekly as long as the corn will allow.

4. To employ the new smoothing harrow, at least once a week, after the corn is planted, until it is ten or twelve inches high. This harrow passes over the corn broadcast, without injuring it, and should be used so often that no crop of weeds can ever reach an inch high.

The first will include the three last, until the thorough extirpation of of weeds is effected. The third is commonly practiced in an imperfect manner. The rows are planted crooked, so that the cultivator cannot be made to cut closely to the rows in all places, and the work is often entirely discontinued after passing two or three times at most. The weeds, which come up later, grow without interruption, go to seed, and leave the land more foul than before. If the plan of making the hills to row both ways, cannot be used for the purpose of effecting a thorough clearing out of weeds, it would be better to resort to a naked summer fallow.

But a summer fallow should not be a crop of weeds. We know farmers who congratulate themselves that their sheep are able to "pick a living" on their fallows. In other words, they appear to be pleased that the plowing has been so badly done that a good deal of grass has been left for these animals, and also that the weeds which come up are interfered with only by their nibblings. Such fallows as these are worthy of the rejection which they meet with from the best farmers, but it is a mistake to denounce all summer fallows on account of such poor specimens as these. We have seen a field that had been copiously encumbered with milk-weeds, Canada thistles and other troublesome weeds, kept all summer as clean and mellow as an ash-heap, at a moderate expense of plowing and harrowing, to the utter annihilation of all the milk-weeds, Canada thistles and the other troublesome weeds, which never again made their appearance. This is the kind of fallow for killing weeds, while it must be admitted that too many seem to be for the special purpose of propagating them. Land may be cleared by summer fallowing at a twentieth part of the cost required for doing it with crops which require chiefly hoeing.

The Smoothing Harrow, an implement invented by the author of this work, consists essentially of teeth driven through pieces of plank hinged together, the teeth slanting backwards about forty degrees, so as neither to become clogged by any litter or rubbish, nor to tear up any plants after they are fairly established. They pulverize the soil, and kill all small weeds that are just peeping at the surface. This harrow runs over the whole surface and keeps it clean, until the corn is about a foot high, after which the cultivator may be used for finishing the dressing. If, therefore, it is passed often enough, it obviates nearly or entirely the labor of hand-hoeing, and by its assistance drill-culture may be adopted without any drawback. Experiments performed both at the east and west, agree with much uniformity in assigning at least 25 per cent. more per acre to corn raised in drills, than in hills rowing both ways. This is the result of our own repeated experiments—

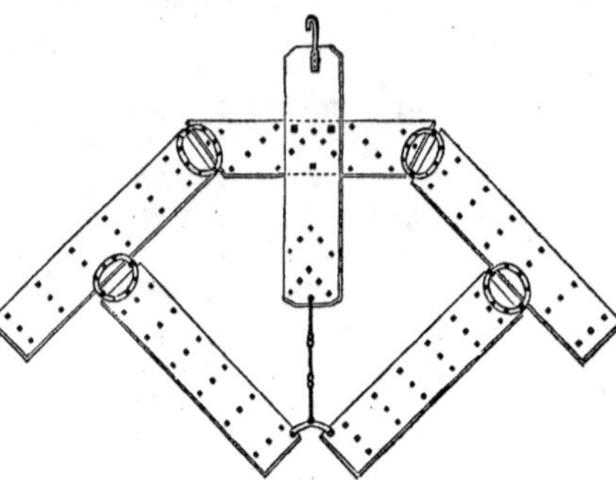

Fig. 3.—*Thomas' Patent Smoothing Harrow, or Broadcast Weeder.*

Fig. 4.—*One of the Sections.*

whether the drills contain hills a foot apart, or single plants at uniform distances. But the objection is the increased labor of hand-hoeing. In weedy soils this objection is a strong one, but on well managed land it is of little weight; and by the method just described it may be entirely obviated. John Johnston informs us that the slight additional expense for hoeing (for he keeps his fields as clean as a floor) is repaid several times over by the twenty-five per cent. of grain, and the nearly doubled amount of corn-fodder. We have already pointed out another advantage of planting in drills or rows but one way—the readiness with which the planting may proceed as fast as each successive land is plowed. Joseph Watson of Clyde, N. Y., plows his sod with a strong team, and while this is going on, a lighter team harrows the freshly turned earth, and the horse-planter follows immediately after. Not a day in waiting for the completion of the field is thus lost. He finds another advantage of the one-row system, besides that of rapid planting by machinery—the hand-hoeing which may be required is done while the soil is fresh after the passing of the horse, and any hills accidentally covered are immediately relieved, without waiting several days until the cross cultivation is finished.

It must be admitted that most of the corn raised in the country is in fields more or less infested with weeds. These are allowed to grow until, by a disputed possession of the field, they have greatly retarded the growth of the crop, (as shown in fig. 5,) instead of allowing the corn full occupancy without any intruder, resulting in the satisfactory crops of seventy to ninety bushels per acre.

If the owners are determined to get rid of the weeds which infest their fields, and cannot devote a year to fallowing or to other special modes of extirpation, they must take another course. They must place corn in rows perfectly straight both ways, so as to obviate work by hand as far as possible. A common way is to mark the rows in one direction by the use

Fig. 5.—*Clean and Weedy Rows.*

of a marker, till the whole field is marked, and then to drop the rows across thê lines by means of stakes, which, being all of a length, equal to the distanee of the rows asunder, the dropper readily measures off the next row as he passes each stake. There is a serious objection to this mode—for although the cross rows may appear generally straight, they will have short jogs or crooks which the dropper cannot well avoid. As a consequence, the cultivator cannot be made to cut closely to all the hills, and the hand labor of hoeing clean is greatly increased. The best way is to use first a good marker, such as the one figured and described on pages 33 and 296 of vol. IV of Rural Affairs, or on page 181 of vol. V, until the whole field is marked distinctly in one direction. Then, before dropping the corn, the cross-marking may be rapidly performed by hanging several chains at equal distances to a long light pole, as shown on page 34 of vol. IV of Rural Affairs. A great advantage in all these modes of marking is that the rows will be equally distant asunder, whether straight or with some curves, and the two-horse cultivator, if it runs closely to one row, must pass equally so to the others.

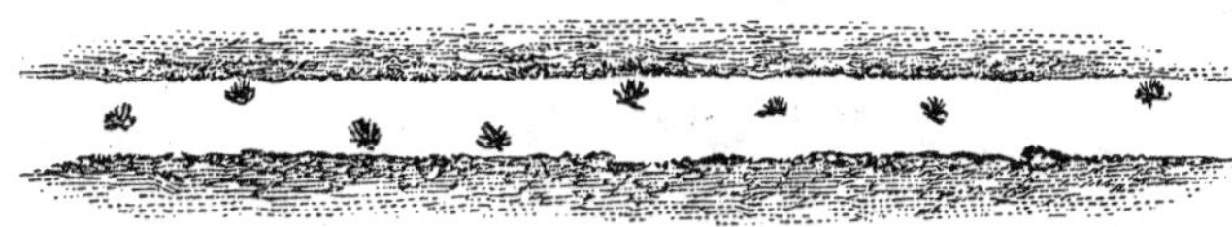

Fig. 6.—*Cultivating Corn in Crooked Rows.*

The importance to perfect and successful cultivation of straight rows, in all modes of planting, whether in hills ranging both ways, in thick rows, or with hills but one way, or in uniform drills, will be apparent from the accompanying cuts, where the close work of the cultivator in straight

Fig. 7.—*Cultivating Corn in Straight Rows.*

rows is shown in figs. 7 and 8; and the half finished work in crooked rows in figs 6 and 8.

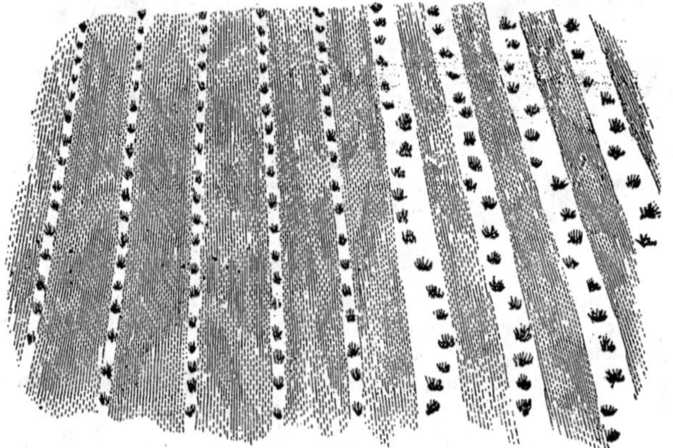

Fig. 8.—*Effects of Planting Corn in Straight and Crooked Rows.*

Many two-horse cultivators for working among the rows of corn are made at the west and east. Those where the driver walks after the implement and guides it by handles, are termed *walking cultivators.* They are cheaper, lighter for the team, and are better adapted to rough or sloping ground. The *Sulky cultivators,* or those with a seat for the driver, require more level or smooth ground, but they may be used by a lame or disabled person, or by the owner who does not desire the hard drudgery of walking all

Fig. 9.—*Phifer Cultivator.*

day; and after full practice the best ones are guided with much precision on smooth land and between even rows. One of these, made at Trenton, N. J., known as the Phifer Cultivator, is shown by fig. 9; levers, at the ready command of the driver, being used to steer the implement accurately. Another sulky cultivator, shown by fig. 10, is the one manufactured by Fords & Howe of Oneonta, N. Y.

Fig. 10.—*Fords & Howe's Cultivator.*

An essential requirement for success in growing corn, is the adaptation of the variety to the character of the climate, or to the length of the season. At the extreme north small early ripening varieties must be planted in order to escape the autumnal frosts. Although the ears are smaller than the large cob southern dent and gourd-seed corn, the plants may be allowed to grow so much nearer together that an equal amount by good management, may be obtained from an acre. The smaller northern varieties are not more than six or seven feet high, and the ears are within a foot of the ground; the corn of the middle States, in southern Ohio and Indiana, and in Kentucky, is twelve or fifteen feet high, the ears five or six feet from the ground, and stalks and ears so much larger as to induce many superficial observers to suppose the crop much greater in amount than the northern small varieties. The southern sorts would fail to ripen at the north, except on the richest and warmest soils. In whatever locality the crop is injured or lost by tardy ripening, this fact points at once to the planting of an earlier and smaller variety.

It is scarcely necessary to attempt to show that the only crops which can be raised to a profit, are those which grow on rich land, and which are cultivated in the best manner. On land that has been hard run, with no manure, every bushel will cost twice as much as the market price; and if a crop of weeds besides is allowed possession of the field, the case will be nearly hopeless. Compared with such a field, a well grown, heavy crop, on well manured land with inverted clover sod, or rich alluvion—the whole kept perfectly clean by frequent horse cultivation and previous good management, presents a contrast in appearance of the strongest character; and the contrast in the net profits cannot fail to be interesting to the farmers who cultivate the two fields, when full accounts of the expenditures and receipts are placed side by side.

THE GREATEST PROFITS FROM DRAINING.

THE PROFITS OF UNDERDRAINING are now admitted by all intelligent farmers, but the question is often asked—How many years are required to repay the cost of draining by the increase of the crops? The time will vary with circumstances, from one to five years, and if the after management of the land is bad, a longer time will be necessary. In our own experience in putting in many miles of tile drain in good upland soil, the increased crops have repaid the expense in an average of about three years. The land was of such a character that superficial observers would say that no draining was necessary. But the old test of digging trial holes two or three feet deep, showed by the water which was retained in these holes for many days during the wet season, that the subsoil contained a large surplus of water; and further proof was given by the long continued discharge of water from the drains which were afterwards cut. The digging was mostly performed by loosening the subsoil and hard-pan by the use of the ditching plow, and throwing the loosened earth out by hand. This method lessened the expense of digging to one-half, and the entire cost of the drains, when completed, to two-thirds. The drains being cut about two rods apart, the whole cost was about thirty-five dollars per acre. Consequently an increase in profit of twelve dollars per acre would repay the expense in three years, the average time as above stated. Where the soil was very rich and decidedly too moist, only two years were required, as but little could be raised before, and a great deal afterwards. If the soil is poor, and the farming bad, the advantages will of course be much less. There is one very important consideration, however, which we have not yet taken into the account. This is the lengthening of the season. Farms in the latitude of forty-three degrees may be virtually carried two or three degrees further south, without the disadvantages of a hotter sun. A well drained field may be plowed without inconvenience almost on the day that the frost disappears from the earth. If the field be not drained, the farmer may very likely have to wait a fortnight, and in wet seasons a month, before he can obtain a finely pulverized soil from his furrows. Cases are not uncommon where half the amount of a barley or oat crop has been lost by such delay, and this loss proved to be fully equal to the entire expense of cultivation. Thus the delay occasioned by water, reduced what would have been a hundred per cent. of profits, to nothing. At the same time the inability to cultivate his fields early, has driven the work of the farmer into a narrow compass, reducing his operations and increasing his expenses. The machinery of the farm has become deranged, and if we had some mode of reducing the vexation which the owner suffers, to an estimate of dollars and cents, it would undoubtedly be found that the profits of draining are considerably greater than our figures already given. Our remarks thus far are intended to apply exclusively to regular underdraining, or where parallel channels, about two rods asunder, extend over the whole surface. There is another kind of draining still more profitable. Or rather a smaller expenditure results in a greater comparative advantage. We refer to its performance on that character of ground made up of knolls and swales, or where upland is traversed irregularly by low and wet portions. It sometimes happens that a few wet streaks across a field, in other respects tolerably dry, have prevented spring plowing for a whole month. Three-fourths of the field perhaps, was quite dry enough, but the mud and water in the remaining portions placed everything at a stand-still. The accompanying figures, 11 and 12, represent a plan and profile of such a field more broken up than usually occurs, where the shaded portions show the wet parts, and the rest, the dry parts of the field.

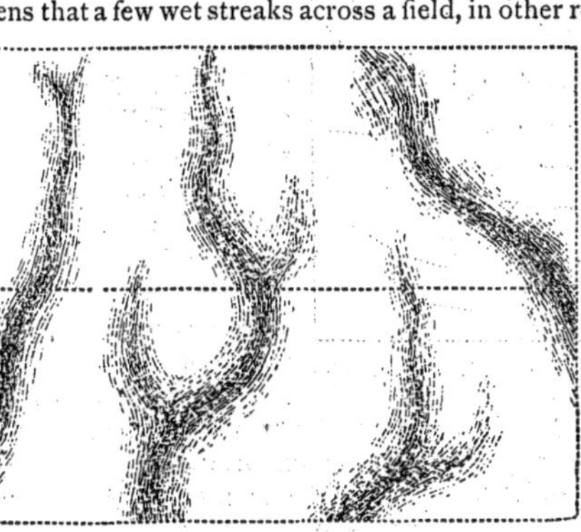

Fig. 11.—*Field before Draining—Dark portions Swales.*

Fig. 12.—*Section of the Field at the Dotted Line of Fig. 11.*

Now the question occurs—what is the best mode of draining such a field? The first thing is to cut underdrains through the centre of every swale or low wet portion of the ground. No difficulty will be found in locating these drains, or in determining the proper places for descent and giving a free flow for the water, if the owner will take the pains to observe where the surface water flows off from melting snows or heavy storms. These ditches should be laid with large tile, say three inches in diameter, which, on account of the greater velocity of the water, will carry off more than three times as much water as a two inch tile. These large channels will serve as main drains in future years, when the owner is able to ditch the whole surface. When these mains are completed, the fields should be measured, and the position of the drains (which of course will be crooked and irregular,) laid down on a map or sketch, as nearly as may be. This will assist in finding the channels when side or branch drains are made on a future occasion. Fig. 13 represents the field when the draining is completed by the construction of all the branches.

As this field is represented as more broken up than most fields, or intersected with more frequent swales, the branches will be much shorter than in a field where a single broad depression extends through the whole ground, in which case the branches will be longer, more nearly parallel, and the whole will be simpler.

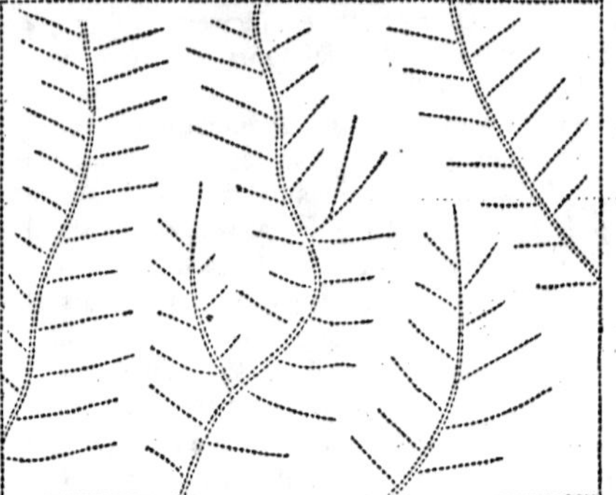

Fig. 13.—*The same Field after Draining—the Mains, double lines—Branches, single dotted lines.*

Many farmers will be quite satisfied when they have drained the low places only. No one can fail to be highly gratified with this result if the work has been well done, for it will make all parts of the land capable of plowing and other tillage at the same period of time, and he will not have to wait several weeks after the majority of the field has become fit for plowing, before he can drive his horses over the wet spots. Another important advantage will be gained by thus bringing the richest parts of the land, which before were useless, into profitable cultivation for all kinds of crops. We have known instances where wheat or corn was so much improved on such land as to repay the entire expense in one year. But if, after this preliminary work of draining the swales has been completed, it should be found by digging the trial holes already mentioned, that the subsoil of the upland needs underdraining, the full extent of the benefit cannot be reached until the whole work is throughly accomplished.

PRACTICAL HINTS ON FENCE MAKING.

DIRECTIONS have been given in former volumes of this work in relation to various details in Fence Making, yet some additional hints may be useful under varying circumstances. The vast expense required to build and keep up farm fences, renders every assistance which can be afforded for lessening the cost, a matter of great importance. In the State of New-York alone, the aggregate expenditure required to build all the fences cannot be less than two hundred million dollars ; and for keeping in repair and for replacing decayed ones, the annual outlay must be as great as fifteen millions. It is well to look into the matter, and see if they cannot be constructed at a cheaper rate in some respects—for if only one-tenth part of the cost of erection could be saved, we should

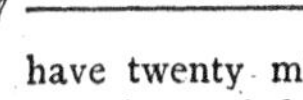

have twenty millions—enough to give a good agricultural school and experimental farm in every county of the State.

BOARD AND RAIL FENCES COMPARED.

The old zig-zag or worm fence is now generally discarded in theory, and adopted on an immense scale in practice. It has its merits and defects; but in heavily timbered regions will be largely built for many years to come. Some of its most prominent defects may be removed without difficulty. The breadth of ground which it occupies, may be easily lessened one-half by properly staking the corners. If the rails are split in a workmanlike manner and well and promptly seasoned, the fences will last at least forty years; they are easily and rapidly built, and they may be removed short distances for the purpose of plowing and clearing up old fence rows and corners, with great comparative facility.

An estimate of the amount of timber required for these fences will show that they need not necessarily consume so much as many have supposed. If the rails are well and evenly split, an average weight equal to three by four inches, or at most four inches square, will be ample. If well laid on good corner stones, and "staked and ridered," seven rails will be high enough for all ordinary purposes, and fourteen rails to the rod all that are required—or rather, each two lengths will make 18 feet in running distance, the rails being 12 feet leng. The timber for the stakes will be about equal to one rail, making fifteen rails for each 18 feet of length. These, at the preceding estimate, would contain timber equivalent to 180 feet, board measure; that is, each rail would be equal to 16 feet, which multiplied by 15, the number of rails, would give 180 feet, board measure, to each double length of 18 feet.

Now, compare this to a good board fence of equal height, as commonly made. If the boards are eighteen feet long, (which length we take for convenience in comparison,) they should be a little stouter than if shorter, but the increased amount will be balanced by the fewer posts, which will be 9 feet apart. Five boards averaging 6 inches wide, with a cap-board of 4 inches, would make 51 feet for each length. The posts, if strong and substantial, or equivalent to 5 inches square, and 7½ feet long, would contain 31 feet board measure—the whole being 82 feet for each 18 feet of length—a little less than one-half the quantity required for the crooked rail fence.

In order to compare the two properly, it is essential that circumstances be taken into the account. If the timber for the rails already stands on the farm to be fenced, the whole labor will consist in splitting, drawing and building; which will be less than cutting and drawing saw-logs several miles to the sawmill, and returning with the boards. If one-half is demanded by the owner of the sawmill for sawing—a rate common in many places—the farmer will be able to fence but little more of his land than if he had retained his logs and split them at home; and all the labor of drawing to sawmill, digging post-holes, setting posts and nailing on the boards, with the added expense of nails, will be avoided by splitting the rails.

On the other hand, if the rails must be purchased and drawn long distances, they may cost as much or more than boards and posts, and require more labor to draw them. Under these circumstances, it will be better to make board fences after the manner described in a subsequent part of this article.

Board fences do not usually last so long as well made rail fences, partly because the stuff is thinner and lighter, and gives way first at the nails; and partly because the more durable kinds of wood are often selected for splitting, while hemlock, pine, &c., which are not so durable, are more commonly employed for sawing into boards.

Importance of Seasoning.

A great loss often occurs in the construction of rail fences by a needless quickness in decay, resulting from imperfect seasoning. To make rails last well, they must be seasoned as quickly as possible after cutting. We have found that basswood rails will last more than twice as long when cut and immediately split at midsummer, than if split in winter. The former dry speedily and become hard like horn; the latter are partly "sap rotten" before the seasoning process is effected. Summer cut timber, however, left in the log, may be more injured by a fermenting decay in summer than if left unsplit in winter. But at whatever time the trees are felled, it is important that the rails be split at once, and placed fully exposed to wind and weather for drying.

Uniformity in Length.

There is another point of importance in procuring either rails or boards for fences. It is to have them cut of proper length. Rails are made of lengths of ten, twelve and of fourteen feet. Sometimes timber that is hard to split is cut shorter than that which is smoother, freer and most easily severed into rails. If these different lengths are kept entirely separate from each other, this difference may answer; but usually it is difficult to do this without considerable inconvenience. When rails become old, they need assorting by throwing out those likely to break, and reserving the soundest. These second-hand rails should be used in a fence together, and new rails kept by themselves; because if the old and new are mixed, they will be like a new patch on an old garment—the old ones continuing to decay and becoming worthless long before the new ones lose any of their soundness and strength. Now if rails have been cut of different lengths on different parts of the farm, as already stated, these old and reduced rails cannot be used together, without the great inconvenience of long ends projecting from the corners, standing in the way of plowing near the boundary, and being liable to catch the traces of the harness; and frequently a rail will be found too short and proving an annoyance to the builder. To make a fence of such mixed rails will cost twice the labor required to build one of good materials of uniform length. It is therefore better to have all the rail-cuts measured carefully aud accurately, and cut

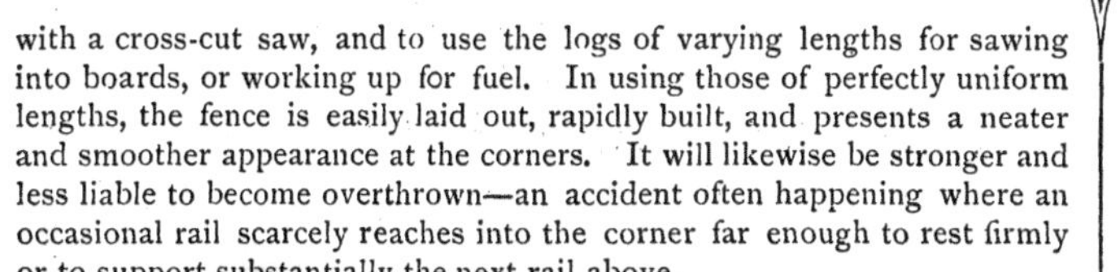

with a cross-cut saw, and to use the logs of varying lengths for sawing into boards, or working up for fuel. In using those of perfectly uniform lengths, the fence is easily laid out, rapidly built, and presents a neater and smoother appearance at the corners. It will likewise be stronger and less liable to become overthrown—an accident often happening where an occasional rail scarcely reaches into the corner far enough to rest firmly or to support substantially the next rail above.

The same inconveniences result from unequal lengths in fence boards, although in a different way. Boards are not, like rails, transferred from one fence to another; hence if a portion of one line of fence happens to be twelve feet long, and another distinct portion fourteen or sixteen feet, no harm results. But if they are intended to be of some definite length, and are carelessly cut and are found to vary all through six or eight inches from each other, continued trouble will arise, unless the boards are sawed off at the ends at a considerable waste. If the attempt is made to vary the distances of the posts, so as to economize, the boards must not only be all previously and laboriously assorted, but they must be distributed with much care in their proper places, and nearly as much more care required in setting the posts in accordance with these lengths. All this trouble will be avoided by carefully cutting the saw-logs of measured uniform length.

System in Splitting.

It is a matter of much importance to do the work of splitting in the best way. A novice might spoil half the timber, and expend double labor in doing it—splitting some rails too large, others too small, and often making rails twice as large at one end as the at the other. Fig. 14 shows the sawed end of a log which is ready to split. Take a piece of red chalk, or a carpenter's black lead pencil, and with the help of a straight-edge, draw the two marks or diameters across it at right angles, as shown by the larger lines. Then draw two others where the dotted marks are seen. This will cut the log in eight pieces; after which it should be again regularly subdivided, according to its size. If two feet or a little more in diameter, it may be laid out for fifty-six rails, as shown in fig. 15; varying with the intended use of the rails. If for a common zig-zag fence, the rails measure about the same in breadth and thickness; but if for post and rail fence, they should be flat and broad. Either of these forms may be easily made by increasing or lessening the number of the cross-connections, between the radiating lines.

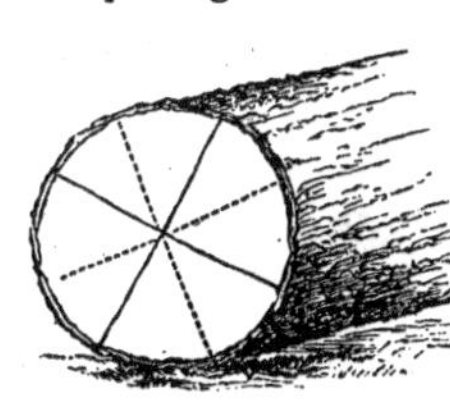

Fig. 14.—*Marking Logs for Splitting.*

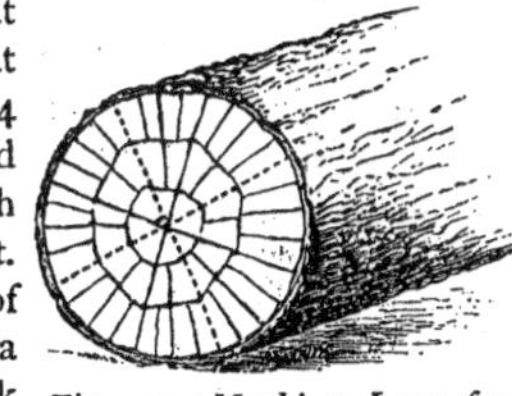

Fig. 15.—*Marking Logs for Splitting.*

Fig. 16 represents a log about a foot and a half in diameter, marked for splitting into twenty-four rails; and fig. 17 another a little larger and straighter, and more free of knots, marked for splitting into thirty-two flat rails, and eight common ones. These different modes of laying out the work will show sufficiently how it is done, and enable any one of moderate intelligence to mark out any log of whatever size, by a little variation. After this is done, the splitter will proceed rapidly with his work, without hesitation, or without losing time in pondering where he shall next set his wedge. He will save both time and timber, and turn out better and more uniform rails.

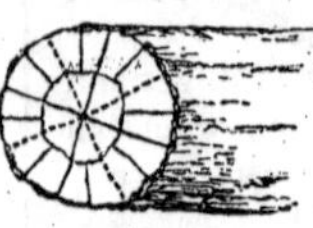

Fig. 16.

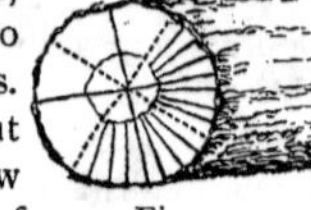

Fig. 17.

Every one who has split wood of any kind, knows that in order to prevent the cleft from running off on one side or the other, it should be nearly in the middle of the stick. Hence the general rule, to keep reducing the size of the sections, not, as some would suppose, by taking off a rail from the side, but by striking directly through the centre. The log should therefore be first split in two, and these halves again into four quarters—one of which is shown in fig. 18. This is to be again split into the two parts, fig. 19; and these again in the manner indicated in fig. 20. In this way, there will be but little tendency to form uneven rails; and if the cleft is controlled somewhat by striking the wedge or axe a little ahead, no trouble whatever will occur.

Fig. 18.

Fig. 19.

Fig. 20.

Laying Out the Fence.

In laying out a zig-zag rail fence, various modes are adopted for fixing the places for the corners; but the most perfect, and perhaps the most rapid, is to stretch a cord or line, (which, if the weather is not windy, may be 20 rods long,) and keeping it to its place by flat stones laid on it, proceed to lay off each corner. Take a rod or pole a foot less in length than the rails, place the middle on the line, and then give it such divergence from the line as will be suitable for a firm fence. Stick in a peg or stick at each intended corner, and so proceed till all is laid off. The work will be greatly facilitated by nailing a piece of lath at right angles to the pole, just long enough to reach from the line to the corner. Where a long line of fence, or many fences are to be built, a still more rapid and equally perfect mode of laying out the fence is to take a cord and mark it off in equal divisions, by sticking a pin through and twisting it round the cord, at each length, and then after stretching it, lay off, right and left from these pins, the proper corners, by the help of a measured stick. In this way two men can lay out with great accuracy several miles in a day.

Such fences occupy too much space, which is a prominent objection to them; the dotted lines, *a a*, fig. 21, showing the full breadth which the fence and stakes must take up, as commonly built—requiring, in order to

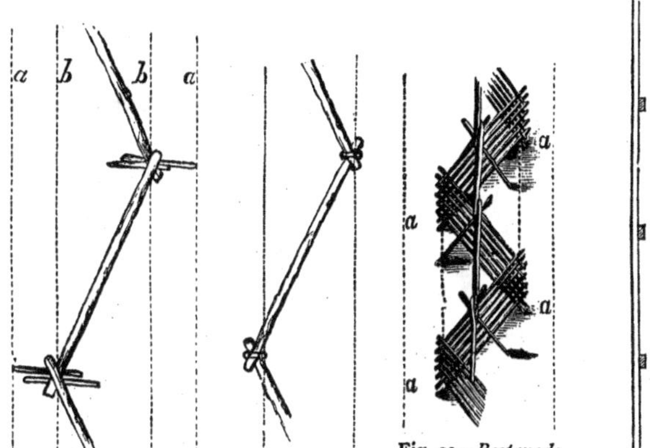

Fig. 21.—*Width of Zig-Zag Fence.*

Fig. 22.—*Zig-Zag Fence made narrow by staking.*

Fig. 23—*Best mode of Staking Common Rail Fences—vertical view.*

Fig. 24.—*Board or Post-and-Rail Fence.*

keep clear of the stakes, which, set in this sloping manner, are easily knocked out, a strip of land nearly a rod wide, which is equivalent to about one acre for every line of fence 160 rods long. More than one-half this space, or from *b* to *b*, is saved by setting the stakes as fig. 22 represents, where they are vertical and wired together; and are less easily thrown out by the plow. The same advantage exists with staking the fence, as shown by fig. 23. But the most perfect fence for convenience, neatness and saving of land which it occupies, is the one made of post and rails, or of boards, fig. 24.

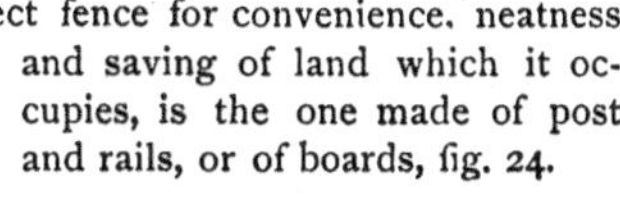

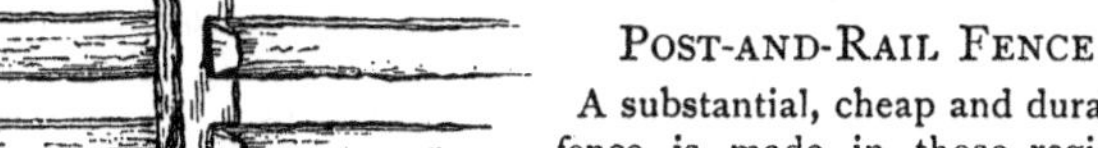

Fig. 25.—*Post-and-Rail Fence.*

POST-AND-RAIL FENCE.

A substantial, cheap and durable fence is made in those regions where chestnut and other good timber grows in abundance, as represented by fig. 25. The rails and posts should be split flat. Two modes are used for perforating the posts—boring and cutting with a two-handed tool made for this purpose. If the former can be done by machinery, the work is cheap and rapid; and cutting the holes may be performed during the leisure of winter. The mode of boring posts is described, with engraved illustrations, on page 24 of vol. IV of Rural Affairs. The bank or ridge of earth obviates the necessity of a bottom rail, saves cutting an additional hole in the post, and by stiffening the post, renders it unnecessary to dig so deep a hole. The rails, being dressed off wedge-form at the ends, pass each other in the holes. It is therefore necessary to insert all the rails as the posts are successively set in, and after once in, they cannot come out or become displaced.

A good cattle fence of posts and common rails is shown by fig. 26. If sheep are not kept on the farm, such a fence is quite sufficient, as no good and neat farmer allows his swine the run of his fields. Good stiff posts, (which may be split ones, as no facing is essential,) are inserted a few inches nearer each other than the length of the rails, and the rails are then secured as shown in the cut, by telegraph or fence wire, inserted through holes bored with a brace-bit, through both rails and

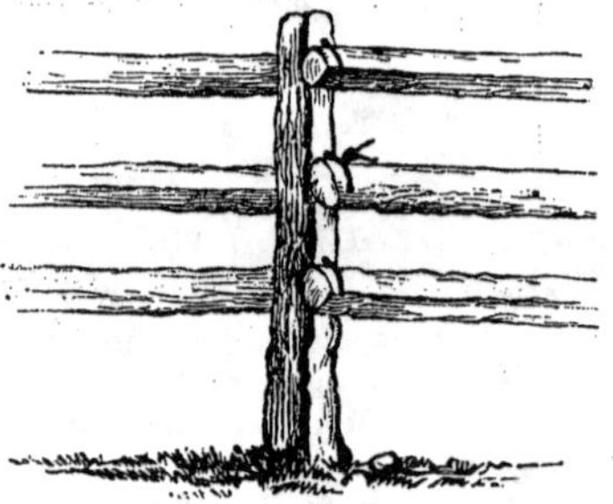

Fig. 26.—*Cattle Fence.*

Fig. 27.—*Two-legged Horse for Supporting Rails—Leaning against Post.*

post, and again through the post alone just above the rails. The best way to hold the rails is to provide wooden horses, each with two legs, as shown in fig. 27, the height of which may be easily regulated by placing them more sloping or erect against the posts, according to the desired height, and on these horses the ends of the rails rest while the boring and wiring is performed. These supports are simple, easily made in a few minutes, and a good supply of them will save much time in building. If there is great difference in the height of the rails, two sizes would be desirable.

Board Fences.

The cheapest and best *board* fence, so far as we know, is the kind represented by fig. 28. It requires only half the number of posts commonly used in making board fences, and digging half the number of holes, and only half the labor of setting—making a difference at present prices of about forty cents per rod, but varying with soil and price of labor and timber. The boards may be 12, 14 or 16 feet in length—the greater the length, the stouter they should be; and the posts should be good and substantial ones. It will save much labor if the boards have been sawed from accurately measured logs, so as to be all of precisely the same length as already shown. The distances of the posts asunder may then be measured accurately with a pole, and every board may be nailed to its place with very little waste in cutting. After having nailed on the boards, (of course without breaking joints,) a small piece of timber or batten is placed upright and midway between the posts, and firmly secured by nailing through the boards into it. These battens may be made of sawed timber 2 by 3 inches; or of split slabs of about the same size, only one straight side being necessary. A cap-board is then placed on the top and secured by nailing into the tops of the posts, the top of the batten, and into the upper edge of the top board. This cap-board is of much importance, stiffening the fence, and, in connection with the batten, rendering the whole firm and substantial. A space of about one foot should be left below the bottom board, to be plowed up to, so as to form a ridge. This saves some lumber, stiffens the posts, forms surface drainage, and prevents young horses from leaning against the fence, as they do not like to stand in the ditch. This ridge should not be made too wide at the top, or these animals will stand upon it and push against the fence.

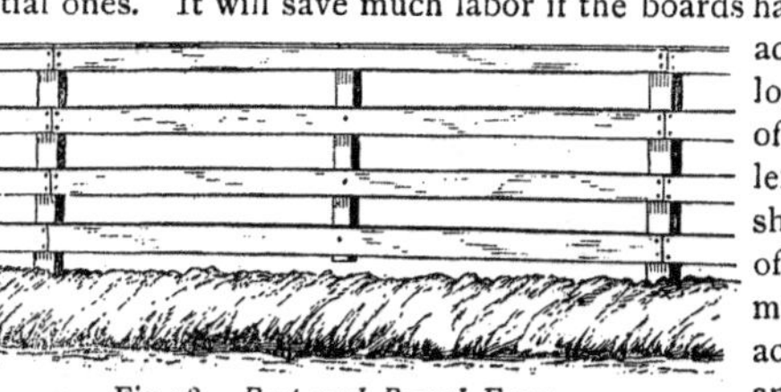

Fig. 28.—*Post and Board Fence.*

If this fence is likely to be much exposed, and if the boards are 16 or 18 feet long, it may be best to place two battens between the posts, as in fig. 29—which will make a fence of great firmness, in connection with a substantial cap-board. If the lower board could be an inch and a half thick, and nailed with twenty-penny nails, it would be still better.

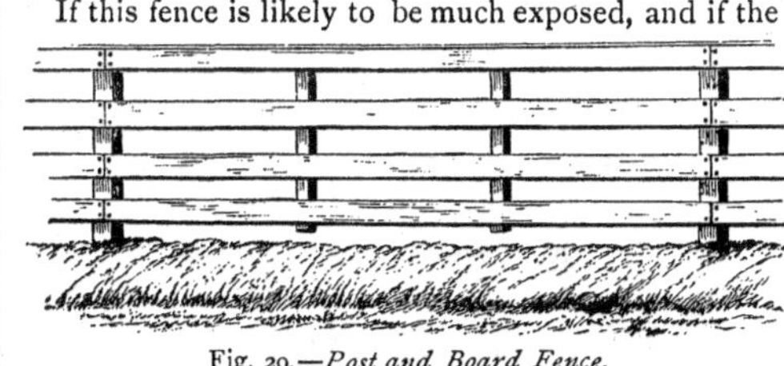

Fig. 29.—*Post and Board Fence.*

The cost of these battens, if sawed out on purpose, or if cut from 2 by 3 inch scantling, at two cents per foot, board measure, would not be four cents each; and the time required for nailing them to their places would not be over two minutes each. This is much more economical than setting another post, including its cost, labor of digging hole, setting the post, pounding the earth, &c.

Board fences are often made of boards the widest and strongest at bottom, and narrowest at top. It would be better if the order were reversed, as the hardest usage comes on the top board, whether by persons climbing over, or by cattle crowding against it. For the same reason it is always a matter of economy to use a cap-board, even if only three inches wide, as it becomes a firm stiffener when nailed to the upper edge of the top board.

Slab Fences.

In lumber regions, and in the immediate neighborhoods of sawmills, good strong durable slabs may be had at a low price, and may be made into cheap and substantial fences. Fig. 30 shows the manner in which the slabs are attached to the posts by means of large annealed or telegraph wire. A single nail driven through the upper part of each end of the slab secures it temporarily to the post, till two holes are bored with a brace-bit through both slabs and the intervening post. The wire is then passed through, drawn tight and firmly twisted. This fence, if the wire is large enough, (and it should not be less than No. 9,) will make a firm barrier against cattle and horses. Fig. 31 shows the structure in section; and fig. 32 the appearance of the whole when completed, with batten between posts for stiffening the slabs, and the ridge of earth beneath, for increasing its efficiency.

Fig. 30.—*Slab Fence.*

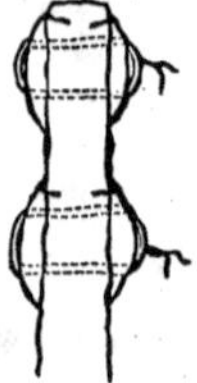

Fig. 31.—*Section of Slab Fence.*

Fig. 32.—*Slab Fence.*

In building fences on uneven ground, awkward workmen often make them unnecessarily deformed in appearance by trying to follow the surface without regard to the angles formed in the line of the boards. Fig. 33 represents a crooked surface, over which a fence is to be built. Fig. 34 shows the kind of fence too often built over such a surface by a bungler. Fig. 35 is a handsome, graceful finished fence as made on the same ground by a skillful hand; room enough being allowed under the bottom board to to make an embanked ridge by plowing a few furrows against the line, and afterwards smoothing with spade or hoe, as described in another part of this article. The curve is first given to the top board, and all the rest follow parallel to it, by using the gauge described on page 273 of vol. II of Rural Affairs. The mode by which the curve is easily and readily made is shown in fig. 36, where by deviating slightly from a straight line, a uni-

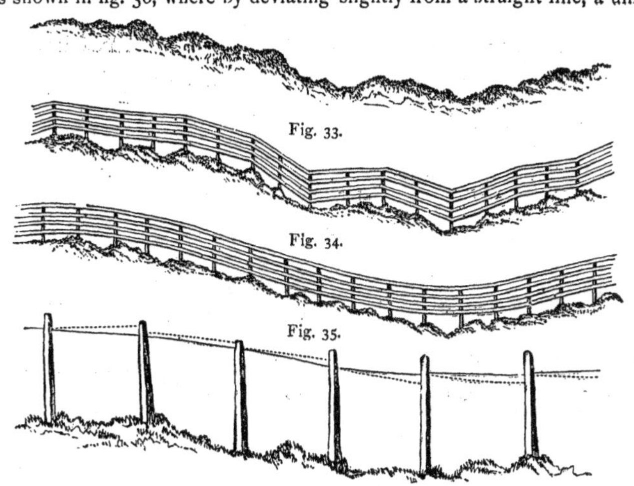

Fig. 33.

Fig. 34.

Fig. 35.

Fig. 36.

form curve is formed. This deviation may vary at each post from the fourth of an inch to an inch or two, according to circumstances; and a long curve may be made to pass by insensible gradations into a short one, by increasing the deviation regularly—or the reverse, by diminishing it. This work may be accomplished by using a straight pole long enough to connect three posts; or more rapidly and easier to the practiced workmen, by slightly driving a nail in each post to range by.

Sharpening posts and making holes with the crowbar in the bottom of spaded holes, for the purpose of saving the labor of digging the holes to full depth, is sometimes recommended and put in practice. It has some advantages—a prominent one being the firmness with which the bottom or end of the post is held by the hard subsoil, into which it is driven by the maul at the top; and as a post never decays below till long after it it is rotted at the surface of the earth, it need not be so large at the lower end. But a serious difficulty results from the impossibility of setting such posts in a straight line, both the crow-bar holes and the sharpening often placing them an inch or two on one side or the other, and rendering the fence zig-zag and distorted. This mode will, however, answer for cheap fences, as, for example, those made of slabs, and which are intended to stand but a few years.

To make a durable, handsome and perfect fence, the earth should be firmly beaten down with an iron-shod rammer, as it is shovelled in by small portions at a time, keeping the post accurately in position against a stretched cord or two cords, as described in a former number of the Register. A post thus set should stand as firmly in the earth as if it stood in solid rock. We have seen a fence thus made which had not twisted three inches from a straight line in twenty-four years—perfect underdrainage having been provided beneath it; while another and adjoining fence, with posts more carelessly set, had become nearly worthless from the distortions to which it had been subjected.

THE MANAGEMENT OF FRACTIOUS COWS.

By a Well-Known Farmer.

DID ANY OF YOUR READERS ever hear of Dame Sniggins, who lived in New-England during the early part of the last century? It is said she was remarkable for her skill in bringing up children, and in the practice of the old, time-honored flagellation system, gave each of her boys a thorough whipping on every Saturday evening, so as to cover any possible offences which they might have committed during the week. I have neve personally witnessed the results of this mode of treatment, but have seen a practice not unlike it adopted by some third-rate farmers in managing their cows. One in particular I well remember, who, as he was about to commence milking, regularly placed his kicking cow in a narrow stall, where he endured for a time her dexterous attempts to kick him over, when at last, patience being exhausted, he commenced an unremitting succession of blows by club or cudgel, until he deemed the demands of vengeance fully satisfied. It will hardly be supposed that this attempt to carry the point by storm was successful on the part of the owner. The poor animal became enraged as a matter of course, and sought in future every chance to express her sentiments towards the owner through her heels. She remained unsubdued—and in fact could have no distinct knowledge what those furious assaults could mean.

Joe and Bob were the two boys of a neighbor, and had charge of milking the old cow, an animal of strength and energy, and generally supposed to be a cow inherently hard to manage. She was milked in one corner of the large pasture, where she was driven, Bob "surrounding" her with whip in hand, while Joe worked vigorously at the dugs. When she kicked or ran away, she was chased around the pasture, and when cornered again, was whipped for her misbehavior. In a short time, by the repeated association of cause and effect, she carefully avoided being driven into the dreaded corner. This cow was of course soon "spoiled," and the boys could not manage her, as the lessons they had given her made her as wild as the roe of the mountains.

The writer, then a boy, had special charge of the cows and milking on his father's farm. Observing that animals could be educated into bad habits by such management as that just mentioned, he concluded that the reverse might as well be reached by the use of suitable means. His experiments were quite successful. The theory adopted was simply this—to make the animal understand distinctly what was wanted, and then make it her interest or inclination to carry out the wish of her manager. It was an indispensable requisite before proceeding, that the milker exercise perfect control of himself, and not allow any provocation to effect his serenity in the slightest degree; he could then apply the remedy with the unvarying precision of a machine. The following is an accurate statement of one of the experiments which was performed in this way on an animal belonging to a widow; the cow, being a notable milker and a still more notable

Fig. 37.—*Management of Fractious Cows.*

kicker, (fig. 37,) had been successively sold from one owner to another, at a considerable discount each time on account of her atrocious character, until she came, at a low price, into the widow's hands. It was necessary to tie this animal's legs every time before milking—and even after this precaution, few were willing to undertake the task, as it was regarded about as much as one's limbs were worth to come near her. The possibility of curing her of this habit, when expressed, was regarded as a chimera; it was however undertaken. Providing myself with a good short whip, I placed it under my left arm, projecting backwards, so as not to be in the way, but so as to be quickly grasped with the right hand. Placing the cow in a yard about twenty-five feet square, the pail was then taken in the left hand, and the dreaded animal firmly approached, with pleasant, decided and soothing words. But the operation of milking was hardly begun before, as was expected, a stroke of the hoof, like a flash of lightning, was made towards me. It was evaded, and a *single* cut of the whip given to the animal. She of course started to run, when a *single* cut of the whip across her nose arrested her at once. With a confident stroking with the hand and a kind word, the milking was resumed. The same kicking and other process was repeated, and continued perhaps a dozen times before the milking was completed—a bystander, the widow's son, repeatedly vociferating the caution, "She'll kill you—she'll kick you over—you better stop!" The cow and the milker, however, soon came to a distinct understanding—she discovering in a short time that every *attempt* to kick was instantly followed by the same inevitable single cut of the whip. It was surprising how soon the cow understood cause and effect, when the two were so closely and invariably connected.

Before the first milking was completed the kicking had been nearly given up. At the next milking the offence was repeated three times; at the third but once. The cow had now learned, in as distinct a manner as a uniform and close connection could possibly convey to her, that every *attempt* to kick brought instantaneously the same terrific and dreaded single blow—two or more blows would have produced a reaction, a spirit of resistance, and the salutary effect would have been lost. She had likewise learned that when she stood quietly, she was treated with kindness and her position made pleasant. Nothing more was wanted—the cure was completed. It was interesting and satisfactory to observe on passing a week

Fig. 38.—*Fractious Cow Cured.*

afterwards that the widow's son, who had so dreaded to approach this animal, was seated quietly on his milking stool drawing the copious streams of new milk from her udder, while the animal stood quietly with that peculiar expression of satisfaction and enjoyment always indicated by chewing the cud with closed eyes, (fig. 38.)

A number of persons to whom this simple secret of management had been imparted, entirely failed in their attempts to carry it out, simply because they omitted essential parts. For example—one who attempted the remedy did not like to disturb the animal at the first kick, because no harm was done and all seemed to be going on smoothly, nor at the second for the same reason; but at the third the pail was upset and the milker's shank bruised, and he then made up the deficiency by an undefined storm of blows with fury and anger. The whole thing was a failure, as a matter of course—as the animal could not possibly understand the intention. Another did better, but could not control himself, and struck two or three times instead of but once, and thus stirred up the anger of the animal.

I have often seen persons who acted entirely in accordance with impulse—in fact very few do otherwise. If a cow attempts a spiteful kick which happens not to do any harm, she is acquitted, and no notice taken of her offence; if she happens, without intending it, to strike her hoof against the pail and upsets it, a storm of blows is poured down upon her. This is the way animals are spoiled and rendered "vicious," and for which the manager, not the animal, should suffer the penalty.

It must be admitted that very few persons possess even the smallest amount of perception and self-possession—they act merely by impulse.

I have tried my mode of treatment on many fractious cows, and never failed in but a single instance. A kind friend, at whose house I made a visit, told me his cow "Gentle," an excellent milker, was an inveterate kicker, and proposed to me to try my skill. I scarcely ever saw a finer looking animal, or, as I soon learned, one that had more sagacity. She saw at a glance, as soon as I commenced milking, that I was prepared for her, and either that I was not to be trifled with, or else being pleased with my manner, did not like to injure me; hence she never attempted to kick while I was at her side. As a necessary consequence, I could not cure her of what she did not practice. She kicked every one else, but never attempted this kind of gymnastics on me.

The author of "My Farm of Edgewood," after describing some instances of the effects of excessive passion on the one hand, and of unmeaning "gentleness" on the other, gravely comes to this wise conclusion: "The moral of the story is—if a cow is an inveterate kicker, tie her legs with a gentle hand, or kill her." It is needless for any one who can control himself, and who is able to teach a cow just what he wants, in the manner already described, to believe a word of this sage morality. It is no wonder, however, that such a conclusion was arrived at, when we read a little further in the book just quoted, "Beating will never cure, whether it come in successive thuds, or in an explosive outbreak of outrageous violence." Of course it will not, and here we entirely agree. The "thuds" spoken of, as we are elsewhere told, were mauling the animal's ribs with the milking stool.

I once tried the same lesson of *cause and effect*, already described as so effectual with kickers, in curing a cow of running about repeatedly over the yard while milking, with entire success, but with more labor and longer time. The animal learned that I wanted her to stand at the exact spot where I commenced the operation, that that was the only place where she could enjoy herself, and that a single touch on the face quickly and surely followed every deviation.

Animals which are wild and frightened by the presence of man, may be soon inured to his presence. It is said that when Rarey, the horse tamer, first takes wild and untractable pastured colts in hand, he walks slowly towards them, and if they run, he stands perfectly still—does not try to "head" them—and when they cease running, again approaches them slowly. They soon become tired of a race in which there is no competitor.

I have tried this mode with entire success. With such animals, not "vicious," time and patience will accomplish almost anything. A man named Cole, well known many years ago as a successful trainer of oxen, adopted this mode to reduce wild steers to submission. He broke oxen as a business, taking about five yoke, or ten animals, at a time. They were placed in a yard, and he passed very slowly among them, or stood still if they showed any alarm at his presence. In the course of half a day they all became quite accustomed to him, and he could then begin to place his hand gently on their backs. By the close of the second day he could handle them freely, and some would allow him to place the ox-bow on their necks. When they became familiar with this part of the yoke, he drew gently upon it, so as to lead them. At first they would not yield to this pressure, but gradually they would take a step or two forward. In this manner he gradually led them on, until at the end of the fifth day he had completed a perfect system of training—they would come or follow him at the mere motion of his hand, and he could drive them with perfect precision without speaking a word, unless occasionally in a low voice—never allowing the noisy vociferations and yells so common in driving oxen.

The perfect manner in which animals may be made to connect cause and effect, I once saw exemplified in a span of horses used to work on the tread-wheel of a ferry boat. When turned loose from the stable, which was ninety feet higher than the dock, they found their own way down the steep hill, walked unattended to the end of the pier, walked on the boat, passed towards the stern, and then stepped down backwards a foot below the floor on the working wheel. What was the reason of this movement, which most horses could not be persuaded to do by whipping? Simply this—they always found a mess of oats ready for them when they placed themselves in position on the wheel. By similar appliances, pleasant or disagreeable, animals may be taught to do almost anything, provided an immediate and invariable connection is made between the object desired by the owner, and the animal's wishes, which the latter can understand. Cruelty and severe punishment are thus avoided, and pleasure and enjoyment for both parties—that is, for owner and owned, attained without trouble.

STONE WALLS FOR HILL-SIDES.

IT IS GENERALLY ADMITTED that stone walls, when well made, form the best of all farm fences, and where the material is abundant they should be adopted in preference to rail and board fence, which decays, or to hedges, which need frequent clipping and cutting, and which often die in patches, and leave gaps or openings. A permanent, well-built stone wall will neither burn down nor rot down; and the only essential requisite is to construct it so that the force of frost shall never throw it over. We sometimes meet with those which have the foundations so well laid below freezing, and the walls so securely laid up, that they promise to stand for centuries without the necessity of any repairs whatever.

In former volumes we have given full directions for the construction of stone fences on level ground. It remains to offer some suggestions for their erection on hill-sides. It frequently happens that the side of a sloping bank may be selected for placing a division line, even if somewhat curved, where a wall may be built more cheaply than on level ground, and which will also tend to give a more level surface to each enclosed field. The border of the higher field will also possess the advantage of never being obstructed by snow drifts—which will be an especial advantage if it is to be occupied by young trees or orchards. It is by no means necessary that the slope should be very steep; a moderate amount of labor in plowing and shovelling will make a good bed for the wall. Fig. 39 represents a slope of this kind; the dotted lines at *a* showing how much earth must be cut away, and at *b* the amount of filling in. The easiest way is to plow down or away from the the bank, throwing the fresh earth towards *c*, where a new ridge is formed, as shown by the dotted lines. Fig. 40 exhibits the same, with the wall in position. It will be seen that but a small quantity of earth at *a* needs removing, and that there is but little to fill in at *b*. It is better to remove less at *a* than is required to fill the space at *b*, because a part of the latter may be supplied from the rising ground at *d*, with less labor than to shovel it up from *a*. Fig. 41 shows the wall as completed. There are a few requisites of importance in constructing such a wall. It should be made sloping towards the bank, to prevent the frost from throwing it over. The degree of slope will vary with circumstances. If built of nearly round stone, like cobble-stone, it will be well to make it recede about one foot in rising three feet. This will be a good barrier against cattle, horses and swine; but sheep will sometimes scale it, and hence one foot in four will be better where sheep are kept. For this purpose, either flatter stone must be employed, or the space between the wall and the earth bank should be filled in with small stone or gravel—which, indeed, should always

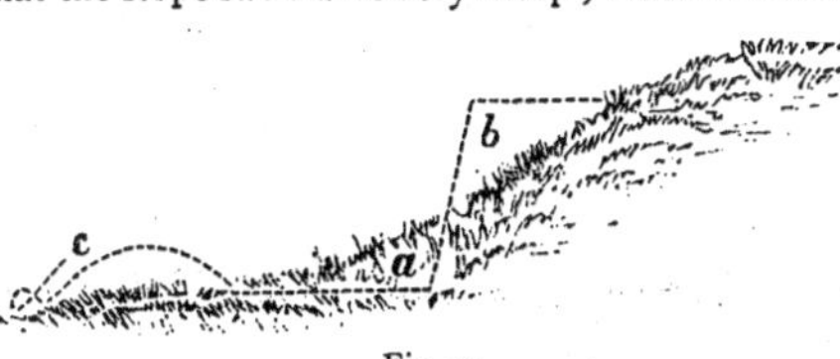

Fig. 39.

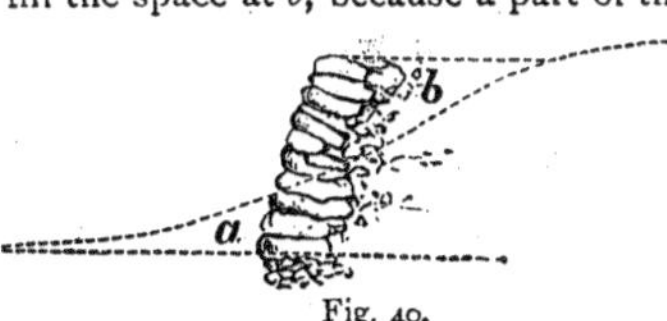

Fig. 40.

Fig. 41.—*Finished Wall against a Bank.*

be done, no matter what kind of stone are used. For if the large stone which constitute the body of the wall, rest against the earth, they are thrown outward by frost a short distance every winter, and will not settle back to their original position; hence in the course of years the face of the structure will become deranged and crooked, and in some places will tumble over. It will also assist in preventing this injury by placing the stone slightly sloping or tipping down towards the earth bank, as distinctly shown in fig. 41; for, if the stone should happen to be thrown outward by freezing, they will be more likely to settle back again than if laid level. Unlike the common wall on level ground, this bank-wall should be thickest at the top, because it is there most exposed to frost and distortion; and for the same reason, the filling in of small stone or gravel should be more copious here than near the bottom. It is less important that such a wall as this be laid on a deep bed of broken or small stone as a foundation, than a common two-faced wall, where such a foundation is so indispensable; but there should nevertheless be a trench of moderate extent thus provided, as seen in section in fig. 40.

The mistake has been sometimes made of building side-hill walls erect or perpendicular, as in fig. 42. Such a wall can never stand permanently, unless several feet thick, and with a large quantity of small stone between wall and earth—which would be expensive and mostly impracticable. If resting against the earth, it will be crowded out a short distance every time the earth freezes; and the stone thus displaced will never regain their original position—the earth settling against them. The next freezing will thrust them a little further, and in the course of time the wall will present the appearance shown in fig. 43.

Fig. 42.—*Vertical Wall.*

Fig. 43.—*The same in after years.*

Where timber is not extremely scarce, and stone is moderately abundant, it may answer well to build a half-wall, fig. 44. Short posts are set in the upper part, and a single board, or two boards, nailed on, according to the height of the wall. There are several advantages of such a half-wall, over those commonly built on a level. The posts, being short, need not be so strong and thick as those of full length; and the smaller ones in a common pile may be picked out and cut into two. Being set in stone, there is a perfect drainage, and such posts will last at least two or three times as long as if set in moist earth. As animals are not likely to crowd against a fence standing thus on the edge of a

Fig. 44.

precipice, the posts need not be nearer than the full length of the boards, if a cap-board is nailed on the top, as the figure shows. If not convenient to use posts and boards, two common rails may be well employed, by first placing one flat on the top of the wall, then a common stake on this

Fig. 45.

Fig. 46.—*Wall on Hill-side.*

towards the lower side, and a short one on the upper, and on these the top rail or rider, fig. 45.

Animals cannot vault over such fences from the lower side, and they are not likely to venture a downward plunge from above.

Fig. 46 represents the mode which should be adopted in building walls, when not made along the margin of a ridge, as already described, but directly up the hill. The principal requisite in order to make a substantial fence in such a position, is to begin at the bottom and to lay the stones, or tiers of stone, on a level. If laid parallel to the surface, the tendency will be constant to tumble downward, instead of standing firm and immovable, as when well laid level.

In laying large and heavy stone, the severe labor of hard lifting may be much lessened by the use of a bench or inclined plane, made of thick, stout plank, fig. 47, up which large stones may be rolled with comparative ease.

Fig. 47.—*Bench for Lifting Stone.*

The thickness of the plank used for making such a bench must depend on the size of the stone likely to be used, but if four inches thick, it will be strong enough for the heaviest masses which can be handled in this way. It will be of more use in building level ground walls, but may also be of much advantage in laying those already described, although a single plank, without a support at the end, will often answer a good purpose, for stone that have been deposited on the upper side.

The use of a *cant-hook*, such as is used in moving logs at sawmills, will greatly facilitate the rolling of the large stone up the plank, and by means of the two, namely, the plank and the hook, combined, one man alone will handle almost any stone not too large to draw on a stone-boat. As the cant-hook (fig. 48) is a simple and very useful implement, and worthy of more general

Fig. 48.—*Cant-Hook.*

use on the farm, we give the dimensions of the different parts of one suitable for handling stone, in order that any farmer may provide one for his own use. Whole length, 5 feet; 3 inches square at larger part, where the mortice or slot is made, tapering to each end, as in the figure; length of iron about 2 feet, an inch and three-fourths wide, and nearly half an inch thick, perforated with several holes to adapt it to large or smaller stone. A strong screw bolt secures it. Sometimes the iron hook is bolted to the side of the lever, instead of being placed in a mortice, but this does not work so firmly. For handling saw logs, cant-hooks are often longer, or about six feet.

In building stone walls, all large stones should be made to lie solid by chinking well with small pieces.

MEASURING AND MAPPING FARMS.

ONE OF THE FIRST THINGS in successful farming is for the owner to know the exact dimensions and areas of his fields. Without this knowledge, much of his work is entirely in the dark. Certain modes of cultivation, such as thick and thin seeding, different ways of manuring, deep and shallow plowing and planting, using different varieties of seed, and early and late cutting, variously affect the expense and amount of product; but as the owner knows only by guessing how much his fields may contain, he is unable to arrive at certain and satisfactory results, or to repeat the modes by which such results are reached. He cannot know how long his team will be occupied in plowing a field, unless he knows its contents, or say how much seed will be needed, how much manure he applies per acre, or whether he obtains thirty, forty or fifty bushels as a crop, which may be affected five or ten bushels to the acre, according to variation in management. It may, in short, be laid down as a certain fact, that the farmer who keeps his lands constantly measured, and regularly weighs or measures their products, will learn more about good paying farming in ten years, than the careless and guessing farmer in forty.

Open weather in winter often affords a good opportunity to measure fields and to map farms. A thin, crusted snow, hard enough to bear the weight, fills up hollows and furrows, and makes a rough surface more easily and accurately measured. With the simple implements we are about to describe, the measuring may be readily accomplished by any farmer. Having taken these measurements in his memorandum book he can lay them down with measure and rule on a sheet of paper within doors, and easily calculate the area by the simple rules here given, if he knows the first rules in arithmetic. He can thus draw his whole farm on a sheet of paper, with the dimensions and contents of every field; and if he has a little skill with pen and pencil, will make a neat and useful map. If he has but little skill, he will nevertheless make a map which, although rougher, will be accurate because its fields have been measured, and will be constantly of great use and value.

Square fields, or those in the shape of a rectangle, are of course most easily calculated; all that is requisite is to multiply the length by the breadth —which, if in *rods*, will be divided by 160 to bring it to acres. Or if the field is measured in *chains* and *links or hundredths*, according to the general practice of land-surveyors, he has the very easy task of dividing these by ten, or in other words, simply pointing off one figure to make the product acres and hundredths of an acre.

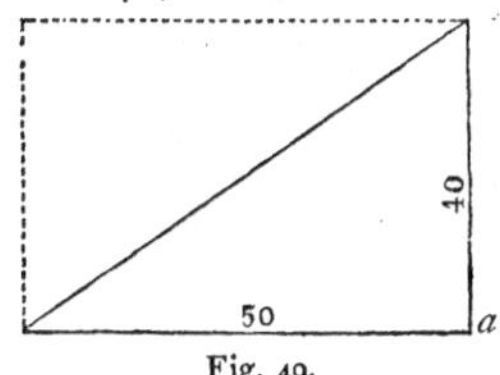

Fig. 49.

But if the fields are three-sided, or with four or more sides in an irregular shape, another mode must be adopted, which is nearly as easy when once understood. If three-sided, with one right angle, as at *a*, fig. 49, this right angled triangle will contain precisely one-half as much as if a square or rectangle, as indicated by the dotted lines. All we have to do in this case, therefore, is to measure the two sides which contain the right angle, multiply them together, and divide by 2. For example—suppose a triangular field measures 40 rods on one side and 50 rods on the other; multiply these and the product is 2,000 square rods—one-half of which, 1,000, is the area—which divided by 160 gives $6\frac{1}{4}$ acres. But more frequently the triangle has no right angle, as in fig. 50—what then? Divide it into two parts, as shown by the dotted line, making two triangles, measure them separately, and add the areas together. The dotted line must be at right angles to the side on which it falls, or nearly so—a slight variation will not affect the result materially. To do this easily, stretch a cord or garden line, or make a straight line in any other way, place a carpenter's square, *a*, on this, moving it along one way or the other until the other arm of the square points to a stake at the corner *b*. Then measure to this corner, and also measure from the square to the two other corners, and you have all the necessary figures to tell readily how much land is in the field.

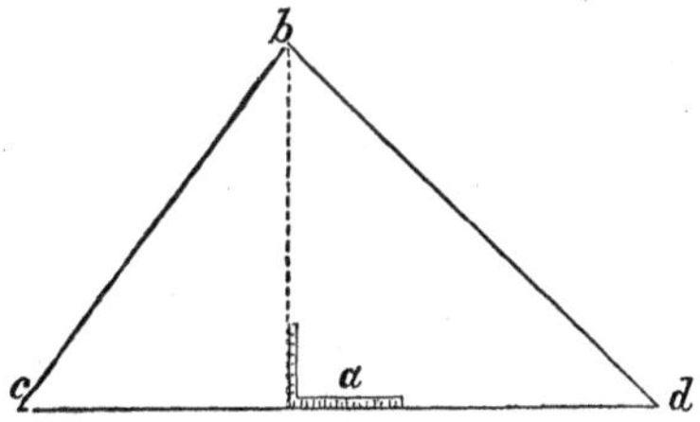

Fig. 50.

Suppose, for example, the dotted line is 40 rods long, and the two parts of the line *c d* are 30 and 50 rods. Multiply, and we get 1,200 and 2,000—add, and the sum is 3,200—divide by 2 and the product is 1,600 square rods, or 10 acres, the contents of the field. The most convenient way of placing this square, is to saw a slit into the side of a stake near the top, drive in the stake and place in the square, fig. 51.

Fig. 51.

A four-sided field with parallel sides, but with oblique angles, fig. 52, requires that a line across it be measured at right angles to one of the other sides, as marked by dots in the figure, and then one of the sides multiplied by the length of this dotted line, which will give the area. If only two

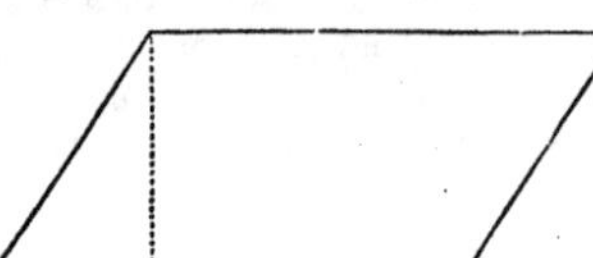

Fig. 52.

sides are parallel, as in fig. 53, add the two unequal sides together and divide the sum by 2; then multiply this quotient by the dotted line for the area.

If the field is four-sided and irregular, fig. 54, cut it up into triangles, as shown by the dotted lines, and measure each of these separately,

Fig. 53.

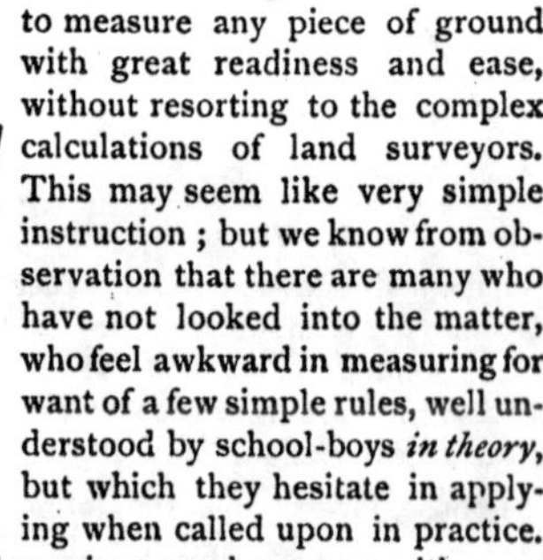

Fig. 54.

as already shown. If five or more sided, fig. 55, pursue the same process. A little practice will enable any man who knows enough to own a field, to measure any piece of ground with great readiness and ease, without resorting to the complex calculations of land surveyors. This may seem like very simple instruction; but we know from observation that there are many who have not looked into the matter, who feel awkward in measuring for want of a few simple rules, well understood by school-boys *in theory*, but which they hesitate in applying when called upon in practice.

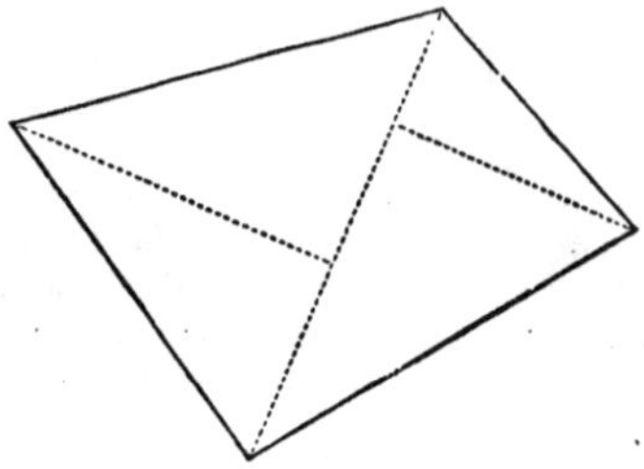

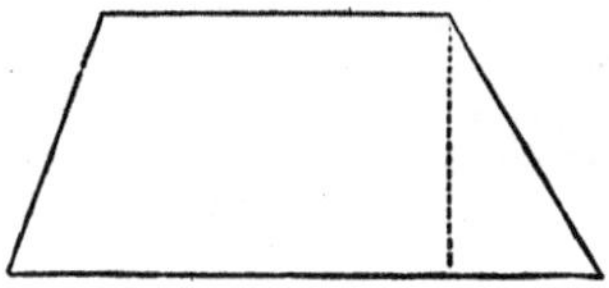

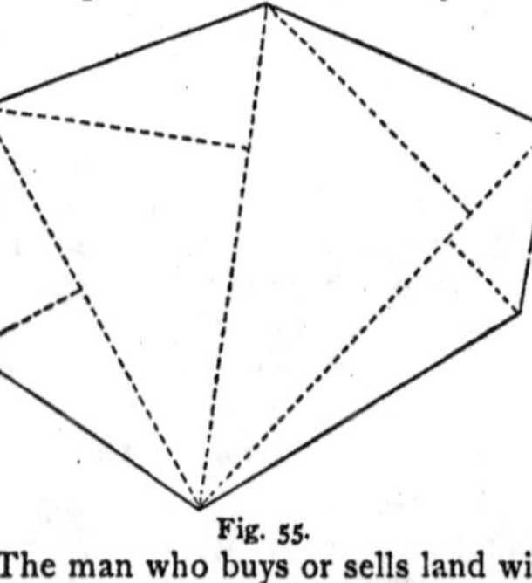

Fig. 55.

The man who buys or sells land will require a good surveyor with compass or theodolite, and a Gunter's chain, corrected by the county standard brass rule in the clerk's office ; but for all ordinary and practical purposes where the farmer keeps the account with himself, this accuracy is unnecessary. Chaining always requires two persons, which is often inconvenient. If the owner can deliberately make his own measurements alone, while his men are at work, he will be much better suited, and will be more likely to enter into the business thoroughly. Pacing is too inaccurate, although some, by long practice, will accomplish it with much uniformity. One of the most rapid and convenient modes is the use of a light angular wheel, which is thrust forward as fast as the measurer walks. Fig. 56 represents

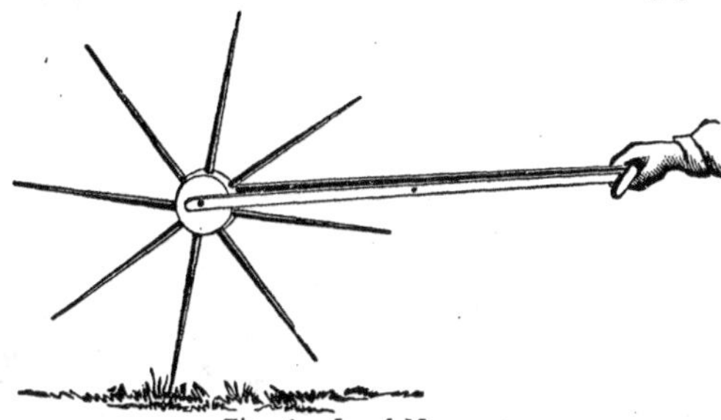

Fig. 56.—*Land Measurer.*

a wheel for this purpose, made of strips of wood a little larger than common lath—lightness being very important, in order to prevent the successive jerking which would take place if the implement were heavy, as each point strikes the earth. The wheel is of such a size as to revolve once at every rod in length. To effect this purpose the strips must be 32⅓ inches long from the centre * (as calculated by trigonometry,) which will give 24¾ inches from point to point. To construct this wheel, take a round piece of board about an inch thick, and saw radiating spaces into it, shaving the wood between the saw cuts out with a sharp chisel, fig. 57 ; then lay in the strips and screw them in. Then screw on another round piece of board and the hub will be complete. The radiating strips or spokes should be fitted with accuracy, so as to be firm, and the points at equal distances. Then measure from point to point, and if all are accurately 24¾ inches apart, the measurements of the land will also be correct, 8 times 24¾ being 16½ feet. It is best to drive a nail lengthwise into the end of each arm or spoke, before whittling it down sharp, as this will prevent the point from wearing down and becoming ultimately too short. A straight, smooth piece of round rod iron, with a screw and nut on one end, is then inserted for an axle ; and two strips of board placed on each side to receive the ends of the axle. A washer made of sole leather may be placed on each side of the wheel and inside the strips of board. These two strips have blocks placed between them to keep them at suitable distances

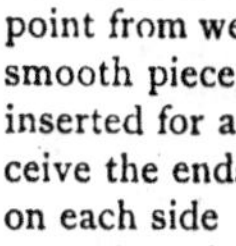

Fig. 57.

* Or with great accuracy, 32.337 inches.

apart, and a cross-bar is passed through the rear end for a handle. For measuring farms of moderate size this will be sufficient, with the addition of a strip of red cloth on one of the spokes, so that each revolution may be easily seen by the operator as he pushes the machine before him. For more extensive work two wheels, for recording, are to be attached, as shown in fig. 58. These may be about six inches in diameter and made of inch board. They are placed in the space in which the wheel revolves, which must be made wide enough for this purpose. The first has twenty small headless nails driven into its circumference at equal distances, projecting half an inch or more. A short tooth projects from the axle, (easily inserted by drilling a hole,) and so situated that at every revolution it comes against one of the nails, thus pushing the wheel on a short distance. At the next revolution it pushes another nail on. In this way the wheel revolves once for every twenty rods. On the axle of this wheel is a similar single tooth, which comes successively against one of the sixteen nails of a second wheel, made like the first. Thus the second wheel revolves once in a mile, or 320 rods, and long distances may be easily counted without much trouble to the observer. It may be used to measure roads, either on foot or by being drawn behind a wagon. The wheels should have enough friction at the axles to prevent any possibility of their slipping, which is easily affected by a spring pressing the axles, or by boring a hole down close beside the bearing, and thrusting in a wooden plug or wedge, so as to press moderately against the axles.

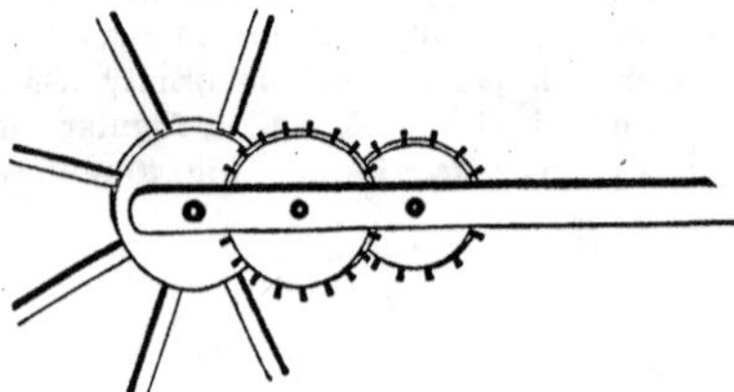

Fig. 58.

This machine may be made of pine, which is light and sufficiently stiff, but the arms or spokes should be of oak or other hard wood. If they are half an inch thick and two inches wide at the hub, tapering to an inch or less at the outer end, we have found them to be quite stiff enough. Any ingenious farmer who has a workshop will readily make one for himself, or a good joiner will do the work well—we had one made (without registering wheels) in a neat and substantial manner for three dollars. It measures land with ease and rapidity, and will soon pay for itself by the increase of knowledge which it will be the means of pouring into the farmer's mind when he measures all his crops.

The measurements will of course be most correct on smooth hard ground. On a freshly plowed field they will be attended with considerable inaccuracy, and should be made after the field is harrowed and settled. Our own experience during the past year shows that on a smooth surface there is rarely a variation of half an inch to a rod, and on ordinary farm ground or grass, not more than an inch, if well made. It is always advisable to prove

the work on a piece of measured ground, to see if the spokes are of the right length. With iron points to the arms, measurements may be rapidly made on hard crusted snow.

For shorter distances, as gardens, village lots, &c., very accurate measuring may be effected by a light eleven-foot pole, three lengths of which make 33 feet or two rods. A blacksmith will make a good handle of round iron, as shown in fig. 59, a screw being cut on the lower end for insertion into the rod. Thin bladed knives, or slits of tin, may be thrust into the ground against the ends, by which one person will do the measuring alone—the handle being placed near one end for this purpose.

Fig. 59.

TOWN AND COUNTRY ROADS.

By Robert Morris Copeland.

A LOVE OF THE BEAUTIFUL, or at least a recognition of its actuality and value, is widely spread among our countrymen, and in every hamlet in the United States there is at least one person who thinks he loves and understands beauty of landscape, and would like in some way to add to the natural advantages of the surrounding country. With most the enthusiasm spends itself in words, and a great deal of talk ends in little advance, for it is very easy to talk, and tiresome, expensive and difficult to create.

In very rude and undeveloped scenery, where the ragged mountain side is blended by wild tracts of forest with the partially subdued country, the introduction of neat houses, trim gardens, and well kept grounds might be actually inharmonious, giving just enough of variety to create a discord. Such cases are, of course, too rare to be considered by us, for in the United States population follows so closely on the steps of the pioneer that the number of clearings and hamlets soon creates distinct features which ought to be improved until they cease to be an injury to the general effect.

How populated places, taken as a whole, may be made agreeable or beautiful, rather than ugly or indifferent, should be a very interesting question, and one to which it would be worth while to devote a good deal of thought and careful planning. There are but few estates in this country extensive enough to give good opportunities to those who would improve the landscape ; but a few farms or country places taken together, or the whole area of a village or town, might be laid out so as to be beautiful in its general effect. Both new and old towns alike are laid out to suit the whims or economies of a few land-holders, or to meet the present daily wants of a handful of people who never think of what may be the future of their homes or towns, but are careful to see that the wants of to-day are satisfied.

In the western country where the land is level or gently rolling it seems as if convenience and economy alike dictate that all roads should be made straight and at right angles, and that if the streets of a new town are wide enough to give good circulation to the air, no further thought need be given to reserving space for health or pleasure. While all would admit the advantage of varying the roads from straight lines in a hilly country,

Fig. 60.—*Straight Road over Hills on the Left—Curving through Level Valley on the Right.*

and the dullest person can be persuaded that it is no farther around than over a hill, but few will understand the expediency of curving roads where the surface is as open to travel in one place as another. This is a serious mistake and stamps all the new towns of the west with such a stereotyped resemblance that a traveler might be excused if, landed on a dark night at the wrong depot, he should go up the main street, turn to right or left, and try his door-key in the same number which marks his own house in his native town.

To create variety in laying out and grouping the streets and houses of a town in a level country, requires more skill and thought than where rolling hills or water courses seem to compel the roads to diverge in well defined directions and the houses to be erected beside the natural highways.

Granting all that can be claimed for the economy and convenience of straight lines and right angles for traffic and travel in cities, we may still ask that the beautiful effects which may be produced by well grouped buildings shall have some recognition. It does not follow that because a straight line is the shortest between two points, therefore it is the best. When we compare the crooked streets of the old part of Boston with the straight ones in New-York and Philadelphia, the argument seems wholly in favor of the latter; but the crookedness is less objectionable than the narrowness of the Boston streets. The most crooked one there, its main avenue, Washington-street, or any of those which are near the harbor, as Devonshire, Broad, Commercial, Sea or Front, are no more inconvenient or crowded than the avenues in New-York which run along the Hudson or East rivers; the latter, though straight, are often impassable for hours.

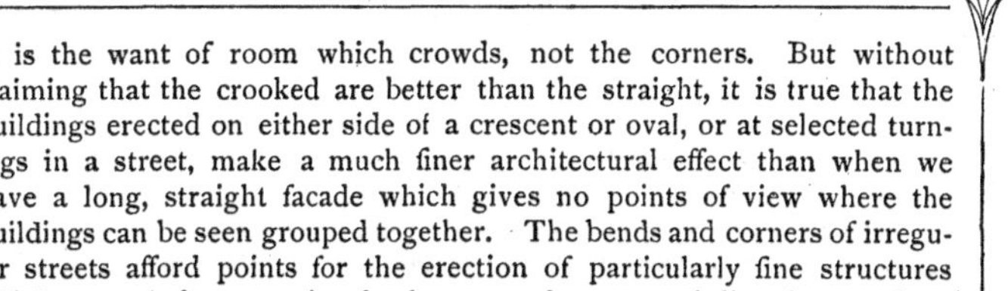

It is the want of room which crowds, not the corners. But without claiming that the crooked are better than the straight, it is true that the buildings erected on either side of a crescent or oval, or at selected turnings in a street, make a much finer architectural effect than when we have a long, straight facade which gives no points of view where the buildings can be seen grouped together. The bends and corners of irregular streets afford points for the erection of particularly fine structures which can only be appreciated when seen from several directions or from a distance. The straight street taxes the invention of the architect and the purse of the builder less, because the buildings erected need have but one front.

We know that a great part of the peculiar charm of European cities lies in the character, grouping and contrasts afforded by the medieval buildings of which they are largely composed. In this country the rawness of our towns and the sameness of much of the landscape, demand that we should make greater efforts to give cities and towns decided architectural beauty, irrespective of the mere questions of cost and convenience.

The necessities of business life compel one who would plan for laying out a city, to consider many things not wholly consistent with what might be most beautiful or picturesque. Gas and water pipes and sewers must traverse the streets, land will become very high priced, and increasing population require more fresh air and ventilation; but these conditions are not inconsistent with a graceful or even picturesque treatment of the surface, if planned for at the outset. It is true that cities which have grown to great size were not foreseen by their progenitors. Most, like Chicago, were the result of a kind of chance; the first settler was attracted by a good spring, the mouth of a stream which gave plenty of fish and access to some large body of water, a hummock covered with wood which promised fuel, and freedom from malaria, a mill-site or some such natural feature.

Whatever was an advantage to the first settler is equally important to others; population and business collect about the favored point, and before the inhabitants themselves realize the fact, they have passed through the stages of a village and town, and find themselves living in a compact city. It would have seemed preposterous to the first settlers to have thought of laying out a city with curved avenues, crescents, squares and parks, and to have provided for the possible elegancies desirable in the future. Many speculators might have been ready to make roads on paper and leave corner lots, and try to sell to new comers eligible sites, but even they would prosecute their plans with some hesitation.

It is hardly to be expected that the few practical men who guide the destinies of a new town should be willing to provide for a city's future greatness; they are too much occupied with private interests. New roads are made wherever accident or convenience chooses to lay them, without any regard to final effect. But while all this is true, and was reasonable in the infancy of our country, we ought to know and do better now; new towns may grow to be a Chicago or St. Louis, and each one should have its maturity foreseen and provided for in its infancy. The old towns and cities which have grown, will increase in the future and ought to look to it that their increasing suburbs are not only convenient, but as beautiful as thoughtful foresight can make them.

Thus far I have reasoned as if cities alone were interested, and as if only masses of people should provide for their future. But if my argument is good for the future or present city, it applies more strongly to towns and villages. Nothing is more refreshing to a traveler, than to come by stage or rail into a quiet country town, where the newness of the houses has been toned by the hand of time; where vines have learned to climb the walls and swing from the rafters; where stately trees spread loving shelter

Fig. 61.—*Street Planted with Trees, with Shrubs in the Door-yards.*

over the streets and houses; the village gardens seeming a part of the country about; the main street, by a pleasant turn, taking us out of the hum of population to the banks of a river or lake, or between farms rich with grain fields and orchards, with farm houses that nestle among trees and shrubs; where a few flowers about the doors show that the inhabitants know the value of beauty as well as money.

Fig. 62.—*Streets without Trees.*

Instead of these pleasing combinations, this blending of town and country, we enter every town between stiff houses, without a vine to decorate or a tree to shade, through the suburbs given up to squalid inhabitants, redolent with bad smells, the pathway disputed by rampant pigs or predacious cows.

The positive difference to the stranger and to the inhabitant, between two such hamlets is apparent, and it ought to be the duty of every village and town corporation to arrange their roads and precincts so as to provide not only for the necessities, but also for the beauties.

There must be some part of every town devoted to the squalor of poverty and careless indifference, as there must be sewers under the streets—why not lay out the town at first recognizing this fact, and instead of thrusting misery and disorder in the face of every one who enters or leaves the place, provide such approaches and exits, subject to civic restrictions in the matter of kind and quality of buildings, as will insure an agreeable feeling on the mind of the coming or departing traveler?

This is no fancy idea; a village, town or city is as readily laid out and improved as a private country place; there is no more need—if the people will act together as a body corporate—of having the agreeable character of a town injured by the way it is presented to the eye, than of entering a fine house through the kitchen, or having the avenue of approach traverse the kitchen garden and barnyard instead of the lawn. It is a more difficult problem to solve satisfactorily than the treatment of a private estate or a public park, but it is also more important and vitally affects a greater number of interests.

This question has been treated rather in relation to the level lands of the new and great West, than to the long settled eastern and middle States. Few would deny the truth of my theory and argument if applied only to the rolling and hilly countries of the East, and many would add facts to mine to show how absurdly many long settled towns are sacrificed to their highways, for the roads are carried up and over the steepest hills, as if some imperative necessity compelled every one to weary out their muscles in daily hard climbing at any rate, and therefore the exercise might as well be got in the village street as in mountain rambles. There are many arguments which I need not repeat in favor of a hill or elevated place as a site for a dwelling-house, which if good for one person, is for all who can find room to build thereon. If one man is benefitted by having his house where it may get the widest views and the purest air, all who can, should imitate his example.

Business necessities, moving heavy merchandise, utilizing water-powers and vicinity to railroads, will always select the level or lower part of towns as the best places for shops and factories. In such places, to accommodate traffic, the roads should be broad, and often will be most convenient when straight, but the high or rolling part of the town should be given up to dwellings. In the old feudal days of Europe this was recognized, and the lord of the town always selected the hill for his tower or castle. It is true that he gained advantages for offence and defence by his elevation, but I do not doubt that, safety apart, the hill would have always been taken as the best site for the palace.

Few towns are destitute of both kinds of surface, but generally the practical habits of our people, their desire to be near their business, church, school and turnpike, induce men to build in the lower and level parts of the town before the uplands are considered serviceable.

Fig. 63.—*Town with Stores and Warehouses on the Level—Dwellings on the Hill.*

The fatigue of climbing a hill at the end of a day's work, and the greater cost of making and keeping roads on hills, are arguments for lowland houses; but every consideration of beauty, picturesqueness and effect is in favor of living on the hills and transacting our business in the valleys. In the old days, when population was too scanty to make good country roads, turnpike corporations were established which, believing only in the shortest road, selected the straightest line between two points, and made their roads over hill and through valley, indifferent to effect or fatigue. Turnpike roads, which are steep and intensely hot in summer days, leading the tired traveler toilsomely up miles of hills, have created in the minds of many persons a disrelish for houses on hills. These old turnpikes have done great injury to the public taste and appreciation by their violation of both convenience and economy, and too often their straight road bisects the most beautiful hill in a country village.

They substitute for smooth, grassy slopes or groves of trees, a long, hard line of road, with fences on both sides, the gravel glaring in the hot sun, the fences like a line of grade stakes, compelling the eye to measure the height of the hill and the steepness of the ascent. We see villages spring up in the most uninteresting situations and gradually expand, while the most inviting and romantic spots will be neglected. It should, therefore, be the duty of the town authorities who want their town to thrive and be selected by strangers for homes, to have their lands carefully examined, their valleys and hills surveyed, their environs studied, and then lay out the future roads which shall define how and where the town is to grow, so as to make the most, in every sense, of the surface. It may be years before such roads are made, but if they are adopted and defined, future improvements of private owners will be in relation to them. This plan requires no interference with vested rights or private interests; no one will be obliged to divide his land into ineligible lots, or to sell against his will, but on the contrary, it will give private property increased value and advantages. The farm of to-day which is to be the suburb of another generation will confine its fences, hedgerows, orchards, ornamental plantations and buildings to the proposed roads, avenue or park, and when the public road is needed there will be people ready to live upon it.

Summon before the imagination some hilly town, the suburb of a city, or so far inland that city fluctuations and interests are unfelt or unknown ; see its hills, some parts farm land, some wooded, some precipitous and almost inaccessible ; such roads as there are, climbing up the steepest slopes, and to be avoided by every one but boys with their sleds in winter. Imagine now the same town in its primitive condition, and then carry a road up the valleys, gently winding round the hills to their very summits, of so easy a grade that neither man or horse would experience unusual fatigue in their ascent. Houses would be built by the side whose windows would look over valleys and meadows, or towards the distant lakes and mountains. In time the land opened by these highways would be sought for the home of those who recognize the advantage of surrounding their houses with all that is beautiful, would gladly build on however small an area, if by the fact of their position they can own the landscape beyond, their lawns will end, not in fence or hedge-row, but gradually blend in the outlines of the valley or distant mountains. This is no dream or impossible theory, but may be realized in every hilly town that now exists, or shall be hereafter built. Wherever the surface is varied, men can improve the opportunities if they will be awake in time to their best interests, and level towns may, by wise planning and judicious planting, create a local beauty for the future which will make them individual in their characteristics and varied in their home attractions.

The authorities representing the people should watch their present as well as future interests, and should lay claim in a general way to such outside lands as must in time become thickly settled, and provide by plans for such drainage, roads and public grounds, such as the wants and increasing tastes of the future population will demand.

FARM BUILDINGS.

SINCE THE INVENTION of horse machinery for elevating hay, and carrying it horizontally through barns to any required distance, the plans of these buildings are likely to be materially modified. There will be less necessity for the three-story barns—into which the crops from the fields were drawn to a high elevation and pitched downwards with great ease—although the filling of granaries from above is quite a convenience. The horse-fork will lift its load to any ordinary required height, from the floor next above the basement. And by the use of Hinman's or Hick's horizontal carriers the load may be run off to any desired distance

Fig. 64.—*Perspective View.*

and dropped at a precise point. Some farmers now carry their hay sixty or seventy feet and fill long lofts, from which it is readily pitched in winter to cattle in the sheds below. Variations in the forms of building will be made by farmers to suit their different conveniences and circumstances. Hence the great varitey of useful plans which may be presented, meeting these various wants; and from time to time we intend to furnish plans as they may be found on the grounds of the best managers.

James Slocum's Barn.

Among the best barns lately erected, intended to supply the wants of general farming, is that of James Slocum of Fayette Co., Pa., (fig. 64.) The plans which we are enabled to give, and which were drawn and engraved directly from the builders' drawings, are the more valuable, as they will enable any one desirous of erecting a similar one, to ascertain readily what all the materials will be, and to construct it without difficulty. This barn furnishes perfect facilities for pitching hay by horse-power into the mows from the second floor, and all its parts are readily accessible. The following is his description:

My farm of three hundred and forty-three acres is devoted mainly to the breeding of Spanish Merino sheep. I have on the farm two other barns, 30 by 60 feet, occupied entirely by sheep.

The barn which I now refer to is 45 by 76 feet. The basement is 10 feet high in the clear. The walls go below the frost, and a narrow trench filled with broken stone, is still below that, for the purpose of drainage. Walls 2½ feet thick. Four feet of the north wall is built against the bank, but the grade is lost in the first fifteen feet. The end center doors open out upon level ground. The basement is aired by sixteen windows,

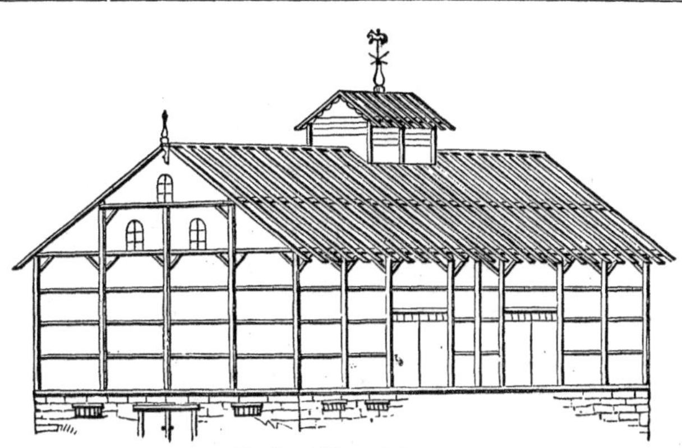

Fig. 65.—*View of Frame.*

six on a side and two at each end, revolving on pivots or axles, and can be held open by means of straps, at any desired angle, the base of the windows projecting outwards ; also in warm weather by the use of slat doors.

The frame (fig. 65) is 20 feet high ; all the posts supporting the purline plates, are 28 feet, (except those of the arch bents ;) by using 28 feet posts for front of mows, no cross-tie is needed between the posts until the purline plate is reached, obviating all obstruction to the use of the horse-fork

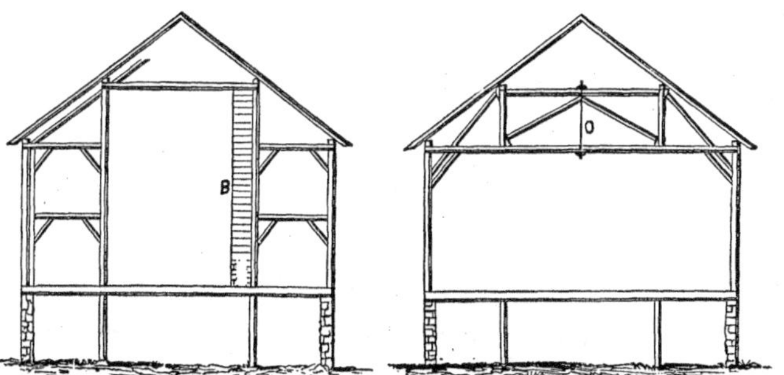

Fig. 66.—*Cross Section between Floor and Bay — B. Ventilator and Straw Shoot.*

Fig. 67.—*Cross Section of Center of Floor—O. 1½ inch Iron Rod, with Nuts and Washers.*

until the hay in the mow is 28 feet high, and allows the load to be backed into the mow. I can drive all over my barn—back into either mow with a load of hay, and fill the mow in thirds or sections—requiring but one man in the mow to take away the hay. By using the full length (28 feet) at the ends, it prevents them from bulging out as they frequently do, where the posts reach only to the cross-tie. The mow and barn floors are laid with planed and grooved pine.

The barn is double-boarded and battened. I used for the first covering 16 feet boards, (one board would cut into the four required short pieces,) 12 inches wide, planed by machinery to one thickness. For the outside covering I had boards sawed 20 feet long, 12 inches wide, breaking the joints of the first covering. The battens were cut from 3 by 3 pine scantling cut across the square, making a three-cornered batten with 4 inch flat surface to cover the outside joints.

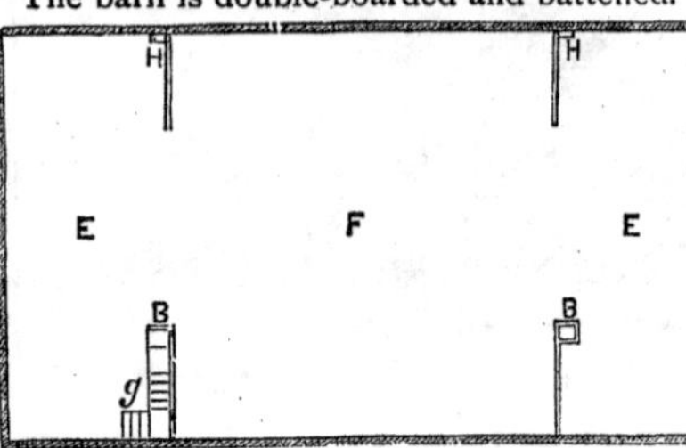

Fig. 68.—*Floor* 36 *by* 45 *feet*—E. E. *Hay Bays* 20 *by* 45 *feet*—B. B. *Ventilators and Hay Shoots* 3 *by* 3½ *feet*—*g. Stairs*—H. H. *Ventilators from Stables* 12 *by* 14 *inches.*

The roof has 18 feet pitch, projecting beyond the building 2 feet at right angles. The cupola is 7 by 12 feet.

In fig. 67 I had a stirrup of two by half inch iron around the post and bolted to the beam, and also around the beam and bolted to the purline posts, which the drawings do not show.

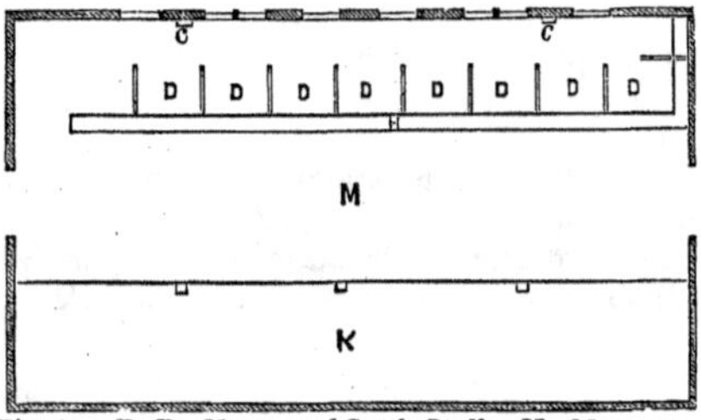

Fig. 69.—D. D. *Horse and Cattle Stalls*—H. *Mangers*—C. C. *Ventilators*—K. *Sheep Avenue*—M. *Sheep Avenue and Cartway when needed—The only divisions in* K. *and* M. *are Sheep Racks.*

The bridge is entirely separate from the barn; the wall of the bridge is three feet from the barn wall.

Plan of a Cow-House.

A correspondent (D. S. B.) has furnished us a sketch of a good plan for a cow-house, adapted to dairy farming, which we have condensed into the accompanying drawings, shown in the annexed cuts. It is built on a slope of ground sufficient to give a basement seven feet high—we think eight feet or more would be better, as contributing to a more perfect ventilation and a purer air. The barn is 30 by 60 feet, with a shed running the whole length on one side. The following is our correspondent's description:

The barn to be built on a slope of ground sufficient to give a basement 7 feet high—the barn to be 30 by 60 feet, with a shed 10 feet wide running the whole length of one side. This may be left off at the desire or fancy of the builder. One objection may be raised to the barn—the width of the bays, 24 feet; but horse-forks are within the reach of all, and the larger a bay the more it will hold in proportion to its size. If the barn is built with 13 feet posts,* each of these bays will hold 18 tons of hay, and the whole barn, including the rooms for straw, 50 tons at least. The floor is too narrow for grain purposes only, but for a grass farm 12 feet is wide enough. There should be three doors in each end—one in the gable for ventilating the hay when first put in; next the floor and straw rooms the bays should be boarded up 4 feet. The basement underneath the straw rooms is intended for the manure from the cows, thus keeping it protected from the weather.

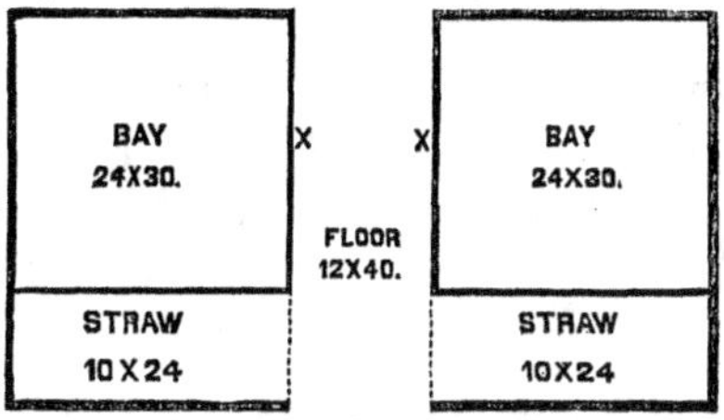

Fig. 70.—X. X. *Doors in the Floor Girt for Putting Down Fodder to Feed Room Directly Below.*

Next comes the cow-stable. Behind the cows there should be a bin made 4 feet wide, to hold muck, leaves, &c., for bedding; this should be arranged so that the partition boards can be taken out as the muck is used, the cows to be kept in stanchions; 3 feet to a cow is my rule; some use only 2½, but 3 is none too much. The stable will hold twenty head. I prefer a dirt floor for cows, having wood for the manure and hind feet only. In front is a feed room 4 feet wide; next a root cellar, an indispensable adjunct to a cow-stable; this should have double walls to prevent freezing. Over the cistern a room will be left 12 by 24 feet; this can be used for a calf-shed and hospital, or sheep-shed, or place for storing wagons in winter and sleighs in summer, &c.

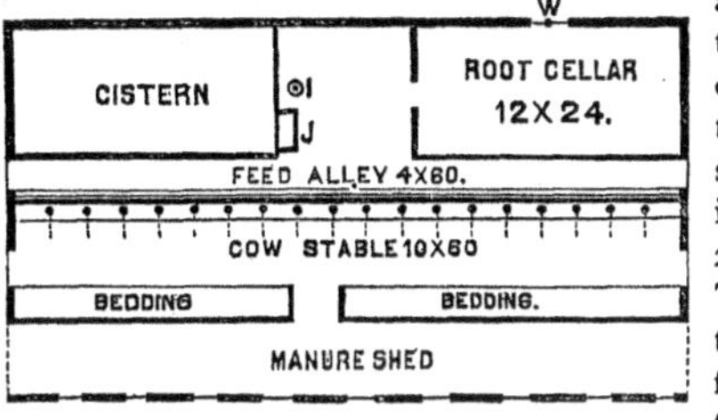

Fig. 71.—W. *Double Window for Filling Root Cellar*—I. *Pump*—J. *Root Washer.*

Of the cost of such a barn I say nothing, because no two sections have the same prices for lumber, &c.

The manure shed can be left open, being supported by posts only, or boarded up as shown in plan. Perhaps the first would be the better plan on some accounts.

Corn Crib.

The accompanying elevation and plan show the design of a neat, compact and convenient corn crib, built by a Long-Island correspondent, and

* If taller, the barn would be more spacious, hold more hay, and cost but little more—the roof being the same.

which after several years trial, he finds to answer the intended purpose to his satisfaction. It is 30 feet long, 4 feet wide at sills and 5 at plates; the middle cross projection is 6 feet long, at which the crib is easily filled

Fig. 72—*Elevation.*

or emptied. The main posts, 4 by 6 inches, are 8 feet from top of sills to top of plates—those of the projection are 6 feet 2 inches. The sills being more exposed, should be of durable wood, and are 4 by 6 inches. The whole stands on seventeen locust posts, set 2½ feet into the ground, and resting on flat stones, 20 inches above ground, capped with inverted milk-pans. These pans, by the way, should have no wire rims, as the rascally rats will sometimes jump, cling to these rims, and throw themselves into the crib. The sides are pine strips 2 inches wide, set just far enough apart to hold the smallest ears. Movable boards keep the corn from the central floor, which is used for shelling and storing the cobs.

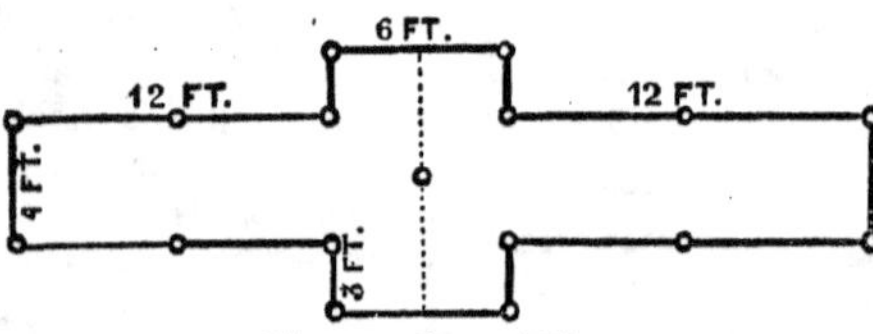

Fig. 73.—*Ground Plan.*

A Good Corn House.

One of the best and most convenient buildings for this purpose, is the one on the farm of George Geddes, near Syracuse, N. Y. It is sixteen feet wide and sixty feet long, and has a capacity for more than four thousand bushels of ears. Fig. 74 is a cross section of the building. A wagon may be driven in, at one end, through the whole length, and out of it at the other. The corn is deposited on each side—the space being five feet wide at top and three at bottom. A strong upright timber is placed for this purpose at every ten feet, leaning toward the wagon way, as the figure indicates. Smaller timbers

Fig. 74.—*Section of Corn-House.*

are placed between. The horizontal slats, for holding the corn, are successively tacked on the inside of these upright timbers, as the spaces are filled from the bottom upwards. Short nails being used for this purpose, the slats are easily knocked off as the corn is discharged. When the spaces become nearly full, the corn is thrown upward with a scoop shovel, over the top of the timbers. The upright timbers merely rest on the floor, to which they are secured by a nail, and may be readily taken out at any time. They are strongly ironed to the timbers above, and iron braces connect these timbers to the posts. No advantage having been found to result from inclined upright strips on the weather side, the posts are made perpendicular. Poles extend the whole length of the roof near the peak, for suspending trussed seed corn. For the purpose of excluding rats, the floor is made of tight two-inch plank, and the underpinning being three feet high, except at the doors, they cannot gain admittance. Iron grates with spaces too small for them to enter, are placed in the underpinning to admit air, and to prevent the decay of the timber. After the house is filled with corn in autumn, both doors are thrown open, and the wind sweeps through freely, causing a rapid drying of the ears—during which time movable gates are placed in the doorways to shut out intruders.

Poultry Houses—Design I.

A correspondent of the Country Gentleman in Maryland, furnishes us the outline of the plan here described, from which we have made the above elevation. It was erected a year ago, at a cost of $62.75, not including the frame, which was supplied from the woods. It gives entire satisfaction.

Length 26 feet—width 10 feet. Height in front 10 feet; in rear, 6 feet. Weather-boarded up and down, with strips 4 inches wide of half inch stuff over joints; underpinning of brick, 10 inches high. The entire floor was dug out one foot deep, then a layer of oyster shells 4 inches thick, put in; then gravel 2 inches, and clear sand 2 inches, and over all gas tar 2 inches, making, in a short time, a hard, solid floor.

It is divided into three compartments by two wooden partitions, beginning

10 inches from the two cross sills, (which are also underpinned,) and running up 3 feet; from there to the roof the partitions are of common plastering laths. The nest boxes are on a wide board that rests on the division sill, and when a hen shows a disposition to set the nest is turned around, and she finds herself within the sitting room, where she is secure from all intrusion. There is a door into each division at the the back of the house. It is kept scrupulously clean, and white-washed frequently.

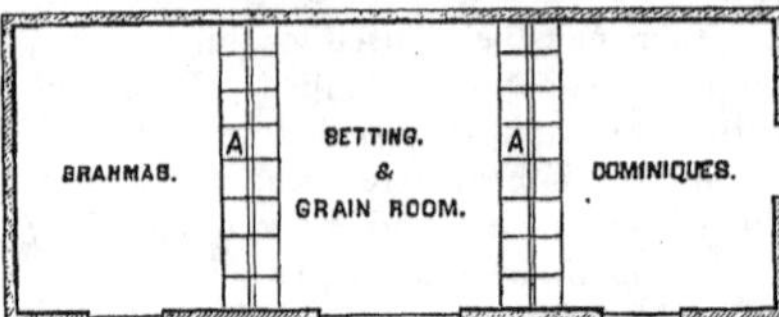

Fig. 76.—*Ground Plan*—A. A. *Nests.*

The roosts are four in number, running across the house, and fit in little grooves made of wood nailed on the side walls and partitions, and are taken down daily, and once a month are coated with gas tar. The yard is in front, the whole width of the house, and 25 feet long, surrounded with a fence 8 feet high, of palings. In a colder climate than ours the house could be lined inside if desirable.

DESIGN II.

The above engraving represents a more substantial and expensive poultry-house, built of stone, although brick or wood may be employed, adopting the same general arrangement. The ground plans were furnished by J. H. DICKERMAN of New-Haven Co., Conn. If built of wood, the posts of the frame may be about 9 feet, the sills resting on walls of concrete about 3 feet above ground, where a slight inclination can be selected. Seven and a half feet will do for the feeding room and manure pit—formed by running a concrete wall three feet high, as shown at the dotted lines in fig. 78. The rear door, (not seen in the view,) admits dry muck or loam, as needed. The windows on the south or front side give light and warmth.

The second floor may be lathed up the roof, to give sufficient height in the centre, which will be 4½ feet under the eaves of the roof. The nests are set in the partition, one foot from the floor, one foot high, and one and

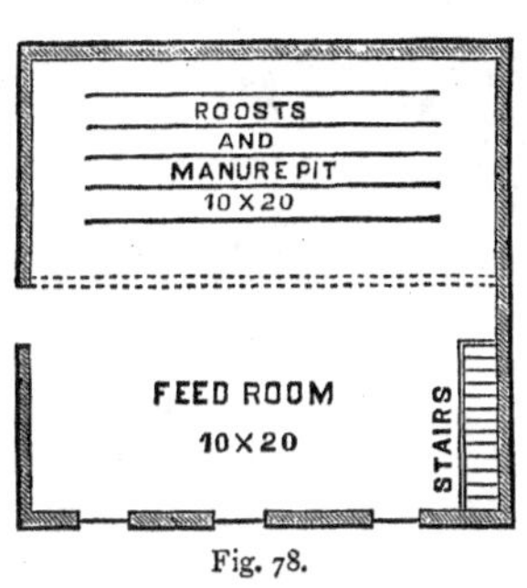

Fig. 78.

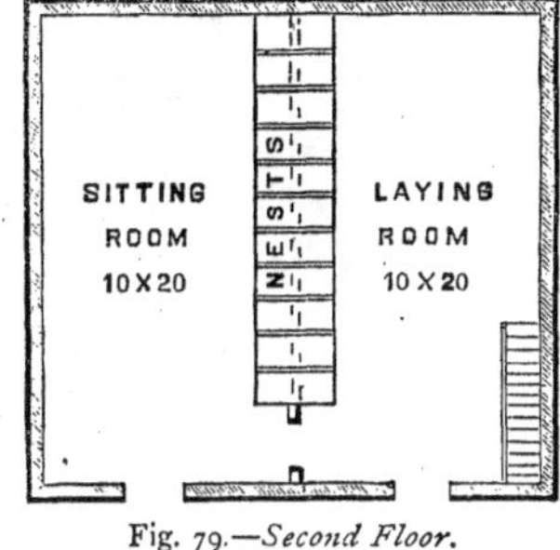

Fig. 79.—*Second Floor.*

a half feet long, open at both ends with a slide door, which is reversed when a hen is sitting, so that she is placed in the opposite or sitting-room, and thus the others never disturb her. A door to communicate between the rooms and windows in the end and south side will give sufficient light and warmth. The whole should be lathed and plastered, or may be constructed of concrete walls, which will make a warmer and more durable building.

This plan will give for 80 feet of outside wall 800 feet floor, and while the lower floor is much the best for roosting and feeding, the hens quite readily ascend to lay and sit, and with equally good success. I find this room sufficient to winter 100 fowls by having liberty to roam at large in pleasant weather, and confined in a yard during the season when they are destructive to growing crops. The cost will vary according to prices of material. One hundred bushels of shell lime would be sufficient to concrete the walls, with small stone and gravel.

DESIGN III.

This design is furnished by J. C. THOMPSON of Staten Island, N. Y., and is the result of thorough experience, and the description is so replete with valuable hints, as well as minute instructions for the erection and furnishing of the building, as amply to compensate for its length.

The foundation is 8 by 20 feet; durable posts set well into the ground, projecting 6 inches above the level of the ground, on which the foundation timbers are laid. The rest of the frame is made of joists or wall strips; planks are sunk a few inches into the ground on the inside and outside of the foundation timbers, and held there by pegs. In summer all may be removed if needed to admit air, or only the inside or outside planks

Fig. 80.—*Thompson's Poultry House.*

removed, which prevents rats from harboring about the house, as they will do in stone foundations.

The building is 8 feet wide by 20 feet long, 6 feet high in front and 6½ in the rear, enclosed with tongued and grooved pine boards; roof, hemlock boards, full length, running out and forming a shed over the front 4 feet wide, the whole covered with tin. The inside is ceiled with boards over head and around the ends and sides, being filled in with sawdust or shavings as the ceiling boards were being put on, except over head, which need not be stuffed.

A sixteen inch square blind ventilator is set in the roof, at the middle and at the highest part of the roof. At the bottom of the ventilator, on the inside of the house, is a shutter hung on hinges; this can be pulled up and closed, or let down by a cord or wire, as required, according to the state of the weather.

In front of the house, on the outside there are two sliding sashes, (set on sash rollers,) each 4½ feet wide and 4 feet high, set in sliding strips at top and bottom. In warm weather they are slid out to and cover the dead wood at the ends of the house. These sashes are made so that the panes slide in from the top, in grooves made in the bars, and meet end to end, which saves the time and labor of setting with putty, so that when a glass is broken it costs no time or labor to slide the glass down to close the broken space by slipping in another at the top; being so easily repaired, it is sure to be done. Sashes made on this plan are 1¼ inches thick at the bottom and sides, and three-fourths at the top, tennoning the bars, (planed on each side, to admit of sliding the glass in.) The bars are tennoned into the bottom rail, and cut out at the top, so as to let the tongue lap well on to the top rail, which is made 4 inches wide, to give strength to the bars, they being nailed to the top of the frame. The bars are set to receive 6 by 8, 7 by 9, or 8 by 10 panes, according to fancy. Small panes are the cheapest, and as more bars are used, there is less danger of breakage, and less cost to repair and keep in order.

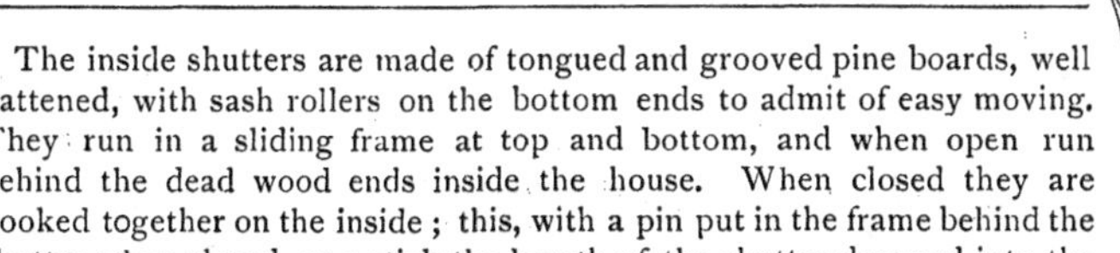

The inside shutters are made of tongued and grooved pine boards, well battened, with sash rollers on the bottom ends to admit of easy moving. They run in a sliding frame at top and bottom, and when open run behind the dead wood ends inside the house. When closed they are hooked together on the inside; this, with a pin put in the frame behind the shutter when closed, or a stick the length of the shutter dropped into the groove behind it when closed, makes a secure fastening.

As the house runs east and west the long way, the nine-foot window faces the sun as it rises, shining directly on the roosts, giving the hens the full benefit of the sun in winter when most required. The roosts are placed in the west end, but not secured to the house. They may be laid on a frame hung on wires from the roof, or secured on four or six posts set in the ground —say three in the rear and three in the front; the front ones 12 to 18 inches high, the rear ones a foot higher; a strip on each from front to rear, for bearers to receive the roosts, which run lengthwise. The roosts used are flat 3 inch fence strips, 12 feet long, laid about a foot apart; a nail tacked in the bearer on the lower side of each roost keeps them from sliding down. The roosts being a little on the incline, the highest are filled first at roosting time, and so on till all are filled. The hens thus roost compactly and keep each other warm in winter; in fact, they form a solid body of fowls—a space 5½ by 12 feet seating a hundred head comfortably. Flat strips are best for roosts, because the hens' toes cannot pass below their feathers, which they will do on small round roosts; and in extremely cold weather, in cold houses, their toes are very apt to be frozen.

In summer the roosts must be spread out level and as wide apart as possible, so as to keep them cool; as the stock is generally reduced as warm weather approaches, my house accommodates one hundred head. Buildings may be made larger or smaller, according to the number desired to be kept.

For nests I use butter or lard tubs, or medium-sized deep cheese boxes, set on the ground around the easterly end of the house; another tier on a shelf over the first.

There are no permanent nests, shelves or roosts; everything is portable, so that each and all can be removed and cleaned and *greased* at pleasure. The house, being new and clean, will be first well white-washed inside in the spring; then oiled with *lamp oil*, (not paint oil, as that dries,) put on with a brush; any kind of pot skimmings, *melted* and put on warm, will do as well as oil. Hen lice cannot live in grease for a moment. Nests can be treated in the same way. Tubs are better than boxes, the corners being nice harbors for vermin, out of which it is hard to drive them. Portable nests are better than permanent ones. When you desire to set a hen, take a clean tub nicely strawed, (do not use hay,) throw in the bottom some flour of sulphur, dry earth or wood ashes, put in the eggs, and set the tub in the place occupied by the setting hen at night; she can be removed to the sitting room next morning by gently placing a hat over her and carefully moving her.

The sitting-room is a place provided for all sitting hens, where the tubs are placed in line along the sides of the room in the order they are taken in—no matter if 10 or 20 tubs are in the room. The hens are not permitted to leave the room until the brood is hatched. Every morning all are carefully removed from their nests, and ample food and drink given them. They eat and quickly return to their nests. At hatching time the chicks are given to the lightest hen, or the one that is the palest about the gills or head, being careful to *remove the empty nest*, for there must never be any more nests than hens in the room, or some stupid hen will take a nest without eggs. If the broods come out small in number, they may at night be put by one hen.

There is a decided advantage in removing each hen every day, as any injury to the eggs can be discovered, the nests made clean, and the chicks cared for as they begin to peep out. Any observer will see at a glance the advantage of a sitting-room over the ordinary way of hatching chicks.

But I must return to the house. Now, it will be seen that with a nine foot window in front, a door in the east end, (which is closed in winter, ceiled and filled in,) a door in the west end, and a ventilator in the roof, the house is about as nearly *turned out of doors* as it well can be for a summer house.

At the west end of the building is a shed 8 feet by 10, tight boarded on the north side; at the west end and in front hot-bed sash are set on their ends, forming an enclosure into which the west door of the hen-house opens. This shed is used as a feeding and drinking place in winter, and in extremely cold and stormy weather the holes in front of the house and the door to the shed are closed; in fine weather all are opened, allowing the largest liberty.

Breeding Coop.

The accompanping figure represents a breeding coop, which will be found convenient under some circumstances. The advantage which it possesses, is that it may be moved from place to place with little trouble, furnishing fresh earth and grass for bottom—giving something of the advantage of a greater range. It may likewise be used for sitting hens. It is 12 feet long, 4 feet high and 4 wide. F. is the nest box, entered inside; H. door for withdrawing eggs; A. windows, each a pane of glass, fastened inside by carpet hooks; E. E. posts 4 feet high, 2 by 4 inches, to nail boards and slats; B. door. The rear, not shown in the cut, is made of inch boards, with door, and with hen-hole. This plan was furnished by a Lockport correspondent.

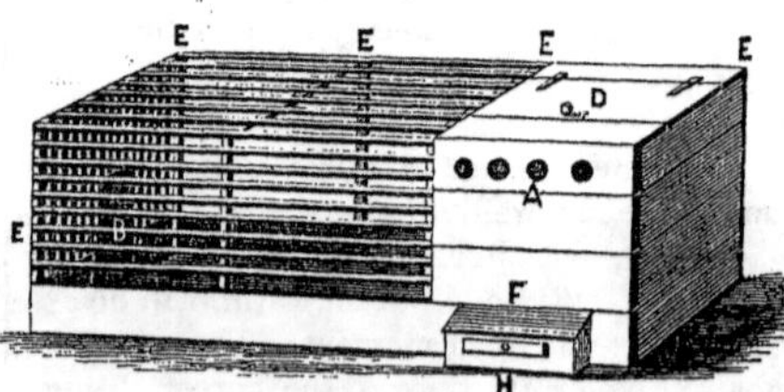

Fig. 81.—*Breeding Coop for Fowls.*

VARIOUS CONTRIVANCES IN RURAL ECONOMY.

CATTLE FASTENINGS.

ON THE GROUNDS of HERENDEEN & JONES of Geneva, we observed a mode of fastening cattle in their stalls, that appeared to possess advantages peculiar to itself, and which circumstances may render just the thing for some farmers. It secures the animal's head in position, without the close confinement incident to stanchions.

The stalls are 3 feet wide from centre to centre; and are separated from each other by vertical partitions at the mangers only. These partitions are nearly 3 feet high, and about 2 feet wide. Fig. 82 represents an end view of one of the stalls as seen from the place where the animal stands—*a. a.* being the partitions made by nailing plank on vertical scantling posts. Upright iron rods, *b. b.*, (about 1 inch in diameter,) bolted on the partitions at the upper ends, and running into the sill at the lower, allow the chain rings to slide freely up and down with the raising or lowering of the heads. The fastening at *c.* is simply a cross-bar and ring, and the two rings admit lengthening or shortening the loop, according to the size of the animal.

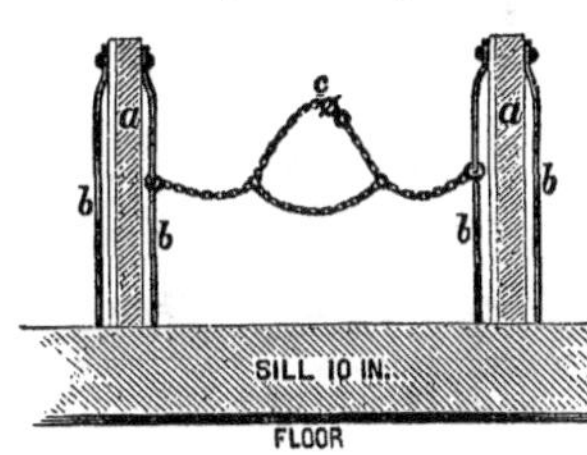

Fig. 82 —*Cross Section of Stall.*

Fig. 83 is a side view of the stable and stall, of which the following are the various dimensions: Whole width of stable about 14 feet inside; width of feeding passage, 4 feet; width of manger, from front to rear, 2 feet; width of floor, from manger to gutter, 4½ feet; width of manure gutter, 1½ feet.

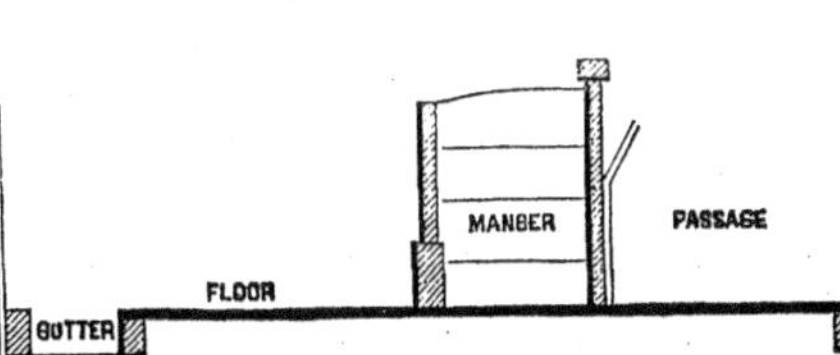

Fig. 83.—*Cross Section of Stable.*

It is not necessary to have any partitions between the stalls, except those which separate the mangers, as the chain loops hold the cattles' necks, and prevent them from hooking each other. The result is quite similar to that of stanchions, with the exception that more liberty is allowed the animals.

WINANS' COW STABLES.

The cow stables of R. WINANS of Maryland, are so peculiar in their arrangement and construction, and possess some important advantages, that we deem them worthy of a description, the substance of which we obtain from the COUNTRY GENTLEMAN. Stables like these, with grated or sparred floors, are better adapted to the warmer winters of the Middle States than those of the extreme north, where the freezing of the manure would tend to clog the openings.

Fig. 84.—*Winan's Cow Stables.*

The stables are either 24 feet in width, with double pitch roof, and a row of stalls on each side a central passage way, or 12 feet in width with single pitch roof and stalls on one side. The stalls are not placed *across* the building, as usual, but in the direction of its length, as shown in the accompanying cut, (fig. 84.) Each cow has a stall to herself, and stands with her right side towards the passage, so as to be accessible for milking. A small wooden bar is suspended along the farther side of the stall; a rope at either end running to the passage way, so that the cow may be gently drawn close to the slats when required for milking or otherwise. The following are the dimensions:

Stalls 5 feet wide at floor; 10½ feet long.

Feeding trough 2½ feet wide; plank floor, for fore feet, 3 feet; iron grate floor, 5 feet—giving total length 10½ feet.

Grate bars of floor, 1 inch wide; ¾ inch deep; 2¼ inches apart each way.

Floor of stall about 2 feet above floor of building; the plank part 1½ inches lower than the grating, and covered with say 2 inches depth of earth.

Guards on the floor a little forward of the grating, to prevent the cow's turning, as she otherwise might.

Vertical rounds between stall and passage, 2 inches in diameter, 9 inches apart, removable one or more at a time—for milking, as shown by the

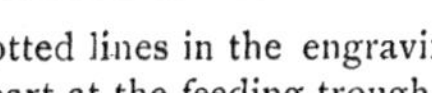

dotted lines in the engraving, and for feeding, as shown by those spread apart at the feeding trough.

The height of the cow above the floor of the passage, is such that the person milking *stands* instead of sitting, and is intended just to suit the milker's convenience in this posture.

The deposits under the stalls fall upon the floor of the stable, which has an outward slope of 6 inches in 12 feet, and are daily removed through a shutter on the outside, not seen in the cut.

A shutter in the weather boarding above, (shown by a cross bar on one of the boards in the cut,) opens for ventilation as may be necessary.

It should be added that these stables were designed after Mr. WINANS' experience had enforced the distinct conviction that cows kept for milk will give a larger and more evenly sustained product, by being stabled the year round, free from sudden changes in diet or mode of life. Hence the objection of inconvenient access that might otherwise arise, is greatly obviated. A short inclined plane is provided, movable on wheels, upon which the cow is led into the stall, and there she remains, under this system, for months, or for the whole year together. A board partition separates the several stalls. The cow is tied by a cord of sufficient length to allow great freedom of movement, and the slack avoided by attaching the end to a small weight, which has room for play between the floor of the building and that of the stall. Nearly the same comfort is provided for the animal as in a loose box, and there is no possibility of its getting its feet into the manger, or injuring itself in any way. The stables here described were built very simply and plainly, with no expenditure for display. Water being easily obtainable, was carried in pipes along the upper frame through which the rounds are inserted in front of the stall, and thence directly into the tub with which each is provided. This tub was alwaws filled, and was provided with plug at bottom, to draw off the contents if desired. The cross-barring in the grate is a preventive against slipping, and by the use of iron instead of wood, the surface of the bars to which manure might adhere, was lessened. The earth in front is gritty enough to aid in this respect, and at the same time the hoofs are kept from growing to too great length. A small blackboard was attached to the partition at the foot of each cow, on which every week or ten days her milk was registered, to ascertain whether the yield of each was well kept up—also any other memoranda of importance. About 200 cows were kept in these stables, and yielded an average of two and one-tenth gallons of milk per day *for the year;* they were almost invariably gaining in flesh meantime, and the air about them, even in very warm weather, was pure and sweet. The cows were nearly as clean as if constantly groomed. Wooden slats would doubtless answer for these stalls, but the iron being found preferable, and the owner having facilities for providing that material, was substituted to advantage. A small car was used for distributing feed up and down the passages.

BARN DOORS ON ROLLERS.

These doors possess some advantages over those hung on hinges, and are also attended with some objections. The advantages are, they are not thrown open or beaten about by the wind; they occupy less room in opening, and snow drifts interfere but slightly with their operation. Their disadvantages are, the greater difficulty of causing them to shut closely, and their liability not to work well, when applied to imperfectly built barns, or to those distorted by sagging or settling. These disadvantages, however, may be obviated by perfect workmanship.

The accompanying figure represents a door thus constructed—the door made smaller in proportion to the rest of the work, in order to show more distinctly the mode of hanging. It is necessary that the doors be placed outside, unless, as rarely happens, there are no timbers within to interfere with opening and shutting. A small, horizontal, wooden rail (*a.*) made of a plank set on edge, is placed over the door or opening, and on its upper edge is a small iron bar. On this the hangings (*b. b.*) roll. A flange on the inside of the wheels, and in the best made rollers on both sides, keeps them to their places. The bottom of the door is kept close to the side of the barn by small hooks of iron. (*c. c.*,) which press against a horizontal iron strip nailed to the lower edge of the outside of the door. Hooks should be placed at proper intervals to hold the door when open as well as when shut. Small cast-iron anti-friction rollers, if placed on these hooks, and also inside the doors against the side of the barn, will cause the doors to run easily and prevent chafing.

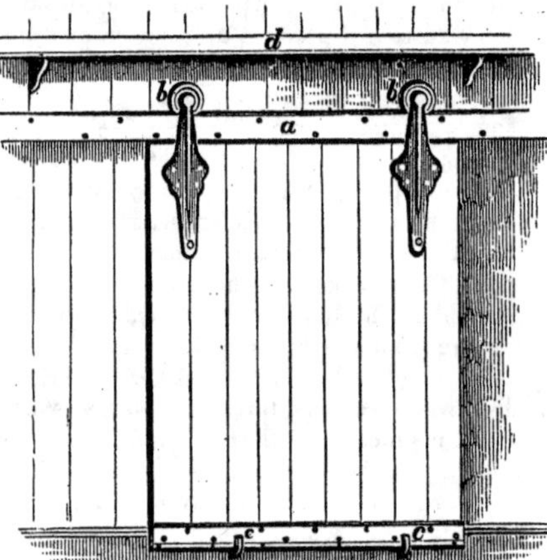

Fig. 85.—*Barn Door on Rollers.*

A vertical strip is to be nailed against the door post outside, and another one on the outer edge of the door inside, so that when it is shut, these two strips, coming together, form a close joint, and exclude the cold air. Rolling doors, here described, are especially adapted to inside partitions, where the accurate exclusion of cold air is not so esssential.

To protect the rollers and rail from rain and snow a horizontal cap (*d.*) a foot or more in width, should extend over the whole.

Door Hasps.

Every door which is often used should be furnished with a good self-fastening latch, but it sometimes happens that on out-buildings, which are less frequently visited, and which must be secured with a padlock at night, a common hasp is regarded as sufficient. Often the single strap is placed over the staple, and during the day-time is fastened to its place by a small wooden pin, and not unfrequently with a corn cob. Several motions must of course be made every time the door is opened, and when it is closed again. Sometimes the pin is lost, and then a search must be made for a stick or a broken limb of a tree to supply its place. To obviate this inconvenience a hasp has been contrived and much introduced into use, like that shown in fig. 86. A small hook is attached, moving on a rivet, so as to supply the place of the pin, and is thus always on hand. Still several motions are required in closing and opening. We have made a still further improvement, as shown in fig. 87, which we find a great convenience; and which may be fastened and unfastened almost as readily as the best latch. A projection is made on the lower side, as distinctly shown by the figure, which is dropped into the staple, and holds the door securely. Another staple is placed on the opposite side of the hinges, by which it is as readily fastened open. At night the loop is slipped on the staple and secured by padlock. This hasp will do for doors that are frequently used or passed many times a day.

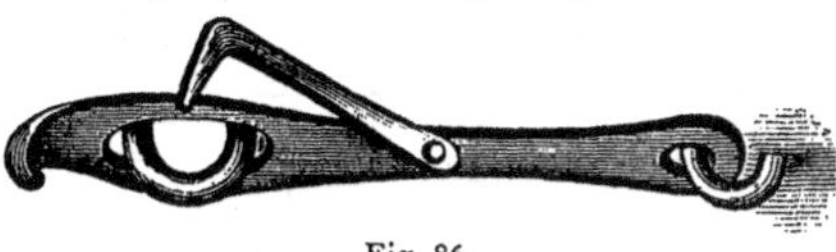

Fig. 86.

Fig. 87.

The one shown in the accompanying figures combines all the advantages of a self-fastening latch and hasp. During the day, when used, the hasp operates like a common straight latch—the loop slipping over the catch quite as readily, and always in position—acting as latch or hasp without trouble or extra motions. The padlock should hang suspended by a small chain, at a convenient distance from the staple, to be used at night—the hasp always adjusting itself by the simple act of shutting the door.

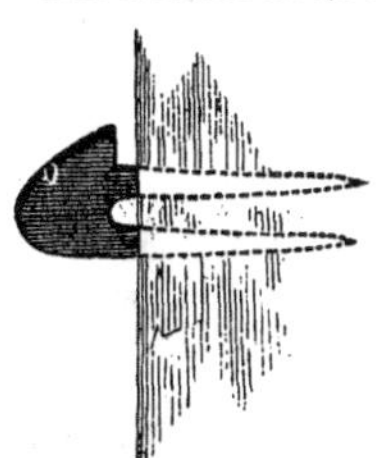

Fig. 88.—*Catch, Driven into the Post.*

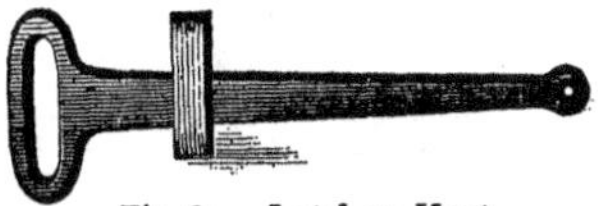

Fig. 89.—*Latch or Hasp.*

Barn-Door Fastenings.

The following is described by L. D. Snook, and is simple, convenient and easily made: A. and B. are portions of barn-doors. K. is a wooden bar 4 feet in length, 2½ in width and three-quarters of an inch thick, loosely

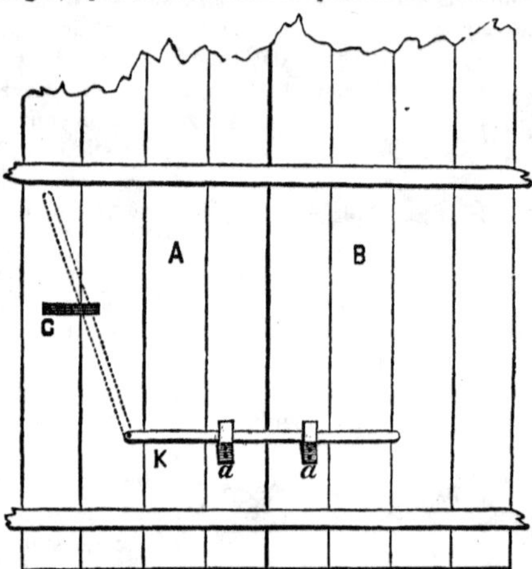

Fig. 90.

bolted at one end, as shown, (fig. 90;) *a. a.* are cleats nailed on each door with a portion sawn away for the resting thereon of the bar K. The dotted lines show the position of the bar, resting against the cleat C. when the doors are to be opened.

Another mode is the following (fig. 91) furnished by M. E. Merchant, of Chenango Co., N. Y. A. B. represent the two doors inside. D.D. two standards attached to the lever C. by tenons fitting sufficiently loose to allow the lever to work up and down. The lever is bolted to the cross-piece of the door midway between the standards D. D. By raising the lever the upper standard drops out of its fastening, and the lower rises from the same. The fastening for the other door is nothing but a block of wood six or eight inches long and one and a half thick, and one of the same on the opposite side of the door, fastened together by a pin at *a.*, and resting on a pin at *b.* The dotted lines show the position of the fastening when the door is open.

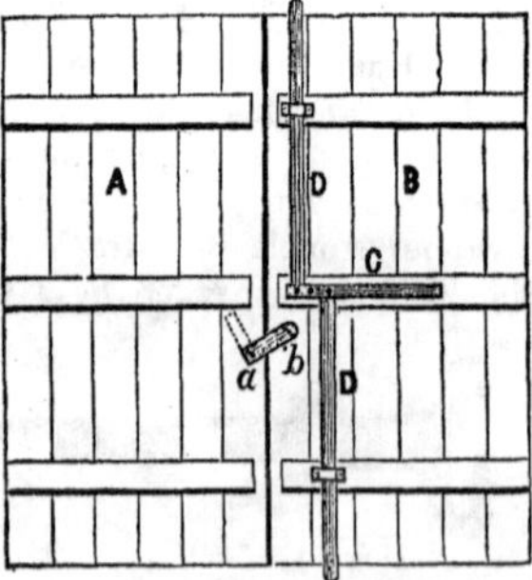

Fig. 91.

Sheep Racks.

C. T. H. of Knowlesville, N. Y., has sent us a description of a mode of constructing sheep-racks, which we illustrate in the accompanying figures. It is a good and cheap contrivance, and may be adapted to racks for other purposes, by adding vertical slats, and by varying the size: Take two boards 16 feet long and 10 inches wide for one side, and three pieces 2½

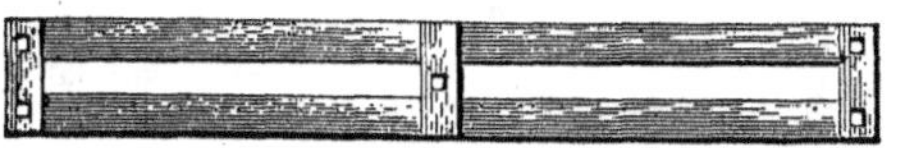

Fig. 92.—*Sheep-Rack—Side Section—16 feet long, Boards 10 inches wide—height 2½ feet.*

feet long and 6 inches wide for cleats; nail one piece on each end and one in the middle, with No. 8 wrought nails; that will form one side, (fig. 92.) Make ends the same, 3 feet long (fig. 93.) Take heavy hoop iron 1½ inches wide, make a staple (fig. 94) by bending it, to be nailed on the end of short section; cut the hoop iron 18 inches long, punch two holes in each end, the same distance, so that when it is bent the holes will be opposite; put a wrought nail through and clinch. Mortice two square holes through each end of side sections, (one in the middle for stay) for staple to go through. Make a half round key tapering, drive them tight, and you have as solid a rack as one could wish. This rack can be taken apart in one minute, and be stored in small compass in a stable or shed.

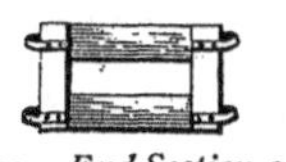

Fig. 93.—*End Section, 3 feet long; Staples project 3½ inches.*

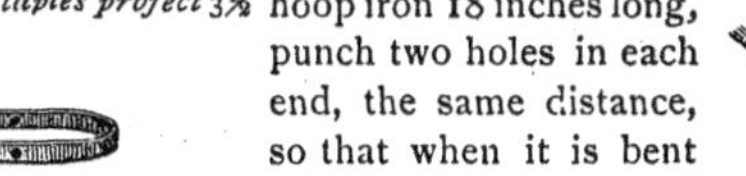

Fig. 94.—*Staple, 9½ inches long.*

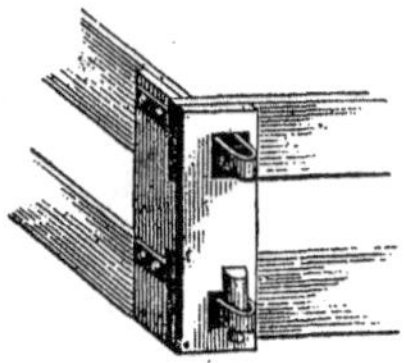

Fig. 95.—*Keying the Corners.*

Cattle Racks for Feeding.

The annexed figure represents a good stout box for feeding cattle. It is about 6 feet square, made of oak plank and corner pieces—*a.* is about 20 inches wide, and *b.* and *c.* about 8 or 10 inches wide—all inch and a quarter thick. Inch boards will do, if very securely nailed and clinched, but they will be more likely to be broken by the cattle. The space between *a.* and *c.* should be about 2 feet for fully grown cattle. If too large, the animals will crawl in and make a bed of the box. Smaller ones for calves should be placed in a separate yard. Four animals feed at once at one of these boxes.

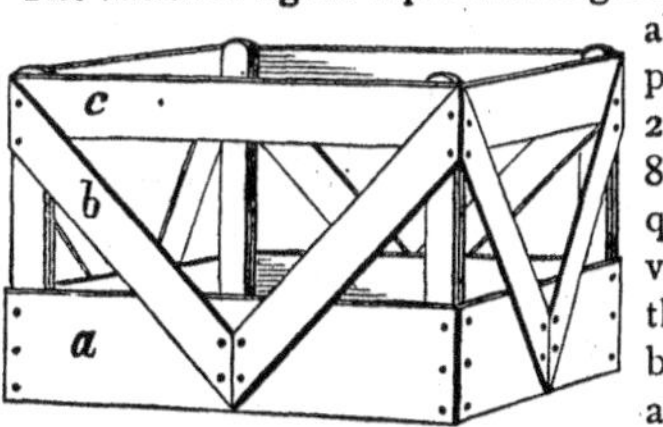

Fig. 96.—*Square Feeding Box.*

Warren Brown of New-Hampshire sends us the following contrivance, which he has used for ten years: The four posts which form the corners are 8 inches square, resting on sills at the bottom, and supporting a beam on which rests the roof. The root projects on each side and at the ends, to cover the animals while eating. The sides and ends of the rack are made from round sticks one foot in diameter, split in two, which are framed into the posts. A floor is laid around the rack for the animals to stand upon, and allow a passage around it. When the ground is dry the floor may be dispensed with, and the posts set in the ground. Some of the advantages of this rack are—1st. It is simple and durable, and almost any one who has the tools can build it. 2d. The food is kept clean and dry, the roof keeping the weather from it. It can be used for horses, cattle or sheep.

Fig. 97.

Another correspondent (H. R. W.) describes a rack intended to prevent animals from chasing each other around it, and it is well adapted to animals not disposed to be quiet and mind their own business. He says:

"For some years I have soiled five or six cows, and during that time have been at my wit's ends to contrive how to keep the strong from over-feeding and the weak from famishing. Small and separate boxes do pretty well—only they are easily overturned and troublesome to fill. The plan upon which I now depend has the experience of two years to recommend it, and seems effectually to baffle the queen of the yard, who now eats her proper allowance in close and peaceful proximity to her feebler associates. You may get the idea of it from the following sketch, adapted to the wants of four animals.

"It is nothing more than a box about 6 feet square, set upon timber or stone, the bottom laid with open joints for ventilation, and the sides boarded up 18 inches. Two partitions, crossing each other, divide it into four equal parts. These partitions are carried up above the sides high enough to blind the cows from each other and

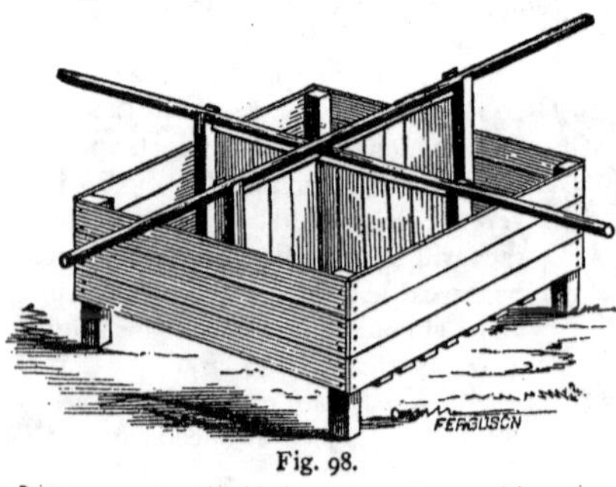

Fig. 98.

prevent their quarrelling. The only important parts of the arrangement are the projecting arms or cross-bars which extend beyond the box and between the cows to a distance of three feet on either side. Each cow now takes possession of a corner and feeds secure from unexpected horns and rear attacks. For should the queen cow, as at the first she undoubtedly will, fancy every other portion better than her own, and conclude to take forcible possession, she must first execute a deliberate and tedious retrograde movement, involving an entire change of base. Her intended victim sees all this, and walks at her leisure out of the open corner, round into the place just made vacant by her belligerent majesty. After a few games of such puss in the corner, it comes to be generally understood that it does not pay, and it is soon abandoned. Any other division of the space would not be so favorable, as not leaving so much room for the retreating cow. As exhibited, it is hardly possible for one animal to confine and worry another. This is all important to one who raises valuable stock. Since adopting this simple contrivance, I have been satisfied with the equal distribution of the good things in my cattle-yard, and the uniform condition and health of the animals.

Farm Gate

A correspondent has sent us a rough sketch and description of a cheap and simple farm gate, which we have re-drawn and put into shape. This gate appears to be well adapted to such openings as do not require to be passed frequently, as for example the entrances to the more remote fields of the farm.

Fig. 99 is a plan—the two posts A. and B. are to be placed in such

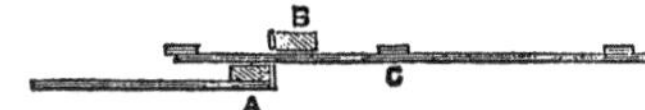

Fig. 99.

position towards each other that the gate C. will slide between them, and also turn 90 degrees, or at right angles to the line of the fence, the cross pieces on which must be nailed on the right side of the post A. and on the left side of B. This cross piece should be an inch and a half thick when the gate lumber is an inch thick. The end pieces of the gate should be on the side opposite the post A., so as to slide by A. and snug up to B. D. is the latch post of the fence.

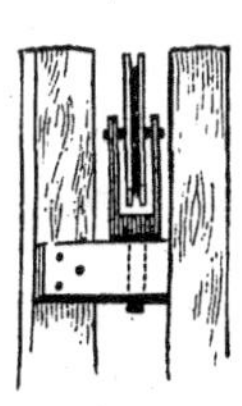

Fig. 100.

Fig. 101.

An improvement is to have a roller with flanges for the upper slot of the gate to run upon, as shown in section in fig. 100, and with a side view in fig. 101. The shank

Fig. 102.

under the roller enters an augur hole in the cross piece, and turns in this hole as the gate is swung open or shut. The roller is of cast-iron. Fig. 102 shows the whole contrivance, with the gate open.

SYPHON.

It often happens that it becomes desirable to drain a pond, cellar, quarry or other excavation without the trouble, time or expense of cutting a deep ditch. J. P. JAY of Kentucky, gives in the annexed figure the mode he has adopted. To start the syphon, stop both ends and fill with water through the vertical branch; then turn the cocks so as to cut off communication with the air; open the upper end of the syphon first, and then the lower end, and the water will run freely. Now when air collects in the bend of the syphon, by opening the cock B. the air will ascend in the pipe at C. Then close B., open A., and pour in water to fill the part C. In this manner the air can be taken out with little trouble. Care should be taken in joining the branch to the syphon that the end does not go inside at all. In case the branch might freeze, it may be joined at D. with a screw so as to be removed. The end where the water escapes should be a little smaller than the rest of the pipe.

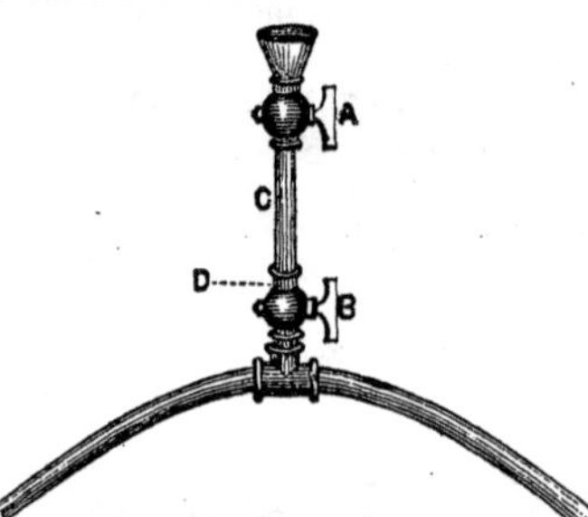

Fig. 103.—*Syphon.*

OSIER BANDS FOR CORNSTALKS AND STRAW.

Nurserymen and others are familiar with the mode of twisting loops on the smaller ends of the long shoots of the osier, in order to use them with facility for bands, in tying up bundles of trees. As the osier is gradually coming into use as bands for straw, cornstalks, and for other farm purposes, we have frequent inquiries as to the mode of using them.

When they are kept cut down yearly, the shoots sent up are six or eight feet long, and the third of an inch or so in diameter. This cutting down must be done in autumn, winter, or early in spring before the buds swell, to prevent the check in growth always resulting from cutting away expanded buds or young leaves. The shoots may be used either in autumn or spring, and work best after a day or two wilting. To use them expeditiously, a loop is formed at the smaller end, through which the larger end is thrust, and then drawn tightly by placing the knee or boot upon the bundle. The two ends are then secured together by a twist.

Before forming the loop, about two feet of the smaller end must be rendered flexible by twisting. This is done by holding it firmly in the left hand, as shown in fig. 104, and whirling the other end rapidly, so as to give it the twist. Then the twisted part is bent around the smaller fingers, and the two parts brought parallel; the fore finger is placed between these parts, and the twisting continued until the loop is made, as shown in fig. 105, and the end tied in a knot. A little practice enables any one to perform this operation rapidly—the bending and looping being done in a moment by the left hand, while the twisting is done with the right hand, without any cessation from the beginning to the end of the work, a few seconds only being required for the whole.

Fig. 104.—*Mode of Twisting the Smaller Ends of Osiers, preparatory to Forming the Loop.*

Fig. 105.—*Mode of Twisting and Looping Osiers, preparatory to Using them as Bands.*

The best osiers, when cut and slightly wilted, are nearly as tough as leathern thongs, and will not only bear any ordinary degree of twisting, but possess sufficient strength for holding a bundle firmly together.

Open Drains Lined with Stone.

Every farmer knows the inconvenience of the washing of surface drains, used for conveying small brooks, and the floods which pour down swales in heavy storms. We have often seen such channels paved with cobble stones, which answered a good purpose, and over this paving the water rushed down without doing any injury. The difficulty is, the outer small stones are apt to be displaced by frost and the plow, when running through fields. A correspondent of the Country Gentleman has succeeded well by adopting the mode of construction shown in the following cut, fig. 106. He cuts a broad ditch, and fills it half full of small and broken stones. He then selects broad or thick flat stones, some twenty inches long, and places them in parallel rows, not quite meeting in the middle, and forming a trough like the letter **V**, as shown in the cut. The thinner ends are placed inward; and if of different thicknesses, the small stones must be hollowed out, so as to make the tops as even and smooth as may be. These ditches must be, so far as possible, in perfectly straight lines across the field, or along a boundary. Neither the plow nor frost will disturb them.

Fig. 106.

Hay Knife.

J. G. Homet of Monroeton, Penn., gives the following statement and description:

Being tired of cutting hay with an axe, I went to the blacksmith, taking about three feet of old buggy tire and one pound of cast steel, and showed how I wished him to make a hay-knife—told him to make two blades, B. B., at about the angle shown, (fig. 107,) with the inner corners rounded and the inner edges thin and sharp. These blades were welded to one end of the tire, and the other end coiled for a handle. At A., (fifteen inches from the bottom,) was welded a step on which the foot is placed to thrust the knife into the hay. At C. the bar is curved out to keep the hand away from the cut hay.

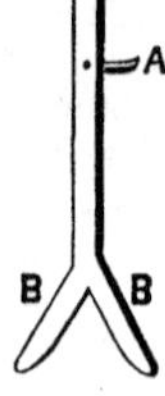

Fig. 107.

The cost of materials is from thirty to fifty cents; cost of making one dollar.

The knife herein described is forty-two inches long, and one and a quarter inches wide.

Whiffletree Hooks.

We have used for some years the whiffletree hooks made and placed on his wagons by J. E. Morgan of Deerfield, N. Y., and find nothing of the kind so convenient for farm wagons.

The hooks for attaching the smaller whiffletrees to the larger ones or eveners, are shown in the accompanying figure. The hooks attached to the ends of the evener are made of iron nearly an inch in diameter, and are consequently so strong as never to be in danger of bending, when subjected to the hardest draught of the team. A slot is made in the ends of the evener, so that these hooks may be turned about on the bolt to which they are attached, through more than half the entire circle. The bolt being nearer the rear part of the evener, the hooks are thrown further off when they are turned backwards, and freely admit the single-trees to be hooked on. When turned forward into position for use, the points of the hook are nearly or quite in contact with the front side of the evener, and they cannot therefore by any possibility allow the single-trees to become detached—they are held as securely as if riveted on. The whole movement of placing on or removing the single-trees is performed in one or two seconds. The ends of the evener are faced with iron plate, so that the whole is strong and substantial, and cannot split or wear. Every one who sees this contrivance is struck with its simplicity and convenience, a single glance showing its character better than this description.

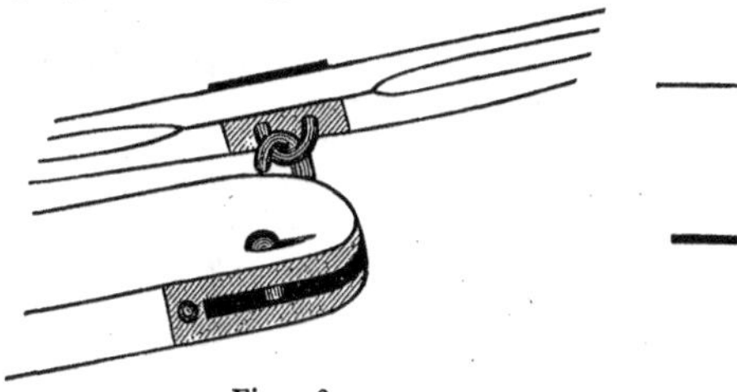

Fig. 108.

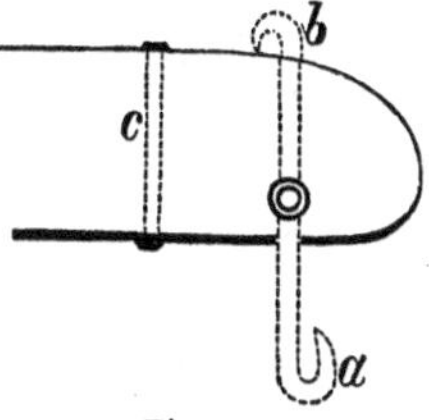

Fig. 109.

Fig. 108 represents a view of this contrivance, showing the end of the evener and of the middle portion of the single-tree. Fig. 109 is a plan, *a.* being the position of the hook when moved backwards to receive the single-tree, and *b.* the position in front when in use; *c.* is a bolt for strengthening the end.

Turning and Mixing Manures.

G. F. P. furnishes us the following description of an implement which he has successfully used for this purpose: I send you a drawing of a contrivance we have used for turning manure in the cattle yards, and found it to be a good thing. In this section we use large quantities of rye straw to litter the yards with during the winter; this remains in the yards during the summer, and is carted out in the fall to the next year's planting ground.

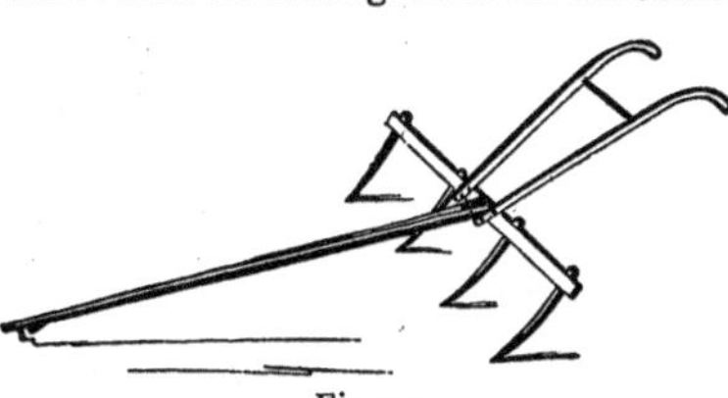

Fig. 110.

It is necessary to turn the manure about mid-summer; otherwise the top will be quite fine, and the bottom scarcely decayed at all. This work was formerly done with potato hooks, and was both slow and laborious. With a concern of this kind, and a good team, it can be done easily and done well. It is worked by pressing the tines into the manure, drawing it forward a few feet and then backing up. I think it will be understood without a description. The one we use has a head about three feet long; the tines are made of three-quarter inch rod, with a shoulder and nut, so they can be screwed tight; they extend a foot below the head, and proiect forward slightly, so as to hold on to the manure.

Stake-Holder for Sleds.

L. D. Snook describes the following contrivance, which will prove convenient for the purpose indicated:

The general method of attaching stakes is to make a hole of the size desired through the rove beam and outer edge of knee. It is needless to state that the making of an inch and a quarter hole through these parts, and that too where the greatest strain comes, cannot do otherwise than weaken these most essential parts of a draft sled. In the cut a hole is made through the rove at A. of any desired size. On the knee, one and a half inches from the runner, is secured an iron band, as shown at B. It is plain to all that a stake can be held firmer by this device than by the old method, and the durability of the sled greatly enhanced. For the drawing of cord wood, I would suggest placing a second iron band between the first one and the rove.

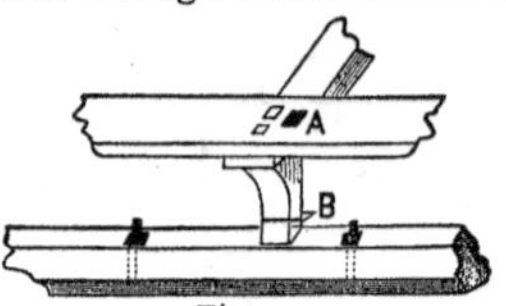

Fig. 111.

Milk Rack.

S. W. Wilcox of South Milford, Mass., says: The following cut represents a rack which will hold forty-five pans of milk of large size, with a a free circulation of air around each pan, which will cause more cream to rise on the milk than if set on the cellar bottom. It takes up but little room, and can be taken apart to wash or move to any part of the house; and with this article it is easy to guard against ants, emmets, mice, or any other kind of vermin, which are a pest in many houses.

Fig. 112.

Scalding Troughs for Hogs.

The following descriptions of a fixed, and of a portable scalding trough were furnished by correspondents. The first is from P. O. Turpin of Kentucky, and he states that he has used it for several years. It consists of a box of two inch plank, made two feet wide and about seven feet long —the bottom being sheet-iron, and all water tight. Set the box over a trench cut a foot deep and a foot wide, and as long as the box, and build a chimney at one end. The fire is built in the trench. Very little fuel is required.—The ladder, fig. 114, is placed within the box, to lift the hog out of the hot water, by means of the windlass shown in fig. 113.

Fig. 113.—*Scalding Trough—not Portable.*

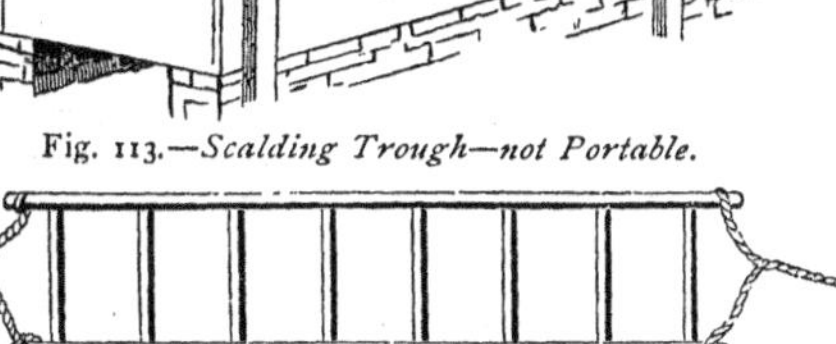

Fig. 114.—*The Ladder.*

Fig. 115 shows a portable scalding vat, the value proved by long use. One man scalded with it ninety-six hogs in a day, and others had them dressed by 2 P. M. The box is about two by five feet. A copper pipe, about twelve inches in diameter, runs in the bottom of the vat from front to rear and returns. Fill the vat with water, and start the fire, and the water is hot in twenty minutes—little wood being required. A rack over the pipe protects it. The roller at the rear end, with with an inclined plane, slides the hog easily in and out. Currents of wind among buildings, require a ventilator on the top of the pipe. This portable trough is in constant use in the neighborhood where used by its itinerant owner.

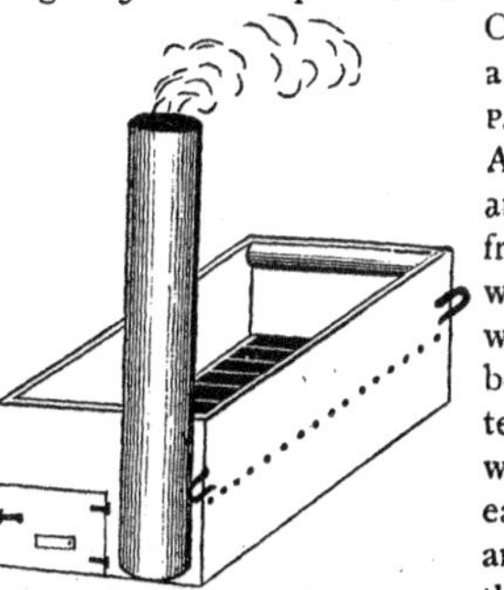

Fig. 115—*Portable Scalding Trough.*

Stacking Hay by Horse-Power.

The following figure (fig. 116) represents the mode adopted in the apparatus made for the use of Raymond's Elevator, by J. H. Chapman of Clayville, N. Y., and which is applicable to any kind of fork capable of running along a rope suspended by means of a wheel and pulley. A similar mode of erecting the poles has been described to us by J. A. Belknap of Yates Co., N. Y. The following explanation will render the whole intelligible: The four pulleys, 1, 2, 3 and 4, hold the rope which works the fork; and *a. b.* and *c.* indicate the inner, middle and outer poles. The pulley marked No. 1 is attached to the middle pole, the horse being hitched to the end at *d*. No. 2 is hung at the top by means of a chain, and

Fig. 116.—*Stacking Hay.*

No. 3 is attached to the fork at the same point where the large rope is fastened—the rope to which the horse is hitched running through this No. 3 pulley around No. 4, and drawing along the loaded fork. The load is thus drawn to any desired place on the stack or rick, and dropped by jerking the cord.

The poles marked *a.* and *b.* are the main poles, and should be thirty feet long; *c.* is the same length as the main poles, and is used as the sheer pole for supporting the others, and for this reason need not be more than four inches in diameter at the top; the others may be a little larger. The two main poles should be bolted together with a three-fourths inch bolt, and the sheer pole placed between them. They should set back from the stack so as to rest on the sheer pole.

This contrivance may be raised in thirty minutes; the pulleys being first placed on the main poles, and the rope drawn through them, a heavy stake is driven, the rope tied, the horse is then hitched, and with a little care and steadiness, the poles are drawn up, resting on the sheer pole, and all is right.

The lower end of each outer pole should be lashed to a strong stake, driven into the ground obliquely, by first making a hole with a crowbar. The two tripods are in most convenient position when placed far enough apart to give room for stack or rick, and to allow the loaded wagon to pass between them. It will be obvious from the engraving that the same motion of the horse, attached to the rope *d.*, which first elevates the fork-load, afterwards carries it along horizontally, until dropped at the precise point desired, by jerking the cord.

A western correspondent gives us a description of another mode of stacking with horse-forks, which, with some modification, has been much adopted at the West. The following is his description:

Get one good stick, say 40 feet long, of some light timber, (we use here pine raft oars;) dress it down so as to leave it strongest about 15 feet from the top. Take a light piece of timber 12 feet long, (4 by 4 pine,) and hinge it with iron, so as to allow it to rise and lower or swing sidewise fully two-thirds the way round, for an arm; run a five-eighths rope from the end of the arm to the top of the pole, through a pulley made fast there, down the pole, and fasten it on a pin for the purpose of raising and lowering the arm from the ground. Fasten three guy ropes to the top of the pole and raise it, first digging a little hole a foot deep to keep the bottom of the pole in its place, and secure it by fastening ropes to stakes, one of which is driven directly behind the pole, and the other two just far enough ahead to keep it from falling backwards. The guy ropes should be 60 feet each. The pole should lean toward the stack, so that when the load comes on the crane it swings of its own accord to the centre. One of the fork-pulleys hangs on the end of the arm, one just under the arm on the pole, and one near the ground. This makes a better rigging in every way, and costs here (fork, ropes and all) about $25.

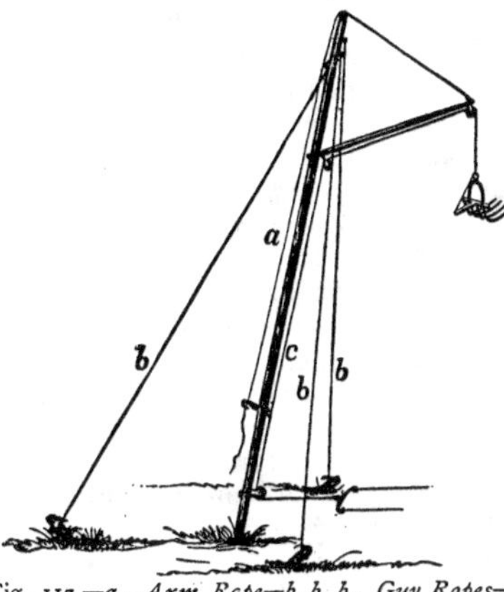

Fig. 117.—*a., Arm Rope—b. b. b., Guy Ropes—c., Fork Rope.*

The advantages are—1st. A much larger rick can be built, and the hay pushed to any part of it easily by the stackers; 2d. All your hay is thrown in the middle of the rick, thereby preventing the settling of one side so as to lean the stack; 3d. It is taken from the ground without dragging against the side of the stack. With such a crane, ricks of thirty or more tons are put up very easily.

FRUIT CULTURE.

THINNING APPLES.—In the summer season orchardists who expect to be successful, should commence their attention to thinning the fruit. A common practice is to let all grow together on the tree, including knurly, scabby, half-grown, curculio-stung, &c. These being mixed up with good fair specimens crowd them, and lessen their size and beauty as well as flavor. Take them in time, thin out all the bad ones, and several decided advantages will result. It is much easier to pick these poor apples now, and cast them to the ground without basketing, than to hand pick them and convey them carefully down to the ground with the general crop. The difference in the labor will be as one to four or five. This subject is very little appreciated; yet when looked at deliberately it will be seen that the farmer might as well let the weeds grow with his corn, to be assorted from it at harvesting, as to allow worthless fruit to occupy through the whole season the place of that which is good.

Rolling Orchards.—The owners of orchards and fruit gardens differ on the practice of allowing their grounds to grow to grass. All agree that for several years, while the trees are young, cultivation is absolutely necessary; and all intelligent cultivators know that a peach orchard, to be of any value, must always be kept mellow. Soils vary in richness, and some good orchardists allow old apple and pear trees to stand in closely cropped grass, but many are tempted to admit a turf merely for the sake of comfort in walking, when, in fact, the trees become stunted in consequence, and the fruit of only second size and quality. By employing the rule which we have recommended of never allowing the annual shoots to become less than a foot in length, (more is better,) they may perceive whether their trees are too feeble. The very common objection that plowing and harrowing make the ground rough and unpleasant to walk on, may be obviated by the use of the two-horse roller. We will suppose an orchard or fruit garden has been kept well cultivated during the first half of summer—a period when none wish to walk in it. Give it a smooth, handsome harrowing, and then pass the roller over it a sufficient number of times to make it as smooth as a walk. If the soil is light and gravelly, it will be in good order at all times; if heavy or clayey, in which case the rolling should be done only when quite dry, it may be freely passed over except in wet or rainy weather. The process may be repeated after some weeks when necessary. Another advantage will result—the rolling, which need not be done until growth is nearly completed, will have a tendency to check luxuriance and to promote a thorough ripening of the young wood.

Apple Trees Overbearing.—We observe in the proceedings of the Alton Horticultural Society, Ill., a statement of one of the members, D. E. Brown, that one great cause of the failure of the apple crop by the dropping of the young fruit was a want of vigor in the trees. Those which bore heavy crops last year, were too much exhausted to do the same again this year. So far as his observation had extended, trees which bore none last year had good crops the present season. He added that last year he "had a good crop—his neighbor Curtis a very poor one. This year I have none, and Curtis has a very good show."

Poor Soil for Blackberries.—Wm. Parry, one of the most intelligent and and successful cultivators of the small fruits, stated before the New-York Fruit-Growers' Club, that a neighbor of his who had about 40 acres devoted to blackberries, purchased a tract of light sandy soil at the low price of $13 per acre, and planted it with them. But desiring to have a model patch, he bought a few acres of very rich land at $300 per acre. The blackberries on this rich land made an enormous growth, but did not

yield as much fruit per acre as the other. It is proper to remark, however, that the poor land had good culture, and we observe in another place that Mr. Parry applies large quantities of black muck along the wide trenches at planting time.

MAHALEB STOCKS.—V. ALDRICH, one of the best Illinois fruit-growers, in allusion to the late rapid growth of the Mahaleb as a cherry stock, and its consequent liability to top-killing in winter, states that he succeeds well with it on high dry ground, not very rich, when there is no top-killing.

FINE EARLY STRAWBERRIES.—The Gardener's Monthly states that at one of the exhibitions of the Pennsylvania Horticultural Society, D. W. HERSTINE of Philadelphia, displayed a large basket of strawberries, "which delighted every one for their size and beauty," produced simply by covering the plants in the open ground with a common hot-bed frame, adding that Mr. H. had an abundance of fruit before the other plants in open ground were in blossom—the strawberries being much finer colored than under ordinary management.

MARKETING PEACHES.—Those who have formerly been in the practice of purchasing peaches gathered when half ripe, and with their flavor less than half developed, will be glad to learn that it has now become fully established that fully ripe peaches (not soft) carry the best, and do not decay as soon as those in a half green state. Dr. HULL of Alton, Ill., an extensive fruit grower, stated at the St. Louis Pomological meeting that he had been shipping fruit to half a dozen different States, and that he has found that when fully mature, and packed tight enough to prevent all friction or rattling, they will "carry six days safely." He uses baskets only, placing oak leaves in the bottom and between layers.

VINEGAR FROM APPLES.—The superior excellence of vinegar made from apples is well known. Dr. CLAGGETT of St. Louis, stated at a recent meeting of the American Pomological Society, that he uses the imperfect fruit too defective for market, and that from his small orchard he thus makes two thousand gallons of vinegar. Mr. NELSON of Fort Wayne, said, at the same meeting, that, after considerable experience, he finds that gathering the imperfect fruit as fast as it falls, for making vinegar, diminishes the ravages of the apple worm annually. He finds it profitable.

CURRANT WORM.—A. J. CAYWOOD of Poughkeepsie, gives in the Horticulturist a good method of applying the well proved remedy for the currant worm, the white hellebore. He uses De La Vergne's bellows, made for applying sulphur to grapevines, &c., and says a slight puff will kill the worms in twenty minutes. In relation to the hellebore being poisonous, he says he has inhaled it for two hours at a time—that it is unpleasant and causes sneezing, which may be avoided by keeping on the windward side—which we advise every one to do. The bellows should not be used when the bushes are wet, as the wires choke with dampened powder.

PEACHES IN ILLINOIS.—W. C. FLAGG, a successful fruit raiser of Alton, gives in the Journal of Horticulture the following list, ripening, we suppose, in regular succession: Hale's Early, Troth's, Large Early York, Morris' Red Rareripe, Crawford's Early, Yellow Rareripe, Oldmixon Free, Reeves' Favorite, Stump the World, Columbia, Late Rareripe, Crawford's Late, Ward's Late, Smock, Delaware White, Heath Cling.

MARKETING STRAWBERRIES.—While many superficial or careless managers cannot send strawberries fifty miles in good salable condition, J. KNOX of Pittsburgh, sends his four hundred miles, and receives double and triple prices for them. The fruit is allowed to ripen before picking: Mr. Knox says, "we allow the fruit to mature enough for our own table before it is gathered for market." It is handled with great care, carefully assorted, and as carefully packed in neat boxes. So large and finely grown are the berries that ten fill a pint box. He has sent the Jucunda to New-York city on Monday, reaching there on Tuesday, and kept it until Friday and Saturday, and sold then at higher prices than other berries brought raised in the immediate vicinity of the city. So much for doing a thing well.

THE CURCULIO.—This insect stings the peaches badly at portions of the West. Mr. EARLE of South Pass, Illinois, stated at the St. Louis Pomological meeting, that an orchard of eleven hundred peach trees was kept free from this insect by jarring, at an expense of two dollars a day during the curculio season—the fruit resulting handsomely paying the cost.

LOOK AT THE INDEX!—We see a great deal of discussion in the papers, on the cultivation of orchards, on manuring the dwarf pear, &c., and a great contrariety of opinion prevails. Some assert that the decline of orchards or the want of success, is owing to the want of fertility in the soil, the absence of manure, and the neglect of high culture; others, that the growth is too rampant, unnatural, and that the trees by this succulent growth become more liable to disease and injury from freezing, and they recommend the growth of grass to retard this unnatural luxuriance. Instances of success and failure are cited on both sides of the question, and no conclusion, satisfactory to all, is reached. The terms "rich soil," "strong growth," "feeble trees," &c., are indefinite, and convey nothing distinct which all may understand alike. We propose that in all discussions of the kind, the writers should make use of figures and accurate measurements—give the length of the annual shoots in feet and inches, and report at what time the terminal buds are formed, that we may know the degree of ripeness or succulence when cold weather comes on. Every cultivator should look at the *index* of growth—that is, the *length of the shoots*—and then we may be able to judge whether to increase or diminish the manuring and cultivation. The whole question would then be rendered comparatively plain and simple.

TOP-GRAFTING OLD APPLE TREES.—A correspondent of the Prairie Farmer thinks it better to regraft old trees (if not enfeebled by age) than to set young ones—although it generally costs about $1.50 to regraft a tree of twenty-four limbs, or to set six young trees. They make a much stronger growth of shoots, come into bearing earlier, and are in less danger of injury by mice, rabbits, sheep, &c. He recommends for this purpose the Fourth of July apple, Ben Davis, Grimes' Golden Pippin, Willow Twig and Stark

—all fair and productive western sorts. He carefully avoids the common practice of cleft-grafting into the limbs some 20 or 30 feet up, but inserts the scions into the sprouts near the centre of the head.

HARDY FRUITS.—In answer to the inquiry of a correspondent who lives in a cold region of New-York State, it may be well to remark at the outset, that for cold regions a selection of the hardiest varieties will save from much disappointment. A few are found to endure the severe winters of Maine, Canada and Wisconsin, where most sorts are badly injured or killed. It is equally important to select dry or well drained upland, not of great fertility, where the trees will make a good, medium, well-ripened growth, and to avoid wet, mucky grounds, which are more liable to sharp frosts, and which, by inducing succulent growth, render the trees more easily winter-killed. Clean and mellow cultivation should be given, which, on soils of moderate fertility, will make a better, hardier and better ripened growth than any manuring without it. The cultivation should not be continued after mid-summer, as if late it might prevent the formation of the terminal buds and cause an unripened second growth. Among the hardiest varieties, the following may be named:

Apples.—Sops of Wine, Red Astrachan, Autumn Strawberry, Fall Orange, Duchess of Oldenburgh, Fameuse, St. Lawrence, Golden Russet (of Western N. Y.,) Northern Spy, Wagener.

Pears.—Buffum, Urbaniste, Anjou, Fulton, Lawrence, Winter Nelis.

Crabs.—Transcendent, Hyslop.

Grapes.—Hartford Prolific, Concord, Delaware, Adirondac (covered in winter.)

Cherries.—Early Richmond, Mayduke, Large Morello.

Plums.—Schenectady Catharine, Lombard, McLaughlin.

Raspberries.—Philadelphia, Black Cap.

Strawberries.—Wilson's.

Blackberries.—Kittatinny.

Currants.—Red Dutch, White Dutch, White Grape, Versailles.

RASPBERRIES AT THE WEST.—At the Horticultural discussions at Alton, Ill., the opinion appeared to be generally admitted that the cultivation of the Raspberry for market, including the Doolittle, did not pay. The Philadelphia has proved more productive, and was thought by others to be profitable.

RAISING GRAPES FROM SEED, NEW VARIETIES.—Procure well ripened grapes, wash the seed from the pulp, and mix them at once with moist sand or leaf-mould. Bury them in open ground till early spring. They should not be allowed at any time to become dry, and care should likewise be taken to prevent their becoming water-soaked. They should, in fact, be treated as cherry stones and pear seeds are managed by nurserymen. Be careful to secure them from mice. Plant in spring, in beds of *deep rich soil*, in drills a foot or two apart, and an inch or two apart in the drills, and about an inch deep. Shade the young plants for a few weeks. Provide small stakes for their support, and mulch the surface with an inch or so of good fine manure. If dry weather occurs, give the ground a thorough soaking as often as once a week. Lay down and cover in winter. The great point is a *deep* and *rich* soil, so as to give the young plants a vigorous start.

THE QUINCE.—In an article on this fruit in the Horticulturist, which urges the importance of good cultivation instead of the common practice of neglect, the writer states that on his own grounds he has trees three years transplanted which yielded this year half a bushel of fruit, selling readily at $4 per bushel. He cites an instance where two hundred baskets were raised from an acre of trees only four years planted—the owner giving the land good cultivation with plentiful annual manurings. As the trees are slow growers, it is recommended to purchase good bushy plants (well pruned of course) about four feet high, as being the cheapest in the end, such trees returning in three years a crop sufficient to pay all expenses.

THERMOMETERS IN FRUIT ROOMS.—The keeping of apples and other fruit depends greatly on the temperature. If the room is too closely shut, from a fear of freezing, the fruit may decay in a few weeks; if kept cold, and with some circulation of air, they will remain sound until spring. The truth is, too much is left to guesswork, and hence, while at some times the temperature may be up to fifty or sixty, it may, on the other hand, run down below freezing on the occurrence of a cold snap, the owner or attendant not always being able to judge by his perception. Thermometers are cheap now-a-days, and such cheap ones will answer the purpose well, not usually varying more than a degree or two at ordinary temperatures. Hang one near the ceiling, and another near the ground. Let the windows of the fruit room be hung on hinges, so that they may be opened to any degree. By means of these windows and the thermometers, the temperature may be kept down to within a few degrees of freezing, if they are examined say twice a day, or night and morning, and the fruit kept sound and fresh, and the owner no longer work in the dark or by guesswork.

ORCHARD CATERPILLAR.—This insect "comes and goes"—is abundant in some years, and in others nearly disappears. Where the millers or moths have left their rings or eggs on the shoots of trees, winter is the time to destroy them. A practiced eye will see almost at a single glance if there are any on a tree, by the swelling or knob which each one gives to a shoot. Select a dark or cloudy day, or else a day when the sky is entirely clear—avoiding thin bright clouds, which will dazzle and hurt the eyes—and cut off every shoot which contains the eggs, and commit them to the fire. A single clip of the orchard shears on a pole will prevent a destructive nest of these depredators next season.

SHARES' HARROW FOR ORCHARDS.—This implement, now so well known, and commonly used for pulverizing the upper surface of inverted sod, appears to be very little used, and its advantages imperfectly comprehended, for working in orchards. The peculiar construction of its teeth causes them to ride over any obstruction, so that they are never caught by roots. When a plow encounters a large root, the point is thrust deeper into the ground beneath it, and the root must be either broken and torn

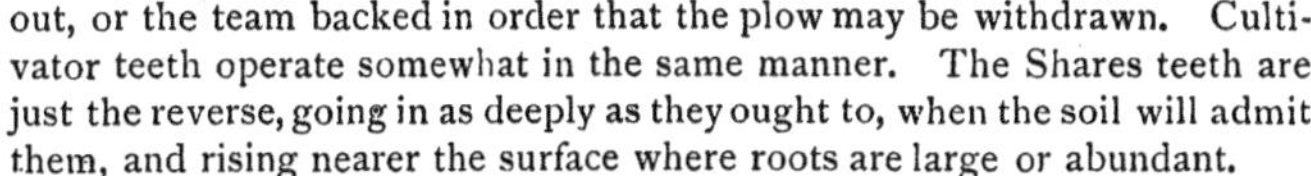

out, or the team backed in order that the plow may be withdrawn. Cultivator teeth operate somewhat in the same manner. The Shares teeth are just the reverse, going in as deeply as they ought to, when the soil will admit them, and rising nearer the surface where roots are large or abundant.

HARDINESS OF PEARS.—There are certain old seedling pear trees, some of them known as "choke pears," remarkable for their hardiness and longevity. The feebleness and tender character of many of the finer varieties have led some cultivators to suppose that the improved sorts must be in these respects inferior to the wildings. This is the case with a large number, simply because they have been selected for their flavor alone, while the wildings have been selected on account of their extreme hardiness—severe winters and absence of culture having destroyed all the feeble ones and left these standing alone. We see no reason why we may not hope to have a list of excellent pears growing on trees possessing all the vigor and endurance of the old wild sorts. We have at least one already—the Buffum. The fruit is not large, but of fair size, and when picked early enough ripens into an excellent flavor, fully equal in quality to some which have a higher name. There is no tree in the nursery superior to it in a handsome, straight growth. It endures the severest winters. Some years ago we saw a tree of this variety growing at Worcester, Mass., grafted 23 years before by the owner, which was then bearing the largest crop of fruit we ever saw on a pear tree. The year before the crop measured 25 bushels, which the owner assured us was evidently smaller.

The Doyenne Boussock, although not equal in vigor to the Buffum, makes up the difference in the size and fine appearance of the fruit. It is uniformly healthy and productive, and appears to possess no drawback.

The Beurre d'Anjou is a slower grower, but is remarkable for its uniform bearing and reliability, while in quality it is not exceeded by many. The Lawrence has not been sufficiently tried in cold regions, but there is strong reason for believing that it will prove one of the most reliable. The growth, it is true, is not handsome, and, on some soils, it is quite feeble; but so far as we know it is uniformly healthy and hardy, and holds its leaves until the fruit is well matured. It was once thought that the Flemish Beauty might be placed in such a list as this, its extreme hardiness having been thoroughly proved. But the natural early dropping of the leaves and fruit, and its more recent liability to scab and cracking, have materially lowered its reputation.

There is another pear, the Howell, which, although not many years tested, gives high promise of value on account of its handsome, uniformly fair fruit, and the fine growth of the tree; although it is represented as being straggling in some places, we find no tree that exceeds it in vigor of growth in the nursery, and but few that are straighter and more upright. We have never seen a scabby or defective specimen. Its flavor is good, although not of the highest quality.

We invite attention to these points in the pear, believing that a more careful selection of those sorts which are vigorous, hardy and reliable, will prevent much of the complaint so generally made of the failure of the pear tree.

Culture of the Raspberry.—L. D. Snook of Yates Co., N. Y., gives the following statement of the mode of management which he adopts, and which refers principally to the Blackcaps :

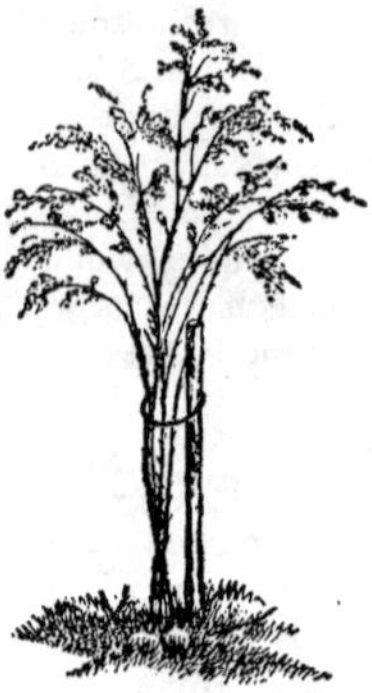

Fig. 118.

Fig. 119.

If no plants are wanted, tie the canes to stakes, as shown in fig. 118, using a strip of cloth or an untwisted strand of rope, but *never* wire or closely twisted twine or cord, as it wounds the cane, causing bleeding, and often the ultimate destruction of the plant. As soon as they are through fruiting, the old wood should be cut out, and all the new wood, except three or four of the strongest, well developed canes, which are to produce fruit the coming season. A very convenient tool for the purpose of removing the surplus canes is a hooked knife of the form shown in fig. 119, which any blacksmith will make for 25 cents ; attach with screws to an old broom or fork handle, 3 feet in length. The hook should be of sufficient dimensions to receive the largest canes, and ground sharp. This is a convenient implement, by the use of which the hands are not lacerated, as by hand pruning with a knife, and it is equally valuable for use among blackberry bushes and sprouts of fruit trees that spring up in undesirable places.

Fig. 120.

Proper mulching is of the utmost importance, and in a dry season it will be the means of producing a good crop, when, without it, a failure would be inevitable. For a mulch use straw or forest leaves, applied after a thorough mixing of the soil in the spring ; it keeps the ground moist, weeds down, fruit clean, and canes that may fall from taking root.

Many cultivators of this fruit support the canes bv tying to stakes, while others use a trellis of the form shown in fig. 120. The parallel horizontal poles are secured to stakes driven in the ground one foot distant and from 8 to 12 feet distant in the row, the length of pole to decide ; connect them together by cleats nailed to stakes or poles, as desired.

ORNAMENTAL PLANTS.

WE PRESENT for this number of the ILLUSTRATED REGISTER, notices of a few fine ornamental plants, for the illustrations of which we are mostly indebted to B. K. BLISS & SON of New-York, and to WASHBURN & CO., of Boston.

Fig. 121.—*The Double Zinnia.*

THE DOUBLE ZINNIA.—This new and showy variety of the old *Zinnia elegans*, has become rapidly introduced of late among florists throughout the country. Many who are not sufficiently careful to preserve it in its purity, are disappointed by finding so many poor or imperfect flowers. This is to be prevented by collecting and saving seed from the very finest flowers only; and as soon as the first bloom appears in summer, all except those of handsome symmetrical shape and full double form, with fine colors, should be immediately pulled out to give place to the best. The finest flowers are remarkable for their brilliant or rich colors, and a symmetry of shape nearly equal to that of a perfect Dahlia.

Fig. 122.—*Pot of Asters.*

THE ASTER.—Although this autumnal flower, unequalled in the floral world in many respects, has been long known, yet the remarkable improvements which have been made of late years in the beauty of many varieties, have rendered it eminently worthy of repeated notice. The dwarf and smaller sorts, with their compact growth and full, rounded, soft and rich flowers, are objects of great interest when grown in pots, as shown in fig. 122.

Fig. 123.—*Salvia Splendens var. Compacta.*

DWARF SALVIA SPLENDENS.—The common Salvia splendens (fig. 123) is well known as a brilliant and showy bedding plant, blooming throughout the latter part of summer and in autumn till frost. The variety Compacta is a new French variety, differing from the old sort by its more tufted, compact and dwarf habit, and by its more numerous spikes of flowers and more dense bloom. Its brilliant scarlet color renders it one of the finest ornamentals for the open ground in summer and autumn.

Fig. 124.—*Ipomea volubilis var. Madame Anne.*

THE IPOMEAS—(fig. 124.)—Nothing excels the genus *Ipomea*, as beautiful and delicate climbing and twining plants, the flowers presenting almost every

Fig. 125.—*Lychnis Haageana.*

imaginable shade of color. They are mostly tender annuals, but if well cared for, will bloom through the latter part of summer in open ground, and may be continued in bloom in sheltered places and in green-houses. The variety represented in the accompanying engraving, is a fine new one, with variegated flowers, striped red on white ground.

LYCHNIS HAAGEANA—(fig. 125.)—This is a hybrid or mixed plant, a beautiful perennial, with large flowers ranging in color from brilliant scarlet to deep blood red, purple, orange, white and flesh color. It grows about one foot in height.

THE KAULFUSSIA is a neat, compact plant, and is a hardy annual. The K. amelloides, (fig. 126,) has bright blue flowers and grows about half a foot high, producing a lively brilliant appearance when growing freely in good soil.

DIANTHUS DENTATUS—(fig. 127.)—This is a beautiful annual pink, flowering the same year as sown, the plant presenting a dwarf

Fig. 126.—*Kaulfussia amelloides.*

Fig. 127.—*Dianthus dentatus.*

appearance, growing in tufts, covered when in bloom with a handsome mass of rose-lilac flowers. It grows well on rock-work, and may be used for edging or for masses where low plants are desired.

THE REDDENED SILENE PENDULA—(fig. 128.)—This is a newly introduced bloomer, the whole plant presenting a remarkable appearance, the flowers of a bright carmine rose, the branches, instead of being green, are of a brownish red, which spreads over the leaves and calyx, giving the whole a striking aspect, quite different from that of the old Silene pendula.

Fig. 128.—*Silene pendula var. ruberrima.*

DOMESTIC MANAGEMENT, RECEIPTS, &c.

Written for the ANNUAL REGISTER by A YOUNG LEARNER.

ECONOMY IN FAMILY EXPENSES.

AT SOME LEISURE TIME, when both are in good humor, let the good-man and his wife deliberately agree upon the sum of money needed by the latter for all expenses in one year. Let the husband then supply it to her in quarterly payments, with the understanding that if she uses her money too fast, she can have no more till the next quarter.

The usual way of asking for cash every time that anything is wanted is a source of mutual irritation. By a regular allowance such conversations as the following would be obviated: "Dear, could you let me have ten cents to buy a paper of pins?" "Mrs. Smith, you waste the most on trifles of all the women I ever saw; besides, I haven't any change to buy your pins. Can't you use carpet-tacks instead?"

Symmetry in expenditure would be also thus secured. An undue portion would not go to this or that, but all could be systematically apportioned. The wife would not feel at liberty to draw upon her husband's purse for superfluities as long as his good nature held out; she would feel the dignity and responsibility of her trust.

2. Let every woman keep an account of every penny she expends. A blank book ruled with separate columns for receipts and expenditures is convenient for this purpose. There are few women who would be extravagant if they were not so ignorant of money affairs. Many wives have no idea of the limits of their husbands' incomes; widows and single women, while knowing how much they receive, often expend their money hit-or-miss, and never know where it all goes. "I always set down every item of my expenses," said a lady to her friend. "Well, I began to do so once," was the reply, "but I got frightened when I saw how much went for mere nothing, and so I gave it up." But a sensible woman will not give it up. The little black figures will teach her conscientious calculation; and the record of early mistakes in outlay will make her judicious in future.

3. John Randolph once rose in Congress with the words, "Mr. Speaker, I have found the Philosopher's stone. It consists of these four words, '*pay as you go.*'" The writer knows a housekeeper who has followed this good advice for many years. Nothing is set against her name in the merchants' ledgers. She has no running account at any store. No matter how much anything is wanted, if the worth of the last "coupon" is laid out, she waits till next payment is due, rather than run in debt. Oh, the blessed independence of this course! A Fifth-Avenue belle, living on borrowed money in a rented house, might well envy her comfort.

Hygiene and Home Remedies.

In order to keep well and grow hardy nothing excels the daily cold bath. A washbowl of pure soft water, a cake of soap, two towels and a wash-cloth are sufficient apparatus when you have no bath-room. Bathe on rising in the morning, washing only a part at a time, and rub vigorously with a towel. An American woman, who has kept up this habit every day for sixteen years, has been confined to bed but once during the time; then it was by a contagious disease. Others of the same family have practiced it with similar results, and though they come of an enfeebled stock, are remarkably healthful. When daily bathing is first practiced, begin cautiously; make only a partial ablution the first day, and gradually increase its extent; do not omit the friction. Your bath will soon become more essential to you than your breakfast. It prevents easy taking cold; keeps the feet warm and circulation even, and promotes a clear skin and good looks.

The privacy of a bath-room can be anywhere secured by enclosing a small space around the washstand with a light screen made like a clothes-frame and covered with paper or chintz. It can be set away when not in use.

All hard-working people need water especially. In one of the best farm-houses I ever saw, a small bath-room opens close to the back door. Here the hot and dusty laborers can enter when they come in from the field, and go thence to the dining-room refreshed and tidy. It is abundantly supplied with coarse towels, toilet and sand soap, and a spotless clean marble basin, with a drain in the bottom, like those in city bath-rooms, but the water comes from a cistern pump just on one side of it.

Farmers' wives, ventilate your bedrooms. Have your chamber windows altered a little, so as to let down from the top. Impure air from the lungs is warmer than the rest, and will rise to the ceiling, near which admitted air ventilates far better, and does not blow about the feet. Eschew the little bed recesses often built in sitting-rooms and kitchens. You might as well sleep in a drum. Also when you build your "new house" make all your bedrooms larger. A room may be large enough to dress in and lie down in, which is too small to breathe in the night long.

Never count the time gained which is subtracted from sleep. Ample and regular hours of rest are necessary for hard-working persons, and "an hour of sleep before midnight is worth two after it."

Children need more food in proportion to their size than adults do, for they have not only to supply the waste of the system, but to furnish besides material for growth. Plain, hearty food, and all they want of it, is right for healthy children. Many mothers of the present day implant disease and a craving for stimulants in their children in weakening their constitutions by insufficient or improper food. There would not be so many "cross" babies if there were not so many half-starved and tightly dressed ones.

Unless you like the heart diseases and consumption, wear no tight clothing. The measures for your dresses should be taken over fully inflated lungs, and you should never wear corsets which do not measure as many inches around as you do.

The sick-room should be quiet, cleanly and well aired. Label all medicines. Poisons should be placed above the reach of children. A good nurse has a steady hand, clear head and kind heart; she is not talkative or nervous. Avoid arguments with the sick; do not tease them with business; do not sit or lean on the bed. If friends call on the patient, their stay should be very short. The practice of visiting the sick on the Sabbath is a very poor one; that day often thus becomes the most trying and fatiguing of all to them. As a general rule do not go into the sick-room unless you go to help and not to talk. Do not deceive the sick; deceit breeds suspicion; they will worry lest you are "keeping something from them." To persuade the dying that they will recover is treason against the interests of the soul.

It soothes and cools a feverish patient to wash him with warm water in which saleratus or soda has been dissolved.

"Composition" tea is good for a sore throat; so is also a piece of salt pork, peppered, wet with liniment, sewed on a strip of flannel, and bound on the neck; it should be laid aside with caution, to prevent taking cold.

Dried rennet or powdered gizzard skin is a great cure for children ill with dysentery. Rennet can be eaten from the hand like dried beef, and some little ones like it as well.

A little cotton batting, wet with sweet oil and laudanum, put in the ear, will cure ear-ache in the beginning of it.

Spirits of ammonia, inhaled, is good for headache, and to help the breathing in bad colds.

To *prevent* chilblains, keep the feet of children always warm and dry. They must wear rubbers out of doors, and woolen hose and thick flannel-lined shoes from fall till spring.

For chapped lips and hands use "camphor ice," or *pure* mutton tallow put on warm. Candles are unsafe for this purpose, as they are sometimes poisoned with arsenic.

Cream cures sunburn on some complexions, lemon juice on others, and cold water suits still others best.

To remove the black spots (erroneously known as "worms") from the face, wash with hot water and press out the oil from every spot with the finger-nails. Then bathe the face with cold water and apply a wash made of twenty drops of tincture of benzoin to one teacupful of water, which is very soothing to an inflamed skin. Observe more cleanliness and eat plainer food.

Bruises are relieved by applying *warm* water.

A persevering application of cold water is the best remedy for common burns.

Cuts are best cured by sewing up with needle and thread—a few stitches being taken through the upper skin to keep the wound from gaping. It does not hurt much if the stitches are not made too deep.

Decorations, &c.

If you want a few house plants and little trouble, the "General Grant" pelargonium will give you satisfaction. Water it often, give it morning sunlight and a big flower pot to spread in. No plant is more long-blooming, vigorous and showy. Hyacinths are easily brought to blossom in winter, each bulb set in a small pot and watered abundantly. When they blossom they fill the whole room with fragrance. Keep the leaves of all house plants free from dust by sprinkling with water.

An old hoop skirt can be made into very pretty hanging baskets. When finished, paint them green.

Fig. 129.—*Wreath of Autumn Leaves.*

The bright leaves of autumn may be pressed and gummed on cardboard, arranged in the form of a wreath. Put a glass over and bind with strips of marble paper. (Fig. 129.) The effect is beautiful.

Neat picture frames for small photographs can be made of straws crossing each other at the corners of the picture, and cut smoothly off at a short distance beyond the intersection. There should be three parallel straws on each side. At the corners they should be sewed or tied together with yellow sewing silk.

Have you an old straight-backed, long-legged chair? Lay it down on the floor and mark across the fore-legs at about fourteen inches from the seat, the back-legs an inch and three-fourths shorter. Saw them off. The chair will then tip back comfortably; cushioned and covered with pretty calico, it will be your favorite when you are sewing.

A small rag-bag, crocheted of white cotton and red yarn, is a very convenient appendage to the side of a sewing-machine.

Feed canaries with mixed hemp and canary seed. Fresh water and an opportunity for them to bathe should be given them every day. A piece of cuttle-fish in the cage is essential. Plantain leaves and seed, chickweed, raw cabbage, young lettuce and sweet apples, given besides the seed, keep them healthy and please them greatly. If they recover slowly after moulting, put a rusty nail in the water-cup. They thrive best in good sized cages. Do not put them out of doors when the weather is windy or cool—say below 60 deg.—they take cold very easily. Keep their surroundings very clean—wash the perches and put a fresh paper on the floor of the cage every day. Our bird is managed thus; he has always been well fed; during the time we have had him, five years, he has never been sick a day.

Sewing and Repairing.

Linen linings are most comfortable for summer dresses. Three-fourths of a yard suffices for one lining. If new linen is considered too expensive for it, an aged linen sheet or pillow-case can be made to do duty.

Dressmakers' wages are high now, and a lady finds it more satisfactory, as well as economical, to do her own sewing. Those who have little ingenuity can make up their own dresses after having them cut, fitted and basted by a dressmaker. Others can fit themselves by a good pattern, or without a pattern, by cutting the lining by a dress already made, being careful to keep the bias the same over the shoulders and to allow the same quantity in the "darts."

Waists and skirts made separately, fold best to pack in a trunk. Separate dress waists should be cut two or three inches longer than others, flaring out corset-fashion on the hips, so as to feel easy. When worn, the belt of the skirt goes on outside, and there will be no gaping between the upper and the lower parts of the dress, or any troublesome basting together to be done.

When you take a journey, do not carry your purse in your pocket to accommodate the thieves. Sew a private pocket into some out-of-the-way part of the underclothing to contain most of your money. You will be then relieved from anxiety and danger.

Old-fashioned circulars of ladies' cloth, cut over by a gored skirt pattern, make good warm skirts, resembling felt. A lady's cloak can be made from an old overcoat. Cut it out economically and clean and press it. Discarded coats and pantaloons can be cut up for little house-jackets, and bound or trimmed with gay braid or ribbon, look well. Gentlemen's silk hats, pressed over, make smart "beavers" for the girls.

Common shirting makes good white aprons, cheap, durable and easily done up. You can ruffle them, or work the edge in scollops with thread, which is pretty.

Pieces of webbing, for making up into stockings, can be had very cheap at the manufactories where merino wrappers are made. They are what is wasted in cutting.

Thin cotton drawers are not enough for winter, and merino underdrawers are expensive; at half their price nice warm ones may be made of white flannel that is not quite all wool—it will not shrink in washing. A yard and a half makes one pair, and no cloth is wasted if cut thus: Mark the middle of the breadth at half the length by sticking in a pin. Then cut across diagonally from two inches above the height of the pin on one selvage to two inches below it on the other, meeting the pin when half way across. The two pieces will now be of equal size, length and slope on the top, and are to be sewed up separately a few inches, gathered each on a band at the lower edges, and both bound on a waistband on the upper or sloped edges, the shortest part of each coming in front. Always tear off the colored edge from flannel, and cross-stitch every seam.

Two strips of swan's down laid together and basted, with ribbon strings to fasten, make a warm tie for a child's neck.

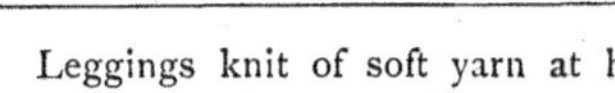

Leggings knit of soft yarn at home are as good, more durable and cheaper than the sales of fancy stores.

Safety from moths for furs consists in leaving them undisturbed through summer in a snug place. Muff boxes are not secure. Taking them out occasionally to air exposes them to the moth. No pepper, camphor or tobacco is needed; after you have worn them for the last time in spring put them into a linen pillow-case, tie up the end in a tight knot, and shut them up in a drawer which will not be often opened. A true and tried prescription.

The warmth of a comfortable is much increased by an interlining of newspaper. To prevent its rattling, cut it into small squares and tack it to the inside of the cover. Put in cotton and knot in the usual way. A very pretty comfortable is made of white paper muslin knotted with scarlet zephyr or red yarn; it keeps clean a great while.

Domestics.

This subject may be likened to that slough "in which have been swallowed up twenty thousand cartloads, yea, millions of wholesome instructions, yet it is the Slough of Despond still."

Knowledge of the cause, however, helps the cure. There is no reason in the world why American women have poor help, except the one for which they alone are to blame. Why is it not just as appropriate and honorable for a woman to cook and wash dishes as it is to teach or sew? Why should it be worse to do in another woman's kitchen for pay the same work she does in her own for nothing? As long as the daughters of a democratic government despise the dignity of labor and respect those who are idly supported like paupers, more than those who win bread by honest toil, just so long do they deserve all the sufferings that inadequate help can occasion. It is the worst of it that the innocent must suffer with the guilty, and those who hold correct opinions are no less victims. Few girls can now be had besides the most degraded and ignorant, and servants, as a class, are constantly deteriorating. The Biddy in your kitchen is not equal to the Nancy that was in your mother's.

There is room for reform. The tone of public opinion must be elevated. The lower classes must be reached and taught and christianized. Training schools, where young women could be taught to bake, wash, iron, churn, &c., in a cleanly and civilized way, would supply us with efficient servants. Public kitchens and laundries, such as exist in Continental Europe, might be most profitably founded in every town and village, and cheaply furnish warm meals and clean linen to overworked and unassisted families.

It is a good way for young housekeepers to take a little girl and bring her up to domestic service. That the plan does not always succeed is often because mistresses are selfish and care little for the lonely child, except to "get all they can out of her." But where mutual relations of confidence and sympathy are established, in after years, when her own family is largest, and her cares most numerous, the mistress will have an invaluable friend and assistant in the girl she has trained, one who will be familiar with all her methods and feel an interest in the household weal.

That a large proportion of our domestics are selfish, dishonest or lazy, is a sad fact; but how few the homes in which the treatment of employers does not confirm them in seeking their own! Disinterested kindness will not always be appreciated or have effect; but the experiment of showing it to servants, if oftener tried, would be often and richly rewarded. The lady who commends the well-done and is patient with the ill-done, and who cares for the physical comfort and moral and religious welfare of her domestics, will find that duty coincides with interest. Even a hired girl ought to have her feelings respected, and be regarded not as a machine, but as a human being and a sister woman. She should not be turned away or neglected when sick. She should not be denied her social rights by forbidding friends, relatives, or even "followers" to see her at her place of service, thus driving her secretly to seek companionship abroad, and in late hours and stolen interviews, often to the poor girl's utter ruin. Mistresses who pursue a system of alternate pouting and petting, tyranny and weak indulgence, are those who "have most to bear" from servants. Scolding and flattery are both unworthy methods. Uniform kindness and uniform firmness is the golden maxim for all those in authority.

Kitchen Conveniences.

A kitchen should not be less than twelve feet square; it should be on the ground floor, be well lighted by opposite windows, and have an outside door. It should contain the cistern pump and the well pump. Some persons arrange the supply of water as if they thought the farther it had to be carried, the better it was. The practice of eating in the kitchen, sitting down amidst the smoke and smell of cookery, and the dirty dishes and refuse of preparation, is fatal to comfort and domestic refinement. Use your dining-room "for every-day;" if muddy feet may enter it, don't carpet it, but have the floor painted dark green and dashed with white. It should be separated from the kitchen by a pantry opening into both. This is the proper place for chinaware and table linen. A small door opening on the shelves saves many steps in replacing dishes when washed. (Fig. 130.) A store-room and closet for iron ware should adjoin the kitchen, and stairs descend from it to the cellar, which should also have an outside door

Fig. 130.—*Kitchen-Opening for Dishes.*

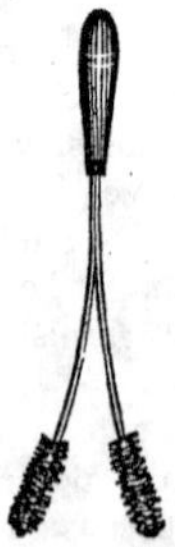

Fig. 131.—*Lamp-Chimney Cleaner.*

and flight of steps, for storing vegetables from the garden. Keep the cellar well drained and clean ; dirt and decaying matter may cause fevers in the family. Wooden sinks for washing dishes are generally uncleanly and ill-odored ; a tin dish-pan with hot suds, a milk-pan with clear rinsing-water, a clean dish-cloth and dry towels will do the business well. Wash tumblers in clear hot water, laying them in edgewise, and rolling them over and over. Wash deep and narrow dishes with a swab on a long stick. Lamp-chimneys should be similarly washed and dried by brushes of carpet yarn fastened to a double wire handle, so as to expand and contract, (fig. 131.) The wires are doubled or looped to hold the brushes.

Cheap Receipts for Cooking,

WHICH HAVE ALL BEEN TRIED.

Brown Bread.—Take 2 quarts of Indian meal, 1 quart of wheat flour, 1½ cups of molasses, 1 tablespoonful of soda, sour milk enough to mix it quite stiff; put it promptly into the oven and bake an hour. It makes six bars.

Buckwheat Cakes are many times better and more wholesome when made light and *thin.* At night mix the flour with milk-warm water, a little salt, and half a teacupful of good yeast into rather a stiff batter and set it in a warm place to rise. In the morning thin the batter with milk and add soda dissolved in hot water. They ought not to be baked up wholesale and pitched into a deep dish—that makes them heavy—but laid in neat piles on a flat plate, and baked as fast as needed at the table.

Scrambled Eggs.—Have a little lard heated in a spider, and break in eggs, one or two for each person ; stir briskly, salt, and take up when they harden. A little chopped parsley and cold ham added to the eggs, makes an agreeable dish.

A dish resembling omelet can be made by beating up eggs and frying them in hot lard. Salt and pepper, and when done invert the spider and turn out its contents on a plate. Send to the table hot.

Cottage Cheese.—When the tea-kettle boils, pour the water into a pan of "loppard" milk ; it will curd at once ; stir it and turn it into a colander, pour a little cold water over it, salt it and break it up.

A better way : Put equal parts of buttermilk and thick milk in a kettle over the fire, heat it almost boiling hot ; pour into a linen bag and let it drain until the next day. Then take it out, salt it, put in a little cream or butter, as it may be thick or not, and make it up into balls the size of an orange.

Potatoes.—It is a good way to pour off the water from boiled potatoes and dry them in the kettle a few minutes before dishing up. Set the kettle again over the fire and shake it back and forth to keep from burning. Daniel Webster recommended this plan.

When mashed potatoes are left from dinner, wet them up next morning into balls with a little milk ; having salted and peppered them, fry in a spider.

Raw potatoes are excellent fried in very thin slices. They must be well salted, fried quickly and crisp to be good.

A GOOD APPLE TART.—Always stew the apples before putting them in pastry. For this tart, make a pie without a bottom crust, and bake. Take off the crust and lay it wrong side up on a plate; put the contents of the pie on top; put on a little sugar, pour in a little cream and grate nutmeg over. An English method.

BREAD-AND-BUTTER PUDDING.—When dry bread is left, spread it with butter and pile up the slices in a pudding dish. Fill in with custard, and add a few raisins. Bake it enough to cook the custard.

BIRDSNEST PUDDING.—Pare about a dozen sweet apples, core them with an apple corer, or by driving through them the tube of a funnel, set them as close as they can stand in a deep pudding dish, fill the cores with sugar, pour in a little water, and bake soft. Make a custard of a scant quart of milk, five eggs, sugar, a little salt and three-quarters of a teaspoonful of vanilla. Fill up the apples again with sugar, pour over the custard and bake quickly.

HOTEL PUDDING.—Heat one quart of milk to a boiling heat; add one-quarter of a cup of butter; then stir in 6 ounces of Indian meal, 2 eggs well beaten, three-quarters of a cup of molasses, one-quarter tablespoonful of cinnamon or ginger. Bake one hour. To be eaten with sauce of butter and sugar.

CORN PUDDING.—Grate the green corn from two dozen ears, then carefully scrape the cobs so as not to get the chaff off; put with the corn about a quart of cold milk, 3 eggs, 2 tablespoonfuls of sugar, 1 teaspoonful of salt; if not sweet enough, add more sugar; if too thick, more milk—the consistency depends on the state of the corn. Pour into buttered plates and bake. A delicious dish for tea.

JACKSON SPONGE CAKE.—One heaping teacupful of flour, one teaspoonful of cream of tartar, one cup of sugar, three eggs, one-quarter teaspoonful of soda dissolved in a tablespoonful of hot water; flavor with extract of lemon; beat together thoroughly and bake 15 or 20 minutes. It makes two bars of cake.

You can always know whether cake is done or not by holding it near the ear. If it "sings," there are little bubbles of water boiling inside, and all is not hardened; if it is silent the cake is done.

POOR MAN'S CAKE.—This receipt, published some time ago in the COUNTRY GENTLEMAN, has been tried and approved: Break two eggs into a teacup, beat, fill up the cup with good sour cream, one cup of sugar, two cups of flour, small teaspoonful of soda or saleratus, a little salt; flavor to the taste, beat well, bake quick.

Happy farmers! Even a "poor man" among you has cream!

APPLE FRUIT CAKE.—Take two cups of dried apples; having soaked them over night, cut them up in small pieces and boil soft; add three cups of molasses and boil three minutes; then add one teacup of shortening, (half fresh lard and half butter is best,) one teacup of sour milk, and one large teaspoonful of soda, two eggs when the apples are cool; spices to taste; put in the oven and bake.

RASPBERRY VINEGAR is good for colds and sore throats, and diluted with cold water, is a most pleasant summer drink. It is thus made: Put a pound of fruit into a bowl and pour on a quart of good vinegar; next day strain the juice on a pound of fresh raspberries, and the following day the same; but do not squeeze the fruit, only drain the liquor as dry as you can. The last time strain it through a coarse bag wet with vinegar to prevent waste. Put it in a stone jar, with a pound of crushed sugar to every pint of juice, stir it when melted, and put the jar into a saucepan of water, or on a hot hearth or stove top; let it simmer and skim it. When cold, bottle it. Use no glazed or metal vessel for it.

Country tables ought to be more abundantly supplied with ripe fruit. In its fresh state, the various kinds, with or without cream and sugar, form the most healthful and delicious food. Dried and canned fruits are good for winter use; but it is strange to see a farmer's wife with much care and pains compounding made dishes which are not half as good as the ripe berries, currants, peaches, &c., left untouched until cooked up and spoiled!

MISCELLANEOUS ITEMS.

A plain sort of extension table is easily made thus. Take a wide smooth plank, cut it so as to be as long as the dining table is wide; fasten under it two long wooden prongs, just far enough apart to slide into the ends of the opening left by the support of the table leaf when drawn out to raise it. This board forms an extra leaf, accommodating four persons; and should have a leg on the outer edge, fastened on by a hinge, to steady it when in use and fold back at other times.

So the legs of ironing tables should be made, and the table itself should be long and high, and attached by hinges to the wall, to double down and save space when not wanted.

A broom is best for the regular family sweeping, and a carpet-sweeper for occasional work. To clean and brighten carpets and rugs, sprinkle wet Indian meal, or snow, over them, and sweep.

Floor matting sometimes will not go down smoothly—the inside being looser than the edge of the breadth; wet the edges and it can be drawn out flat and straight.

Do not buy dark or very white kerosene—the light yellow is safest. Lamps with "sun-burners" economize the light most. There is a patent hanging lamp fastened to the ceiling by a strong coiled brass spring, which can be pushed up or pulled down to any height at pleasure. The apparatus is durable. Such a lamp is nearly as convenient as gas, and is cheerful over the tea-table or work-table.

A light in the room at night is bad for the eyes and the nerves; a kerosene lamp turned down low gives out pernicious gases. Where a light *must* be used, a home-made taper is good and costs nothing. Pinch up the middle of a bit of unsized paper between thumb and finger, and twist it spirally till it has a point; tie it tightly with thread around the neck of this point, and cut a circle around it about 1¼ inches in diameter, and flatten out the taper all around the point till it stands up at right angles to the rest. When the taper is to be burned set it on a thin slice of cork on a saucer of lard; touch the tip of the taper with lard and light it.

Paper sometimes will not stick to an old wall. Where there is danger of this, painters use vinegar in the paste, instead of water.

Whitewashed walls will not rub off if the lime is thus prepared: Mix half a pailful of lime and water; take half a pint of flour, make a starch of it and pour it hot into the whitewash; stir it well.

Bricks, covered with carpeting, are good to set behind doors to keep them from going back against the wall.

A convenient bag for shoes may be made of gingham or delaine, and hung inside a closet door; the back of it straight, nearly the width of the door, and a good deal longer than a pair of shoes; attach loops to it to hang on nails driven into the door; the front of it is gathered full, sewed by a seam to the back at the bottom and sides, and stitched fast to the back at regular intervals, forming three pockets; hem and shirr the top and draw it up loosely with tape. (Fig. 132.) Others recommend a low square box on casters, cushioned on the top, and fitted with two drawers, one for hose and one for shoes.

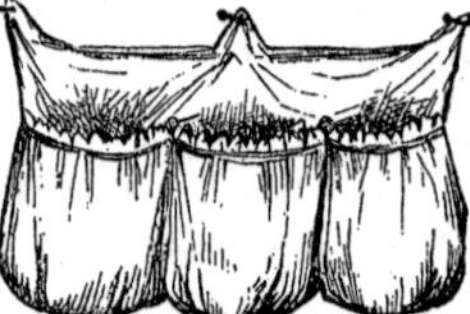

Fig. 132.—*Shoe-Bag.*

A fold of cotton wadding laid across the shoulder blades within the vest or dress, is a protection to the lungs in a long, cold ride. A newspaper is also good to keep out the wind.

Miss Beecher says: "In packing household furniture for moving, have each box numbered, and then have a book in which, as each box is packed, note down the number of the box and the order in which its contents are packed, as this will save much labor and perplexity in unpacking."

White lead paint will sometimes mend broken china very nicely. After applying it to the fractured part, let it remain undisturbed two or three months to dry. Quick-lime mixed with white of egg is a perfect cement.

These methods of restoring unfortunate knife-handles to blades which have come off are worth a trial: "Take a small portion of goose-quill and put into the handle of the knife, warm the blade and when it is hot put it into the quill in the handle, and press it in very firmly. Or brick-dust stirred into melted resin makes a composition that will fix knives and forks in their handles. The tang should be thrust in warm. N. B.—Knives and forks that are not fastened to the handles by rivets should never be put in hot water."

Iron rust is removed by salt mixed with lemon juice. Mildew, by dipping in sour buttermilk and laying in the sun. Ink stains may be sometimes taken out by smearing with hot tallow, left on when the stained article goes to the wash. Freezing will take out old fruit stains, and scalding with boiling water will remove those that have never been through the wash. Sizer's "Magic Lightning or Grease Eradicator," is, despite its pompous name, a most useful family servant in the way of removing all grease from cloth, kid, silk, ribbons, carpets, &c. Wet a little cotton batting with it and rub on the spot. If you conveniently can, put a cloth under to absorb the departing grease. It leaves no stain or odor. Keep it corked tightly. We know of nothing better.

When a hoop-skirt is soiled, put it into a tub of hot suds, and rub it with a clean scrubbing-brush or whisk broom. Rinse it and hang it in the sun. It will be whitened, but a little stiffer than before.

It is advised that calicoes be stiffened with starch made of coffee-water, to prevent any whitish appearance. Colored articles should never be hung to dry in the sun, which is sure to fade them.

Dry white woolen stockings on shingles cut the right shape and size. Each member of the family should have a pair or more of these stocking boards. Pin the hose over the upper edges and hang on the line by strings, to dry. They cannot shrink and need no ironing.

Old ribbons will look quite renewed if washed in cool suds made of fine soap and ironed when damp. Cover the ribbon with a clean cloth and pass the iron over that. If you wish to stiffen the ribbon, dip it, while drying, into gum arabic water. White silk gloves wash well, and should be dried on the hands. Never dampen bonnet ribbon and iron it wet—it makes it stiff as horn.

Nottingham lace curtains may be done up to look quite as good as new by the following process: Make a thin starch and add, for each pair of curtains, three cents' worth of gum arabic, six cents' worth of white glue, a tablespoonful of crushed or granulated sugar, and butter the size of a small plum. After the curtains are washed and dried, dip them in this starch; spread them out on the line; when dry dip and dry again, and then dip a third time. Then when they are partly dry, set large tables out of doors in the bright sun, cover with sheets, pin on the curtains and keep stretching them and changing the position of the pins till they are quite dry. Be careful to draw out every mesh to its original form, and to get the curtains even in length and breadth. If the sun is bright, this part of the work will last about an hour.

This selected receipt for clear starching is a good one: "Collars, undersleeves and handkerchiefs of very fine muslin or lace will not bear much squeezing or rubbing when washed. They can be made perfectly white and clean without either. Rinse them carefully through clear water; then soap them well with white soap; place flat in a dish or saucer, and cover with water; place them in the sun. Let them remain two or three days, changing them frequently and turning them. Once every day take them out, rinse carefully, soap and place in fresh water. When they are white, rinse and starch in the usual way." Starched laces, &c., look best pressed between towels in a large book, and not ironed. Embroideries, daisy edgings, Marseilles goods, &c., look much better when ironed upon the wrong side.

SWEET CORN.

GREAT IMPROVEMENT has been made in the varieties of sweet corn, so that by planting the earliest sorts, followed by the medium varieties, and ending with those partaking of the character of those termed the "evergreen" sorts, a succession may be kept up for some months.

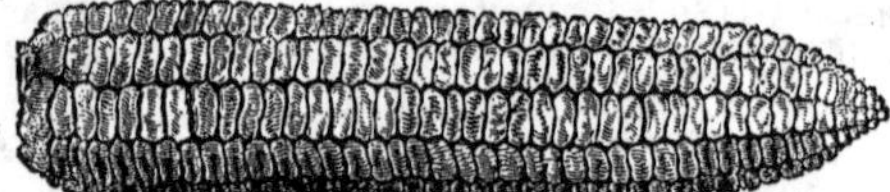

Fig. 133.—*Darling's Early.*

The accompanying cuts represent three valuable varieties. Fig. 133 is Darling's Extra Early, the earliest of the tall sweet kinds, the ears small,

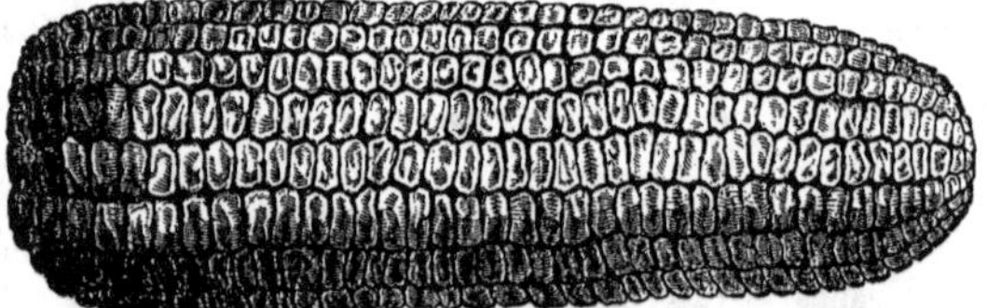

Fig. 134.—*Burr's Improved Sweet.*

well-formed and very sweet. Fig. 134 is Burr's Improved Sweet, larger, later and more productive; and fig. 135 is the Red Cob, of medium earliness, with

Fig. 135.—*Red Cob Sweet.*

a red cob, kernels very large, deep, and in straight rows; the ears are long and well filled. This is one of the best varieties for the main crop.

CULTURE OF THE WATERMELON.

THE WATERMELON, being a fruit from hot climates, needs a warm, dry soil, and warm summers. For this reason it succeeds admirably at the south, and next in the southwest and western States. It grows well in the sandy soils of New-Jersey, and supplies the New-York and Philadelphia markets in abundance. Further north it does not succeed so well; although on warm, rich, light soils, large crops are sometimes grown as far north as forty-three degrees of latitude. Where the soils are heavy more care is required to produce fruit; there must be a perfect under-drainage, and the warmest aspect must be selected. Where the summers

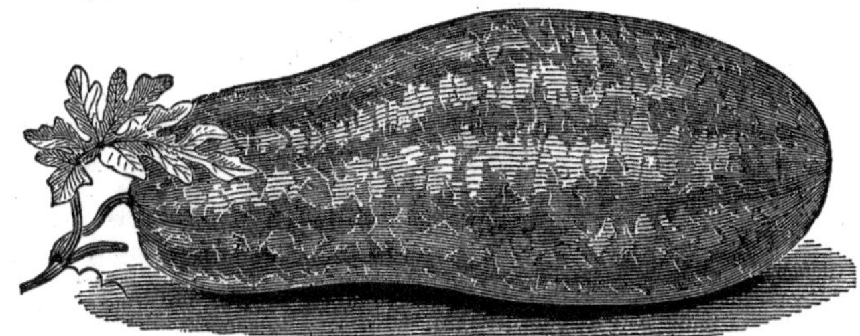

Fig. 136.—*Mountain Sweet Watermelon.*

are warm and long, as in the Middle and Southern States, it does best, and runs less to stems and leaves if the soil is not rich; but at the north, more fertility is absolutely necessary, in order to compensate for the lack of hot sun. We have known fine watermelons to be grown here on compost heaps where there was necessarily a perfect drainage and a large amount of fertility. Sometimes a turf bottom, with good light, rich soil above, has yielded good melons.

TRELLIS FOR FLOWER POTS.

A NEAT TRELLIS for flower pots contributes to the neat appearance and to the ornamental effect of the plants. The accompanying cuts represent a form made by B. K. Bliss & Son of New-York, and manufactured of reeds instead of wire, being lighter, equally durable, and retaining the shape better. They are usually painted green, but perhaps a rich brown would appear better for many plants. Fig. 137 has a 30-inch stick, is 10 inches wide. Fig. 138 has a 24-inch stick and is 14 inches wide. Fig. 139 is taller, and is suitable for placing near the centre of a group of pots. The stick is 3½ feet high, and is 14 inches wide. These supports are sold at from thirty to sixty cents each.

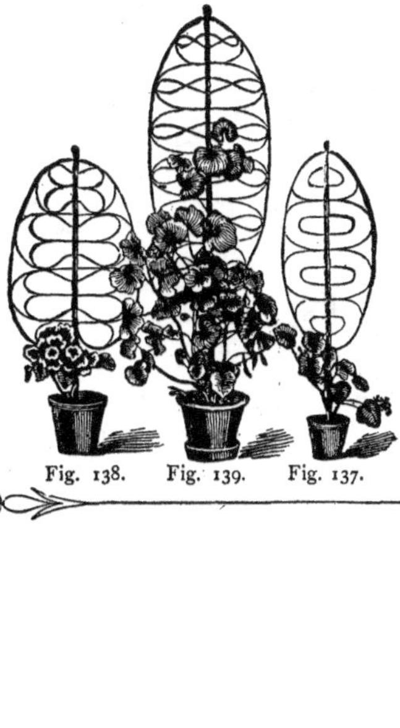

Fig. 138. Fig. 139. Fig. 137.

DITCHING IN WINTER.

FARMERS are generally crowded with work throughout the season. There are many who are days and often weeks behindhand in various jobs which they promised themselves should be done. Some of these jobs are crowded out of the list, and thrust over for another year. But we hardly ever meet with a man who keeps ahead of his work, and does more than he promises. It is therefore well to make up, so far as practicable, in winter, what has been omitted in summer and autumn.

There is another reason for making arangements for work in winter. Hired men want steady employment, and can well afford to hire at lower wages in summer, provided they can be promised continued employment through the winter. Farmers—head men—have no trouble in finding plenty of business for themselves, in the innumerable jobs that present in the shape of repairs of tools, arranging buildings, planning work, overhauling the divisions of their farms, buying or selling stock, preparing grain for seed, &c.; but hired men cannot always do these things, and must have simpler and more continued labor.

One of the most essential of all improvements, often postponed till too late, is underdraining. Very few suppose it may be performed when the ground is frozen hard, and as soon, therefore, as sharp frost commences, the work of cutting drains ceases. This is not at all necessary; but, on the other hand, it may be carried on through a considerable portion of the winter months, if properly conducted. We have on former occasions described the process by which ditches are cut with the ditching plow, the loosened earth being thrown out by hand. This process specially admits the performance of the work in winter. The following mode has been adopted where several hands were employed. Late in autumn, before the ground had become permanently frozen, the drains were laid out and the work was commenced by plowing furrows on the lines. These were deepened by repeating the process and throwing out the loose earth by hand, with pointed shovels turned up at the sides, (fig. 140.) The shovels were such as are sold commonly in the market, and the work of turning up an inch or two of the sides readily done by any common blacksmith. When the ditches become a foot or more

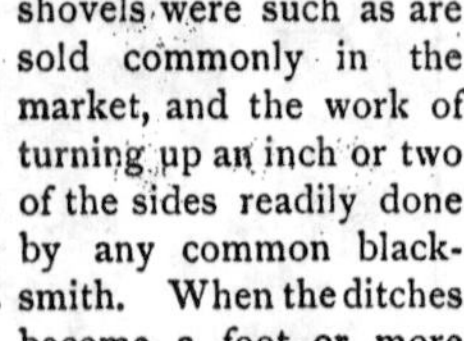

Fig. 140.—*Shovel with Sides turned up, to Throw Out Earth.*

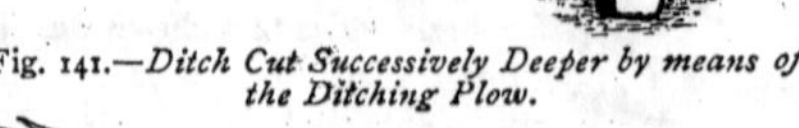

Fig. 141.—*Ditch Cut Successively Deeper by means of the Ditching Plow.*

deep (being as narrow as may be thus made) it will require a hard freeze to affect at all the earth in their bottoms. About at this point the ditching plow is brought into requisition; and loosening up the subsoil, the shovelling out is continued by hand until the required depth of two and a half or three feet is reached, (fig. 141.) Fig. 142 represents the form or construction of the ditching plow, showing the mode by which it is successively adjusted so as to run down to a depth of three feet or more, for loosening up the hard subsoil, and obviating the slow and laborious use of the pick; and fig. 143 shows the mode of attaching the horses to each end of a long evener, for drawing this plow.

Fig. 142.—*Adjustable Ditching Plow.*

Fig. 143.—*Mode of loosening up the hard Subsoil in Ditches, by means of the Ditching Plow.*

If the cold is quite sharp, the motion of the plow and of the shovels through the day will keep the earth open; and on the approach of a cold night, the ditching plow is passed all along the ditches, so as to leave several inches of mellow earth in the bottom. This loosened earth being full of air cavities, is a poor conductor of cold, and will prevent the subsoil from freezing below, at the same time that it is easily broken up again the next morning, if somewhat frozen itself.—Hence the work can go on without hindrance or difficulty. A little snow, if it happens to fall into the ditches, entirely prevents freezing, and is easily shovelled out. If very hard frost is apprehended, a load of cornstalks, well bound in bundles, will go a good way in protecting the ditches, by being dropped lengthwise along them, usually remaining at the top without falling in, and affording efficient protection.

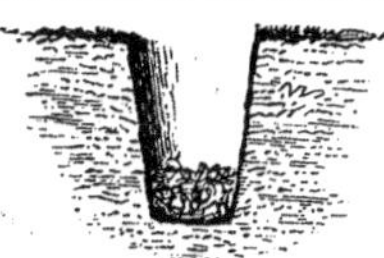

Fig. 144.—*Ditch with mellow earth left in, unfrozen next morning.*

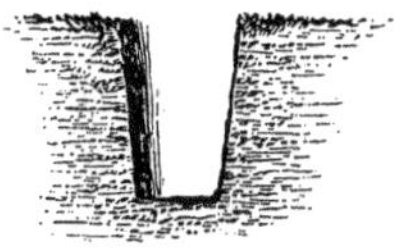

Fig. 145.—*Ditch well cleaned out, and frozen hard next morning.*

The tile should be laid and slightly covered before the thawing of spring, or much of the labor will have to be done over again by the falling in of the thawed earth. This process will have to be varied somewhat in different places, according to the severity of winters and the amount of snow which falls—which the previous directions will enable any farmer of common judgment to perform.

MARKING SHEEP.

THERE ARE TWO KINDS OF MARKING—one with paint on the wool, applied annually, to preserve the owner's name; and the other in the ears, to give the numbers and the age. The paint is often applied with a brush by the workman, in a bungling manner, "in characters uncouth," as Cowper says, if not "spelled amiss." Different modes are adopted to do the work neatly. One is to cut the initials of the name through pasteboard, sheet lead or tin plate, and paint through with a brush; but this is not so convenient or rapid as to put the letter on with a stamp. The stamp may be made in different ways. One is to cut the letters out of a piece of thin board of some kind of wood not easily split, or to cut them of stiff and hard sole leather, and then to screw them on the end of a convenient handle, as shown in the cut, (fig. 146.) A better and more substantial stamp may be made by a blacksmith, forming the letter and handle all in one piece of iron. Or the farmer may make his own stamp of a piece of small iron rod, either giving his initials, if they are of simple shaped letters, or bending the rod so as to make a circle, a triangle, or any other figure which he and others will understand.

Fig. 146.

The paint may be made of lamp-black and linseed oil—the lamp-black being previously mixed with a little turpentine. Or red ochre may be used. Sometimes hog's lard is used instead of oil, which answers well if used in warm weather. The stamp is dipped in or touched to the paint, and a single pressure on the animal just after the fleece is removed by shearing, fixes the name till the next shearing. Some sheep-owners put the letters on the shoulders, but as sheep usually stand in the pen with the back to the manager when they are selected from the flock, it is most convenient to mark them on the back or rump. The wethers may be lettered on the right side and the ewes on the left, which readily enables one to distinguish them; and if there are several rams, these may be marked on the top. An additional mark may be applied on the shoulder at the proper season, to designate the ram that has been employed.

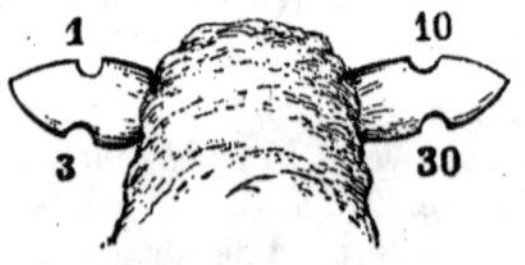

Fig. 147.

The numbers and ages are shown by marks on the ear, and these should be made when the lambs are quite young, or a day or two old, when the dams are more readily known than after the lapse of some weeks. The mode of numbering adopted by the celebrated Von Thaer has been generally adopted, by which the numbers may be readily carried up to 1,000 or more. It is as follows:

Fig. 147—One notch cut in the left ear, *at the top*, is 1.
do. do. do. under side, 3.
do. do. right ear, at top, 10.
do. do. do. under side, 30.

A combination of these notches easily makes any number up to 99.

Fig. 148—One notch cut in the left ear, at the end, is 100.
do. do. right ear, at the end, 200.

Fig. 149—The point of the left ear cut straight off is 400.
do. do. right ear do. 500.

The figures furnish examples of these markings, to which are added the holes punched through to show the age. As no owner would make a mistake of ten years in the age, these marks are much simpler:

Fig. 148.

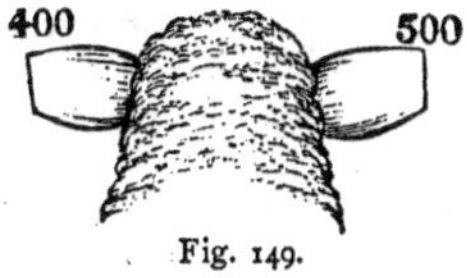

Fig. 150—One hole in the left ear is 1.
do. right ear, 3.

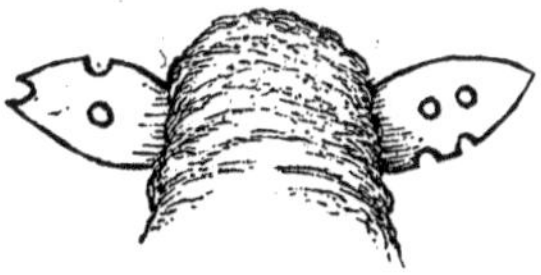

In order to explain more fully these different marks, the following references to the figures are added:

Fig. 149.

Fig. 147 is 1, 3, 10 and 30=44.
Fig. 148 is 100 and 200=300.
Fig. 149 is 400 and 500=900.

Fig. 150, giving the age, is 1 and 3=4; which means that the lamb came in 1854 or 1864, as the case may be—no hole indicating a year as 1850 or 1860, a mistake of ten years in the age not being possible.

Fig. 150.

Fig. 151 is an example showing a combination of these marks, as follows:—1+30+30+100=161; and the lamb belonging to the year 1867.

The numbers being marked every year. and the age marked besides, there is no possibility of making any mistake in a single individual. By a book register the number of the dam may be kept, the date or day of lambing, the ram, and any additional remarks.

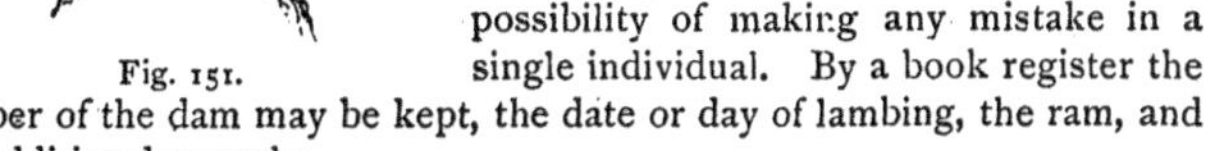

Fig. 151.

The best marker is a saddler's spring punch, which may be used for cutting the notches by placing it at the edge of the ear; or for puncturing the holes in the middle. The holes should be about a fifth of an inch in diameter. If too small, they will grow up when the wound heals.

WARTS ON ANIMALS.—A correspondent of the Germantown Telegraph says that bathing the wart two or three times a week with turpentine and sweet oil mixed, will certainly effect a cure.

RURAL ECONOMY.

WATERING HORSES.—A correspondent of the COUNTRY GENTLEMAN says: Horses should never be kept so long without water that they will drink largely when they get it. Give it to them often, and they will never injure themselves with it. Nothing is more common than to hitch a team to the plow, and make them work half a day without a drop. What man would submit to such treatment? If the plow is started at seven in the morning, water should be given again before ten; and again in the afternoon by four o'clock. Even if half an hour is thus consumed, more work will be done in a day. The objection that horses on the road should not be "loaded with water," is not valid. A horse weighing 1,200 will not be much encumbered additionally by twenty pounds of water, while the distention will give him additional strength. Every farmer knows that when he himself undertakes to lift a large log or heavy stone, he can do more by first inflating himself with air, and not unfrequently he loses a button or two from his pantaloons in the operation. Some degree of inflation by water will add to a horse's strength in a similar manner. In driving a horse on the road at a natural gait of nine or ten miles an hour, I have frequently had occasion to observe that he was laboring with perspiration until I let him drink freely, when he ceased to sweat and evidently travelled more freely. Don't be afraid to give your horses water; the danger is in making them abstain too long, in which case care is needed.

OILING AND GREASING WAGONS.—There is a matter connected with oiling machinery, of no little importance. This is keeping all journals and gudgeons free from *grit*. The farmer who thoroughly cleans frequently all the rubbing parts of his mowing machine, so that when he applies new oil it is the clear transparent fluid, and not oil mixed with sand and dirt, will keep his machine in good working order long after his careless neighbor has ground off with grit the journals of his wheels. It is so with wagons and carriages. If the sand of the soil is allowed to work into the space around the axle, it will convert the two surfaces into something like a grindstone or emery wheel, and these vehicles will need repairing in a very few years. On the other hand, keep them frequently washed throughout, and there will be less chance for the grit to enter; and if the axle is scraped and rubbed clean of the old grease or oil, every time a new application is made, the effect will be still better. It is easy to clean the axle thoroughly, but the hole should have a piece of wood made to fit it, with occasional grooves cut lengthwise in it. Thrust this into the hub and turn it about, and the old grease or oil will collect in the grooves. If this attention is carefully given, wagons will last much longer and not become so frequently out of order; and carriages and buggies will run evenly and track well for a long time after neglected ones have become swinging and ricketty.

THE

ILLUSTRATED ANNUAL REGISTER

OF

RURAL AFFAIRS.

FARM BUILDINGS.

THE TWO GREAT AIDS of improved husbandry are farm machines and farm buildings. Machinery plants, cultivates and harvests the crops, and fits them for market and consumption. Buildings secure from waste all the products of the farm, and protect from suffering, disease and death, all domestic animals. The former have received a vast amount of attention, and the inventive genius of thousands has been brought to bear for the improvement and perfection of plows, harrows, cultivators, mowers and reapers, threshers and grinders. Much less of hard thinking has been devoted to the construction of the various buildings of the farm, for securing cheapness, durability, convenience in erection, and for a saving of labor in filling, transferring and removing the many tons of their contents. With the consciousness that every contribution to a collection of desirable plans may prove acceptable and valuable, we present a few in addition to the several that have already appeared in the ILLUSTRATED ANNUAL REGISTER OF RURAL AFFAIRS.

A Curb-Roof Barn.

The accompanying view represents a curb-roof barn, known also by the terms gambrel or mansard roof, and distinguished by the angle half way up the rafters. It has an important advantage in giving more capacity

Fig. 2.—*Curb Roof Barn.*

for a given amount of siding and shingles; and a greater height above the eaves or posts, which presents no difficulty in filling, now that the work of pitching from the wagon load is wholly performed by means of the horse-fork. The cross timbers above the cross beam being entirely omitted, (except at the ends or outside,) the horse-fork has room to work freely. This barn was planned and built by W. D. Herendeen of Wayne Co., N. Y. The main portion is 40 by 56 feet, and it is well adapted to a farm of 100 to 150 acres.

Beginning with the plan, fig. 3, the construction is briefly explained. The beam between the bay and threshing floor is 12 feet high, and the girth being left out, the whole floor is accessible for the storage of wagons, farming tools, &c., after the grain is threshed or the hay fed out. Broad

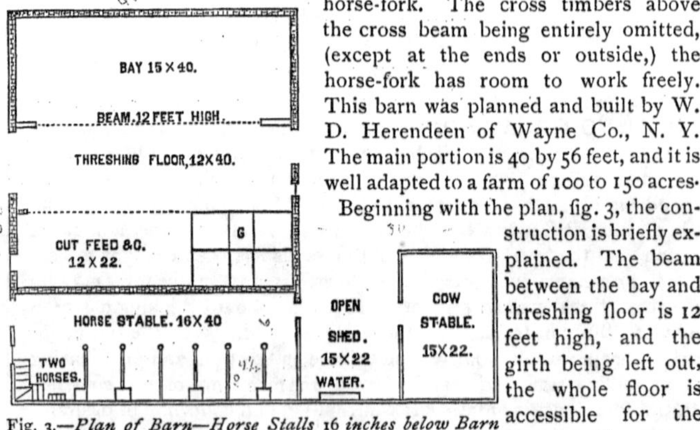

Fig. 3.—*Plan of Barn—Horse Stalls* 16 *inches below Barn Floor*, 8 *feet high in the clear.*

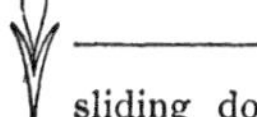

sliding doors open from the outside (on the left) to this floor, and a narrower door at the opposite side. The whole floor is covered with two-inch pine plank, planed and matched.

The cellar is under this floor, 40 feet square, (occupying all the barn except the horse stables,) and 8 feet high in the clear. It consists of an open shed on the side next the cow stable, 10 by 40 feet, with mangers next to the walls. The remaining space, 30 by 40 feet, is enclosed, and has likewise mangers next the wall. The building stands about four feet above the level of the ground in front, and the earth is filled in so as to make an easy ascent. The granary, marked G, affords ample storage for threshed grain.

The horse-stable, 16 feet wide, 16 inches below the floor, and 8 feet high, is entered by a sliding door on the left wide enough for two horses to pass. There is room for eight horses. The larger stall for two horses is often convenient for feeding a harnessed team. The others are 4½ by 10 feet in the clear, and the box hay feeders are those now commonly employed, extending upwards about 20 feet, with doors at the sides at different heights to throw in the hay. The space over the horse stables is occupied with hay for feeding them, the floor of this bay being about seven feet above the floor of the barn.

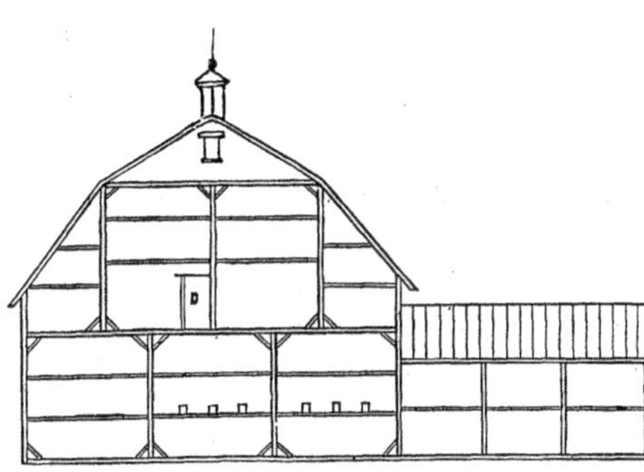

Fig. 4.—*End view of Frame. There are no Purline Girths, Beams or Braces, except at the ends, and all open from big beams up for using the Horse-Fork.*

The cow shed and stables, which are 22 by 30 feet, have a space overhead for the storage of fodder.

The end-view of the frame, fig. 4, needs but little explanation. There are no purline girths, beams or braces, (except at the ends,) to interfere with the free working of the horse-fork. Although the sills of the horse stable are lower than the others, the plates are on the same level. The ventilator is 5½ feet square and 7 feet high, with blinds.

A corn and wagon house is adjacent, 20 by 24 feet, with grain bins overhead, and places for the storage of corn in the ear on each; the wagon passage below; with a hog-pen underneath, in a walled basement seven feet high.

The cost of this barn (which was built several years ago) at the present time would vary much with the price of materials and with various facilities, from $1,800 to $2,500.

A Cattle Barn.

The accompanying design is of a barn on a farm devoted chiefly to the sale of milk, and to the dairy. The plans are taken from the barn of James Wood of Westchester Co., N. Y. The perspective view is varied somewhat from the original. The main building is 30 by 75 feet.

Fig. 5.—*View of Cattle Barn from the Southeast.*

The plans furnish nearly their own explanation. The basement is placed beneath the surface on the west, north and east sides, and, fronting the south, affords a warm aspect for the animals in the yard. The basement walls are of stone, the rest of the building wood. The cow stables in the basement are under the right hand floor above; those above are in a separate lean-to, built at the east end, with room for storing straw overhead, easily accessible from the threshing floor. The cows in the upper stable stand in yokes, three feet being allowed to each, and they face the barn floor from which they are fed—"the same arrangement," writes the owner, "as Whittier describes in his *Snow Bound*—

> "Impatient down the stanchion rows
> The cattle shake their walnut bows."

In the basement the cows are chained and separated by short partitions, and are fed from the passage-way in front of them. Chains are regarded as preferable to stanchions, being nearly as convenient, and allowing the animal greater freedom of head, so that she can lick herself, &c. The loose boxes for cows in the basement, are 5½ by 11 feet including manger. The horse-stalls are five feet wide. The shoot S, (fig. 6,) on the right hand floor, is for throwing hay down to the feeding apartment below, and the shoot S on the left floor is for roots to the turnip bin. T is a trap door for passing oats down to the bin under the stairs, and at the corner of the

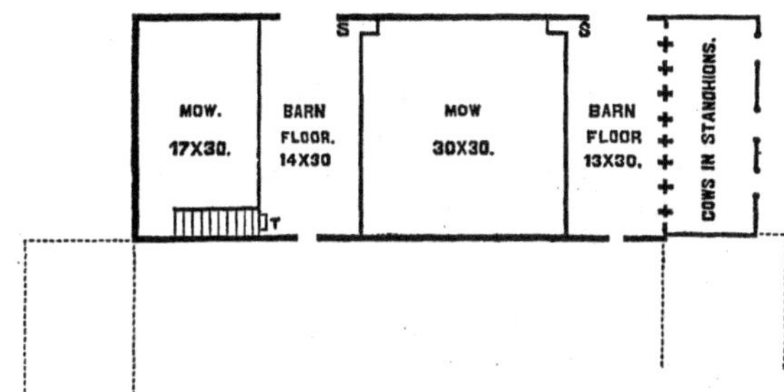

Fig. 6.—*Upper Floors.*

horse stable, from which they are easily conveyed to the animals in front. The root cellar is filled from the outside. Over each end of both barn floors are ample platforms for storing unthreshed grain and straw. The barn floors are entered from the ground at the north side or rear, where it is nearly level with these floors.

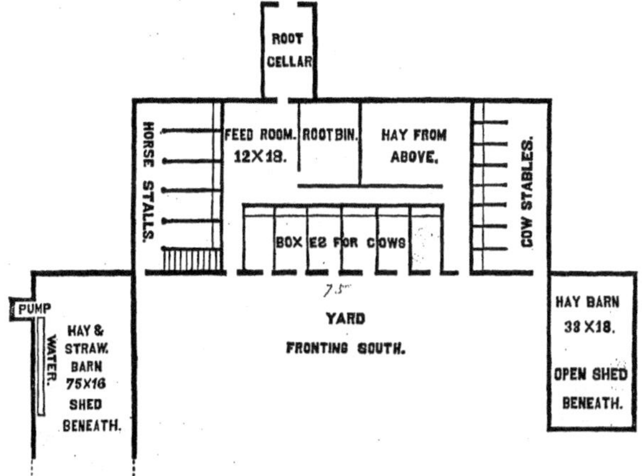

Fig. 7.—*Basement and Stables—three sides with Stone Walls under ground.*

The following is a more particular description of the fastenings in the cattle stalls, where the stanchions are used in this barn, furnished by the owner:

The posts are securely fastened to the timbers below and above, (fig. 8.) The two inch plank *a. a.* is a foot wide, with the edge to the floor, and fastened to posts, one on each side, with space of 2½ inches between; *a' a'* are similar planks, though narrower, which run above, with same space between; *b. b. b.* are the stanchions, of oak, 2½ by 3 inches, working between the

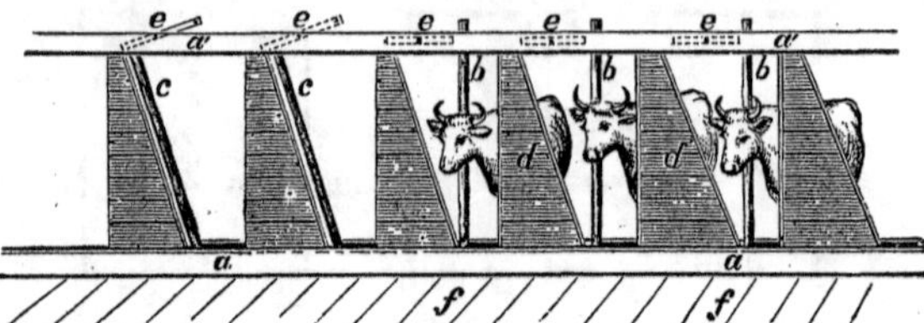

Fig. 8.—*Stanchions for Cows—a. a., Plank with its edge on the Floor—a' a', Plank overhead—b. b., Stanchions—c. c., Stanchions thrown back to let the cows enter—f. f., Floor.*

planks above and below, and turning upon a pin below. When thrown back as *c.*, they leave a wide space for the cattle to put their heads through, or back out, usually about two feet open at top. When closed, as *b.*, they leave a space of 6½ or 7 inches for the animal's neck. They are fastened in place (*b.*) by keys, *e. e. e.*, which work upon pins between the upper planks, and fall behind *b.* when in place. To unfasten the animal this is raised, and *b.* falls back under it, as *c. c.*; *d. d.* are the spaces between the upright, stationary against the animal's neck, (which should be 2½ by 3 inches, oak,) on one side, and the open stanchion on the other; *f. f.* is the floor in front of the animals, upon which they are fed. Sometimes it is left open between the animals, but it is usually better to have a narrow board partition to separate their heads, to prevent their reaching after each other's food. Sometimes separate mangers are built in front to place each animal's hay and other feed in; but this increases the expense and is inconvenient in feeding cornstalks and other coarse food; *a' a'* should be about 4 feet 6 inches high from the floor. For cows from 3 to 3½ feet should be allowed to each.

The spaces *d. d.* should be boarded up to prevent the animals working the hay through under their feet, but they are often left open.

By this arrangement a large number of cattle can be conveniently and quickly fed, requiring less labor and time than when in separate stalls. A little attention is required to place the food within their reach when pushed too far away.

The cattle should stand upon a floor, with trough for manure behind. For cows this floor should be 4 feet 6 inches in width. Then their droppings will fall directly into the trough behind them, and they will lie upon the floor in cleanliness. The stanchions do not permit the animal to move backward or forward more than the length of neck, and thus the droppings fall in one place. The manure is easily removed, and the stables kept in order, and a greater number of cattle taken care of by a given amount of labor than in any other way. The animal has entire

freedom of head for vertical motion; and, where cattle are turned out a portion of each day this mode of fastening answers well, but where they are kept up all the time chains are preferable, giving the animals more liberty of movement, allowing them to lick themselves anywhere.

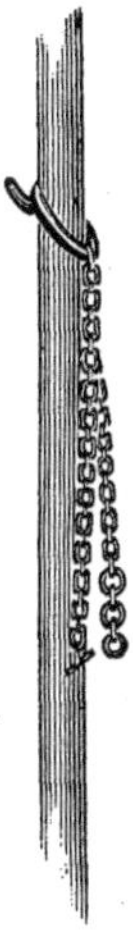

Fastening with chains is a simple operation. An upright, (fig. 9,) a post or a long staple in a post, (fig. 10,) by the side of the animal's neck, holds a large ring at the end of the chain; the ring sliding up or down, gives free vertical motion. At the other end of the chain is a spring snap, which fastens into a small ring in the chain where the snap reaches when the chain is loosely around the animal's neck, (fig. 10.) A pin in the upright to hang the ring on when not in use, is a convenience. A foot of chain between the neck and post is ample.

In chain fastenings there should be a partition between the animals, running back at least two feet from the heads, to separate them when feeding.

Fig. 9.

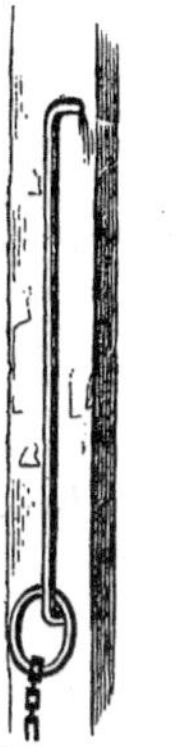

Fig. 10.

The preceding modes of securing cattle by stanchions and by chains are familiar to many; but it often happens that they are not well understood by others, and imperfect, inconvenient, or awkward styles of construction are employed, and these may assist in making some improvement.

Fig. 11.

Carriage and Horse Barn.

The accompanying view and plan represent the barn recently erected by A. B. Allen, Esq., of Tom's River, New-Jersey, and were kindly furnished by him at our request. It will be readily understood that this is not intended as a general farm barn, but for one, as Mr. Allen remarks, " who wants to

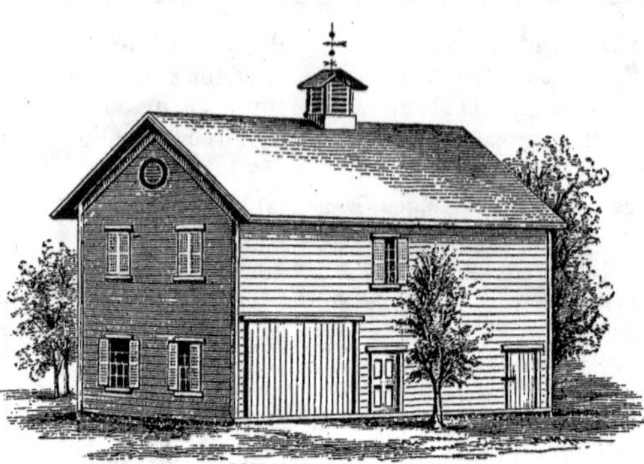

Fig. 12.—*Carriage-House and Barn.*

keep from four to eight animals, and house four to six carriages or wagons." He adds, "The great fault in stables is not depth enough from the mangers behind the horses to the walls. There ought to be room enough here to hang up rough, common harness, and lead the animals out by each other, and also to back them out easily from their stalls and turn in leading them out. In my plan there is not an inch to spare for this; and if I build another, I think I would make it one or two feet more, although this does well. Many stables are two feet less than mine. I find constant mistakes and inconveniences wherever I go, that might have been avoided as well as not. A neighbor has just completed stables with no good ventilation, although I took him to see mine (which are admirably ventilated) before he commenced building."

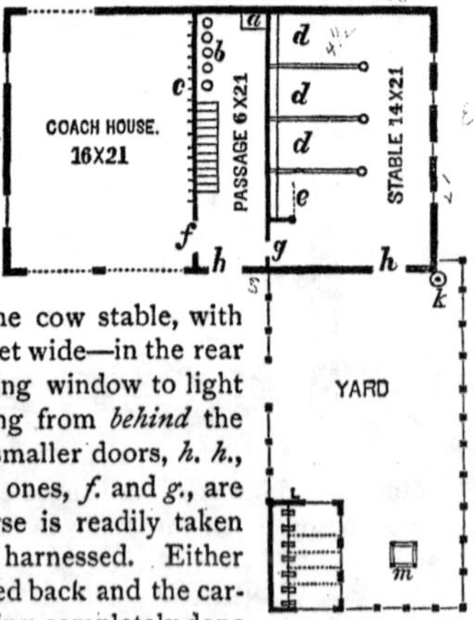

In the plan, (which is 21 by 36 feet, and with 18 feet posts,) *e.* is the cow stable, with stanchion; *d. d. d.*, horse stalls 4½ feet wide—in the rear of each of which is a one-pane sliding window to light and air the stable—the light coming from *behind* the animals, as it always should. The smaller doors, *h. h.*, are 3½ feet wide; and the two inner ones, *f.* and *g.*, are 3 feet wide, through which the horse is readily taken from the stall to the coach room and harnessed. Either of the larger doors may be then rolled back and the carriage driven out, the whole work being completely done indoors, and protected from the weather. All the doors are hung on iron rollers overhead, which run on iron rails. They can be thus opened from a hair's breadth to the full width as desired, and they

COACH HOUSE. 16X21 — PASSAGE 6X21 — STABLE 14X21 — YARD

Fig. 13.—*Plan of Barn and Yard.*

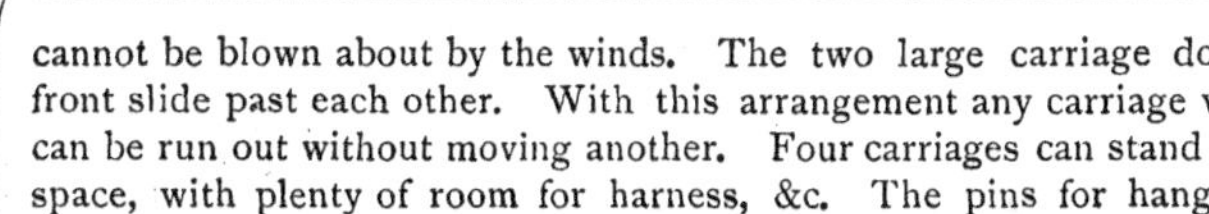

cannot be blown about by the winds. The two large carriage doors in front slide past each other. With this arrangement any carriage wanted can be run out without moving another. Four carriages can stand in this space, with plenty of room for harness, &c. The pins for hanging up harness, are shown at *c*. In the rear of the stairs in the passage are bins or barrels for meal and grain ; *a.* is a vertical box 2½ by 3 feet, of inch matched boards, which comes down within four feet of the floor, and runs up three feet above the floor in the second story, to receive the hay thrown down for the horses. The dust and seed are thus kept from scattering about.

The windows are of two sash, each with six lights, or twelve lights for each window, 10 by 12 glass. If the upper story should be filled wholly with hay, the windows would not be needed there—or should have bars inside to protect them until the hay is thrown out. The upper window at the side is to lighten the stairway, passage, &c.

The floor of the first story is two inch plank. The partitions of the horse stalls are one and a half inch oak plank, *planed smooth*. The mangers are of the same oak plank. The coach-room partition is inch pine matched boards, shutting out all stable odors from the carriages and harness. The floor of the second story is made of matched inch boards, laid as close as in a house—keeping the dust from sifting down on the carriages. Small doors are placed at suitable points to pitch in the hay, not visible in the view, being on the back side.

The stanchion is preferred for the cow stall. Her position is regarded as comfortable as if tied up, and she cannot step into her droppings—her udder and haunches are thus kept clean. She can be turned out every day for some hours into the well protected yard, and exercise and rub herself against a post to her heart's content.

The ventilator is exactly over the centre of the building.

The yard is surrounded by a high and tight board fence. The poultry house, L., is cheaply made by using the yard fence for the south and west sides. On the west side is a double row of nest boxes. The cistern, with non-freezing pump, is at *k.*, and a box hay rack at *m.*

This view of the building is taken from the south side, and the gable, showing the four windows, faces the west.

This plan may be enlarged for more animals or carriages, as the builder may desire.

Corn-House and Shop.

The accompanying plan was prepared for the Country Gentleman in response to an inquiry as regards the arrangement of a corn-crib, with a work-shop above, intended for a farm of moderate size. To secure accommodations suited to a large farm, the arrangement proposed would necessarily be different. Fig. 14 is a perspective view—the building being made of as light frame as will accord with sufficient strength, and placed on short durable posts, for the purpose of excluding rats. These posts, which should be round, and of some durable timber, should be four feet or more in length, and be set at least two feet into the ground, and stand nearly two feet above. Each one should stand on a flat stone at the bottom, and gravel or fine broken stone rammed very compactly about them. If the ground has not a perfect natural drainage, drains should be cut so as to carry off all the surplus water very promptly, as the building will not only stand better on hard well-drained soil, but the posts will last much longer. The posts are cased with tin, and an inverted tin pan placed on the top of each. The wire rim must be cut off from these pans, to prevent the rats leaping and clinging to them.

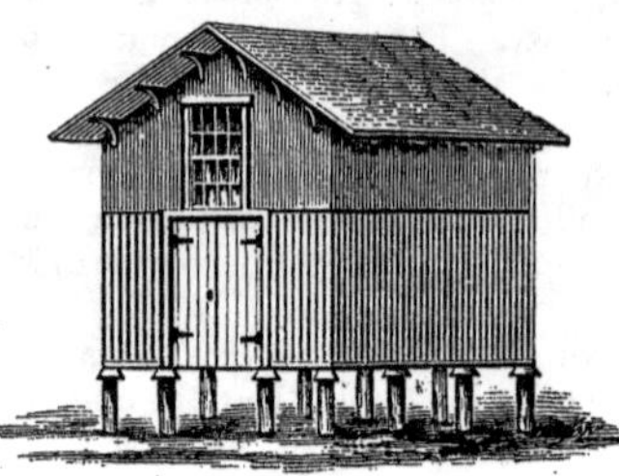

Fig. 14.—*Corn-House with Shop above.*

The doors at each end admit the entrance of a wagon, and in a great measure exclude rain and snow, although not perfectly, as the whole stands on posts. A cross-bar may be used for securing them. The floor of the driveway may be of gravel, flagging, or in the absence of these, of plank, like a plank road. On this the farm wagon may stand when not in use; and this floor will also be convenient for the shelling of corn with a hand machine on rainy days—a small opening being made to withdraw the ears. When the cribs are to be filled, the wagon is driven about half-way in, the large sliding window opened, and the corn thrown on the shop floor above with a scoop shovel. From this floor it is easily shoved into the two cribs through trap doors.

In the elevation, fig. 15, the dotted lines above show the places of the two portions of the window, when slid open. The dotted lines below show the slanting sides of the cribs inside. There is no absolute necessity for this, and none at all for slanting sides without, especially with broad projecting eaves. The stairs to the shop above may occupy the further end of one of the cribs, and a movable step be placed at the bottom.

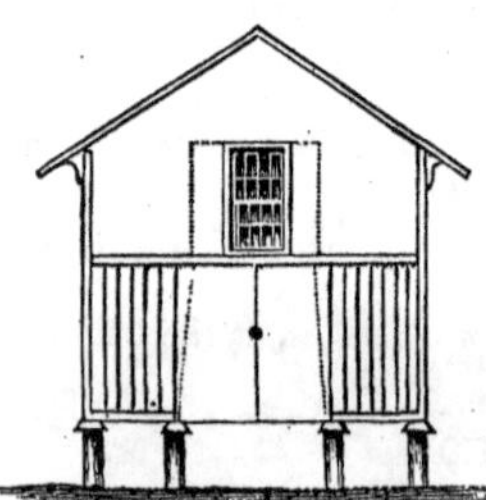

Fig. 15.—*Elevation and Section.*

When more room for corn is needed, it may be provided at the expense of shop room, by running cribs up to the roof—a second shoveling from the shop floor being required for filling the part above this floor.

A modification of this design, for a smaller building, and possessing some advantages, is made by omitting the wagon drive, and throwing the corn from the wagon, which is driven across the end of the building.

Instead of a wagon-way is an alley four feet wide between the two cribs, furnished with a floor, and with stairs at the farther end. A single door enters this alley in place of the double door in the view, fig. 14; and a plank with slats for steps is readily placed in position for going in and out, and removed when not in use, to prevent the ingress of rats.

A good, neat and moderately cheap corn-house may be built like this, twelve feet square, with twelve feet posts above the short posts, the lower part seven feet from the floor, leaving four and a half feet above at the eaves, and eight feet at the middle, with one-third pitch to the roof. The cribs will be four feet wide on each side, and the alleys, both above and below, four feet wide, which will be sufficient for filling the cribs, shelling and doing some other kinds of indoor work, besides room for smaller tools if kept neatly hung in their places. Or, the whole of the upper part may be devoted to a work and tool-shop. Fig. 16 is a perspective view of this building, with movable steps under the door, and fig. 17 a plan of the same. Such a building will hold about 500 bushels of ears up to the second floor, and 900 if filled to the roof.

Fig. 16.—*View of Small Corn-House.*

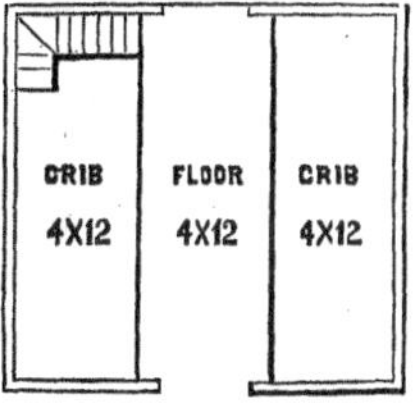

Fig. 17.—*Plan.*

Sometimes such a floor is made for the larger plan, a movable plank bridge being provided to drive the loaded wagon up. But to be strong enough, this bridge is too heavy to be easily handled, and heavier timbers are required for the floor.

The better way, where a large building is needed, is to place it on a smooth stone underpinning three feet above the ground, with an earth roadway for entering. Make the floor tight with two-inch plank, and let everything else be made substantially and fitting closely, and rats will find it difficult to get possession. They will not like to climb up a smooth wall and do the gnawing from the outside. The general arrangement may be the same as that already described, but on a larger scale.

It is most convenient to place the slats which form the inside walls of the cribs in a horizontal position, and put them in as the cribs are filled. This will enable the workmen to fill a large part of those shown in figs. 14 and 15, from the wagon way inside the building. In order to place and take out readily these horizontal

Fig. 18.

pieces, a part of the studs or supports which hold them should have cuts made in their faces, as shown in fig. 18, to receive the slats or rails, the pressure of the corn holding them in their places, or a very slight tack of a nail will secure them perfectly.

Three Designs for Barns.

A correspondent lately applied to us for plans on which to construct "a stable 35 by 25 feet, containing stalls for four horses and two cows, carriage house, harness room, &c., to be built on the slope of a hill which will permit a cellar at small cost; loft to hold six or seven tons of hay—also estimate of cost where rough boards and battens are used." As there are many inquiries of the kind, we have given some attention to the wants expressed, and present below different designs, each possessing some peculiar merit.

Fig. 19.—*Carriage and Horse Barn.*

Fig. 19 is a view of a carriage and horse barn, built of wood, with rough vertical and battened boards, but designed with a view to symmetry and ornamental exterior. It is 25 by 35 feet. Fig. 20 is the plan. The carriage room is 15 by 25 feet, which admits two carriages to stand side by side, with ample room to pass around them, and four can be placed within it without inconvenience. This apartment is separated from the stable by a matched board partition, with a door between them, which should be kept shut, so as to exclude any odor from the stable from entering the carriage room, which has a tendency to corrode the varnish. Pins for harness are placed in this partition, obviating the necessity of a separate harness room, there being ample space in this apartment between the carriages and the wall. Over each pin should be painted or printed the name of each harness, so that every one may be always hung in its

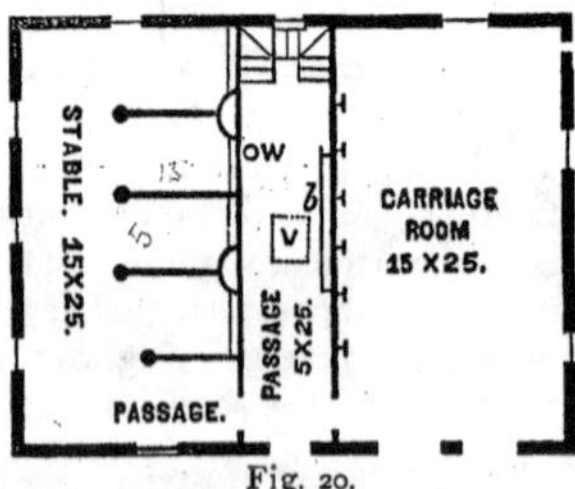

Fig. 20.

appropriate place, preventing confusion, and enabling the attendant to find readily the harness he wants.

The horse stable is 15 by 25 feet, giving space for four stalls, each 5 feet wide and 15 feet long, which is the shortest length admissible with free passage around the rear of the horses for leading them out, removing manure, &c. The light being admitted behind them, the animals' eyes are not injured by light, as they would be if their faces were in front of the openings, and the ventilation from without is more perfect. The mangers are only breast high, admitting free circulation of the air between stalls and passage. The arrangement is such that any horse can be taken from the carriage room to the stable, or the reverse, with only a few steps, and all within doors, so that there is no exposure in time of rain or snow storms. A pump at *w.*, from the rain water cistern below, is convenient for watering. The place of the ventilator is shown by the dotted lines at *v.* It reaches down to the floor above, and passes up through the roof. It has a large board valve near the top, which may be opened or closed at pleasure by means of a rod extending down within reach. This ventilator is simply a square board box or tube, two feet or more square, allowing the immediate escape of all bad odors from the stable, and preventing the diseases so common in horses that have to breathe rank exhalations. In the side of this ventilator, in the hay loft, is one or more large board doors, hung above and swinging inwards, through which the hay is thrust for the horses below. As soon as the forkful has dropped, the door falls shut, and excludes any possible vapors from the stable from entering the loft. The stairs at the end of the passage are boarded up, and a door shuts the passage between the stable and stored hay. Under the higher part of the stairs, is a closet for brushes, curry comb, soap, oils, &c.; and *b.* is a bin for horse-feed.

Fig. 21 shows a proposed plan of the basement, which may be varied with circumstances. Loose boxes or pens are intended for the cows, which, if well littered and kept clean, give more comfortable quarters than small stalls with chains or stanchions. The cistern and roots are in the most sheltered part of the basement. The roots should be kept covered above with two or three feet of straw, to prevent freezing, and the cistern should be similarly protected with six or eight inches of chaff or sawdust, in the space between the board coverings. The pump should be one of those especially intended for cisterns, and so constructed that the water will sink after using, and prevent freezing.

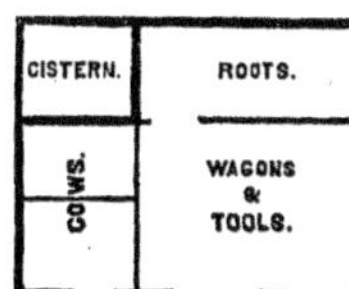

Fig. 21.—*Basement.*

Cisterns are commonly made much too small for the roofs that feed them. With three feet of annual rain-fall a roof the size of the barn we have described, would yield over six hundred barrels of water a year, and the cistern should hold enough to last three months, or at least one hundred and fifty barrels—which would require one nine feet square and six feet deep, or of equivalent dimensions. It is hardly necessary for us to remark that the inner walls of such a cistern as this, to be strong enough to withstand the pressure of the water, should be at least two feet thick, of heavy block stone.

The lower or main story of this barn should be about nine feet high in the clear. With a good ventilator, this is better than twelve feet without a ventilator. With fourteen feet posts, we should have a hay-loft over four feet high at the eaves, and with one-third pitch of the roof; there would be space in the loft (allowing more than a foot vacancy under the roof,) of about six thousand cubic feet—which would hold from eight to ten tons of hay where, as here, it could not settle as in a deep bay.

In the view, fig. 19, the lower or descending ground is seen in the rear. If the ground is not high enough at the end to pitch the hay into the loft through the large upper door, it may be put through the door under the small gable at the side.

Fig. 22.—*View of Carriage and Horse Barn.*

Fig. 22 is a view of a modification of this barn, so arranged that it may be made 5 feet shorter, the dimensions being 25 by 30 feet. Fig. 23 is the plan. The horses' heads are placed towards the outer wall, each stall being lighted by a large single pane window. In order to exclude the light from the horses' eyes, these windows are set seven or eight feet high; and immediately under them on the inside, is a narrow plank hood or shelf, entirely covering the light at the animals' heads. The hay is thrown down from above through the square boxes, from which the horses obtain their supply through the side openings at the bottom. A separate harness room is provided. Pins may be placed in the partition in the carriage apartment, in addition, if desired, as described in the preceding design. The rest of the plan explains itself.

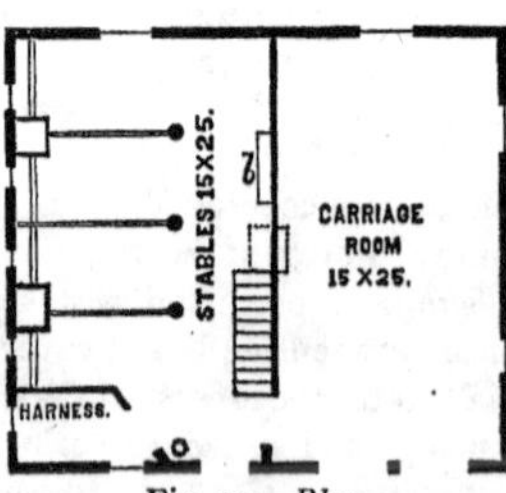

Fig. 23.—*Plan.*

The plan of a third design is shown in fig. 24, the object being to place both cattle and horses above. The dimensions are like those of the first plan, 25 by 35 feet. There are four horse stalls, each 5 feet wide, and two cow stalls, each 4 feet, with a post between. The ample width of the passage between them, prevents danger of the horses kicking the cows—especially if the more quiet animals are placed opposite. The harness room is 5 by 8 feet. A small entrance door is needed at the side, the animals being led out at the end; and a door in the elose partition between stable and carriage house, admits ready passage from one to the other. A large box or bin for feed is shown at *b*.

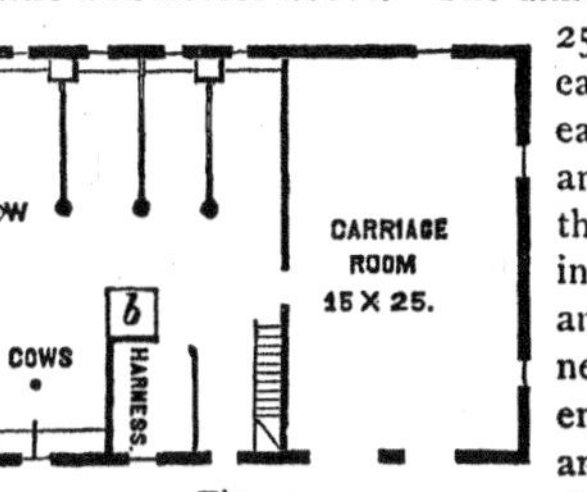

Fig. 24.

The cost of such barns will depend greatly on the price of materials and labor where they are erected, and the degree of finish given them, from $800 to $1500. White oak plank, planed and matched, for the divisions of the stalls and for the floors, will be better than hemlock or pine, but cost more. Good pine siding will be better, but more expensive, than hemlock. The painting may be cheap or costly. We have painted similar buildings at a moderate expense, and in a perfect and durable manner, as follows: First apply with a whitewash brush a heavy coat of crude light petroleum. After a month or two paint the whole with a good coating of Averill paint, which is one of the best and cheapest paints we have tried. Two barrels of petroleum would go over one of these barns, roof and all; and $30 worth of Averill paint, more or less, complete the job.

AMPLE VERANDAHS.—Woodward's little work on Country Homes remarks: The place designed simply for a summer residence for the citizen who is obliged to be at his office or counting room daily, bating the few weeks of summer vacation, need not be so complete in its appointments and arrangements as the permanent country residence. One essential condition, however, in this case is, that there shall be *room enough*, with ample verandahs, and shaded gravel walks, which will afford opportunities for open air exercise in all states of the weather. There is nothing, perhaps, that interferes so essentially with the citizen's enjoyment of the country, as the want of facilities for out-door exercise. It is too hot or too dusty to ride or walk before the shower, and after its refreshment has come it is too wet and muddy. Spacious verandahs, shaded with vines and well made walks, always firm and dry, bordered with shrubbery, or overhung with trees, will give us "ample scope and verge enough."

CULTURE OF THE CARROT.

THE VALUE OF THE CARROT for the winter feeding of horses, and for the increase in the flow and richness of the milk of cows, renders it important to reach the easiest mode of raising it. We sometimes see bad managers obtain the roots at a cost of a dollar a bushel, while the actual cost need not be a tenth part under proper culture.

SOIL.—Three requisites are essential—it must be deep, dry and loose. A shallow soil, with a hard-pan immediately below, will never raise good carrots. It will not allow the roots to extend to their full length, and it will fail to furnish the constant and uniform supply of moisture essential to their growth. It must be dry, not holding stagnant water at any time; but while it allows a proper retention of natural moisture, permit all surplus water to flow freely through it, to which end looseness is essential. The best soil is a deep sandy loam—with enough clay in its composition to absorb and retain the enriching portions of manure. A heavy clay soil is unsuitable, but may be much improved by subsoiling and trench plowing, and still more by the addition of a heavy dressing of sand. We have raised heavy crops of carrots on clayey soils, to which a coat of two inches of building sand had been added and worked in. The invertion of a heavy clover crop to the depth of a foot, by means of a large double Michigan plow, makes an excellent preparation for these roots.

RICHNESS.—On nearly all soils, the application of a heavy coat of manure is essential to a heavy crop. It will, however, be of little use unless well intermixed. If carelessly spread and plowed under in large lumps, it may do more harm than good. The best way is to work in the manure the previous year, by plowing and harrowing repeatedly, breaking it up thoroughly. Old or rotted manure will intermix best, but fresh manure will answer a good purpose if thoroughly pulverized with the soil. We have met with a few farmers who have found too much manure cause the carrots to run too much to stalks and leaves, at the expense of the roots; but failure is more apt to result from too poor a soil.

CLEAN LAND ESSENTIAL.—Failure more commonly results from the overgrowth of weeds than from any other cause. Carrots germinate slowly, and at first are small and delicate. If the soil is full of the seeds

Fig 25.—*Carrots Allowed to be Encumbered by Weeds.*

of weeds, they render early cleaning difficult, and smother more or less the young plants. (Figs. 25 and 26.) It is therefore absolutely essential to have a clean piece of ground. It should be prepared the previous year, working out all the weeds by repeated plowing and harrowing, which may

Fig 26.—*Carrot Crop (end view) Kept Clean and Mellow from the first.*

be done to advantage at the same time that the manure is worked in. It is better to fallow it one whole summer by horse labor, than to be compelled afterwards to go over the whole field with finger labor, pulling out the weeds. A successful farmer, who often raises a thousand bushels per acre, informs us that he made it a *first requisite* to work out completely all the weeds in his carrot land; and that since he has done so, he thinks, from careful estimate, that not more than *one-eighth* the labor is now required than when he first commenced.

PLANTING THE SEED.—As but few cultivators have the weeds totally eradicated, it is desirable to give the young plants as early a start as possible. Germination should therefore be accelerated by steeping in water. The best way is to enclose them in a bag and place them in water for some 24 hours. Then spread them on a shelf or floor, several inches in thickness, where they will continue moist for some days. When just ready to burst, the seed should be planted in freshly worked soil. It may be mixed with several times its bulk of sand, and planted with a drill. If not soaked, it may be well rubbed so as to render the seed smoother, and sown without mixing with sand. Two or three pounds are enough for an acre and a depth of an inch is enough, if the soil is moist.

THE DRILLS.—There are two modes of planting—on the flat surface, and on ridges. A convenient distance asunder is twenty-eight inches, and the plants should be thinned out to about three and a half inches in the

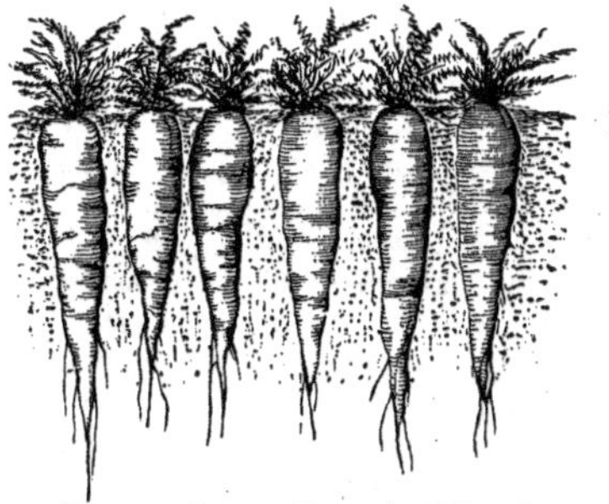

Fig. 27.—*Carrots Properly Thinned.*

Fig. 28.—*Carrots not Thinned.*

row. (Fig. 27.) If thicker, they will crowd and reduce each other in size. (Fig. 28.) If more remote, the increase of size will not compensate for the smaller number. A difference in richness of soil will cause some variation in the best distance, but 3½ inches is about the average. In our own

Fig. 29.—*Young Carrot Crop just coming up on Ridges.*

experience we have effected a saving of labor by planting on small ridges, made by throwing up opposite furrows by a one horse plow, (figs. 29 and 30.) These ridges serve as a guide and protection for the row of young plants when the harrow is passed the first time, which leaves a narrow ridge about two inches high, (fig. 31,) and this renders the hoeing more easy and rapid. The hoe, having a narrow blade set nearly at right angles, is used both for a thrusting and drawing motion—working with double speed, (fig. 32.) The objection to ridging, that it makes the soil too dry, does not hold if the soil is deep and loose, and the crop is planted quite early, as it always should be for the purpose of finding a moist soil at the surface.

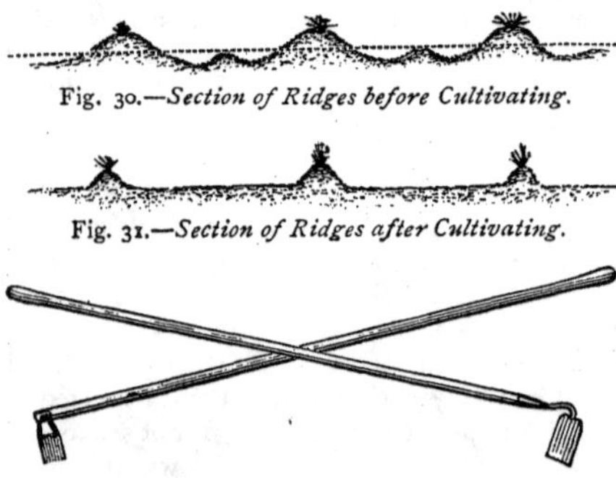

Fig. 30.—*Section of Ridges before Cultivating.*

Fig. 31.—*Section of Ridges after Cultivating.*

Fig. 32.—*Carrot Hoes.*

CULTIVATION.—This should be commenced at the very moment the plants appear. If in ridges, there will be no difficulty in seeing where the rows are. The few weeds are more easily destroyed if taken when the first green points appear at the surface, and the carrots are pushed on by the early mellowing of the earth. A narrow harrow with fine teeth, drawn by one horse, is the most useful implement for this purpose. If the weeds are allowed to get a foot high, it will of course be useless—and so will any other implement. A perfectly clean and mellow surface is more important to carrots than to any other farm crop; and if well attended to for a few weeks, the growth will be so rapid as nearly to shade under all weeds afterwards.

HARVESTING.—A good mode is to plow away from each row on one side, running as near the roots as possible, and then draw them out by hand. Use a side-hill plow for this purpose, so as to throw the earth always in the same direction, keeping the plowed earth all on one side, and the hard or unplowed portion on the other, for receiving and drawing off the roots. A strong team should be employed, and the plow run down to a depth of nearly one foot. One man should drive, and the other hold the plow, so that he may keep his eye constantly on the row. Another mode of harvesting, especially where the Long Orange is planted, is first to cut off the tops by means of a common hoe ground as sharp as a knife, and then to run a subsoil plow as deep as practicable close on each side the row, so that the blade may cut under the row. This loosens the roots, and they are readily thrown out by hand.

Carrots may be dug by hand, with considerable expedition, in the following manner: Take a good steel spade and loosen the roots along the first row, and grasping the tops in one hand throw them about three feet back from the row, with the tops towards you, shaking the roots well. Proceed in the same way with the next row, placing the roots on the top of the

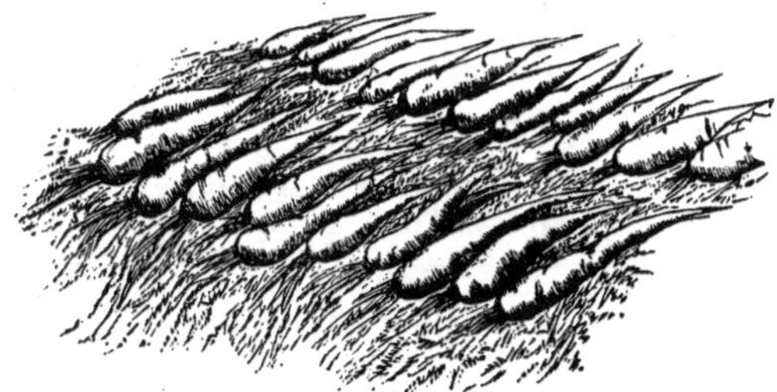

Fig. 33.—*Mode of Drying in the Sun at Harvesting—Roots up, Tops down.*

previous row, (fig. 33.) Proceed in this way until as many are thrown out as can be secured the same day. The tops on which each row rests, keep them from the ground, and they dry rapidly; and the dirt is easily removed and the tops readily cut. This mode of drying is adapted to harvesting with the plow.

KEEPING IN WINTER.—A cellar is most convenient. But if there is none, dig a shallow trench of any desired length, about 3 feet wide and 8 or 10 inches deep, where there is a perfect underdrainage. Place the roots here, heaping them up evenly, like a two sided steep roof. Cover them thickly with straw. so that when packed it may be nine inches to a foot thick, according to the latitude and winters. Cover this with three or four inches of earth beaten smooth. (Fig. 34.) The carrots will keep

Fig. 34.—*Cross Section of Carrot Heap.*

much better with plenty of straw and but little earth, than under more earth and less straw. Make air-holes with a crow-bar for each four or five feet of length, and when winter comes, plug these with straw wisps.

A cellar is more convenient, accessible in winter, and attended with less labor. A house cellar is usually too warm, and if kept as neat as every good farmer wishes, ought not to be encumbered with cattle feed. Every barn should have a cellar under it; the sills will be more durable, and the only cost is excavation and walls. A cellar large enough to hold five thousand bushels, need not cost over $200 to dig and build walls—or at the rate of four cehts per bushel once for all time. Filled with roots it would supply sixty cattle for five months with a peck each night and morning. Such a supply, in connection with dry fodder, would contribute greatly to their health and thrift. The bottom should be paved and covered with water-lime cement, so that the earth which drops from the roots may be easily and neatly swept out. A root cellar should be kept in perfect order, and no filth or decaying vegetables ever allowed in it a day. It should have windows on opposite sides, and free ventilation allowed whenever the weather will admit. If the thermometer is not much below freezing, let the wind blow through, and lessen the current as

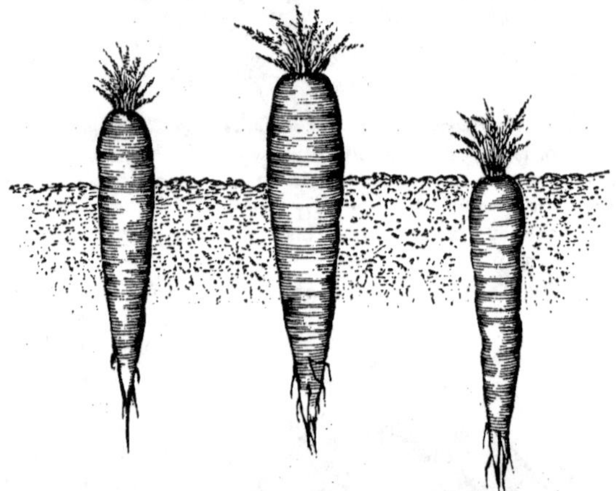

Fig. 35. *Altringham.* Fig. 36. *White Belgian.* Fig. 37. *Long Orange.*

the frost increases. The more uniformly cool the roots are kept, the better and fresher will be their condition. Many would be surprised at the amount of air which might be admitted with safety. If the tight plank floor above is insufficient to exclude frost during the severest weather, a foot of straw over the tops of the roots will be an additional protection.

When barn cellars are large, and hold many thousand bushels of roots, the ventilation will be imperfect unless a sparred floor is made a few inches above the cement bottom. It may be made of scantling of any convenient size, with inch spans between. The passages between the bins for receiving the roots should be in the direction of the prevailing winds, so as to effect a free circulation; the wind being made to blow under the sparred floor. Without free ventilation, both the roots and the timbers of the barn will decay.

VARIETIES.—Formerly the Long Orange (fig. 37) was chiefly planted for field crops, and is still an excellent sort. As the root is wholly in the soil, it is less affected by late autumn frosts than some others, and may be left later before harvesting. The White Belgian (fig. 36) is very large, and projecting several inches out of the ground, is rapidly harvested. It is one of the best varieties for general farm crops. The White Altringham (fig. 35) is similar but not so large. The Yellow Belgian is preferred by some on account of its color.

RAISING AND CURING CORN FODDER.

RAISING CORN thickly in drills or broadcast for fodder exclusively, is becoming more common, and is found to possess some important advantages. Among these are, the large quantity of fodder for soiling or for winter food which may be thus raised on a given piece of land; equalizing farm labor by sowing the fodder after the season of common corn planting, and harvesting it after common grain harvests; and the ease which it affords for extirpating weeds and leaving a clean surface. The crop may be often sown to advantage on spare pieces of land, which the owner has either not had time to prepare for a spring crop, or which may have been unsuitable for spring working.

Three different modes have been adopted for sowing the seed. When raising this crop was first introduced, sowing the seed broadcast at the rate of about four bushels per acre, and harrowing in, was the most common. But it had two objections. As it could not be cultivated, it grew with less vigor and yielded a smaller crop than when raised in drills and worked with a horse. For the same reason the field was not left clean, but was more or less encumbered with weeds, (fig. 38.) It was therefore found best to sow in drills, which may be kept clean and compact, (fig. 39,) a practice which the writer has pursued for about twenty-five years with advantage and success. The best way is first to plow and harrow the ground, as for planting potatoes or corn. Then furrow it out

Fig. 39.

Fig. 38.

with a plow and one horse, so that the furrows will be about three feet apart. The nearer they are together, provided room is allowed for the passage of horse and cultivator, the larger will be the crop. Next take a basket with a convenient bow handle, and holding about half a bushel, and from this proceed to sow the corn along the furrows by scattering it from the basket with the right hand while the basket is held by the left, or on the left arm. Any one will soon learn to do it evenly, and as fast as a man will walk. It is well for a new hand to measure a foot in length, and to count the grains, in order to determine the right quantity of seed. There should not be much less than forty grains to the foot, which will be about two and a half to three bushels to the acre, of the smaller northern varieties. If only fifteen or twenty grains to the foot, there will be only about two-thirds as much fodder per acre, and it will be coarser and be less liable to be all eaten by cattle. The thinner it is, however, the taller it will grow, and appear larger and heavier to a novice, who will at once pronounce it the best crop. The writer has tried the experiment fully by weighing the product, and finds the thickly sown corn, however short and insignificant it may appear, to yield much the heaviest.

As fast as the corn is sown it is readily covered by running a common harrow or a one-horse cultivator along the furrows, which is all that is needed.

After the corn is up, all the working required is to pass the cultivator between the rows two or three times till the crop is too large to admit it. The dense growth of the corn in the rows will shade and kill all the smaller weeds. When, therefore, the crop is harvested, the ground will be left nearly as clean as a floor. In this way the crop becomes one of the best to clean foul land; and waste pieces not in good order may thus be brought to a clean, mellow condition for crops that are to follow.

Sown corn-fodder has one great advantage over the common crop when grain is the principal object. If sown thick enough, it will bear no ears, except small ones without seed. It will consequently exhaust the land far less than the grain-producing crop, and hence also be better to precede in a rotation any that is to follow. We have taken four crops of densely sown corn-fodder, in as many years, from one piece of ground, and each succeeding crop was larger than its predecessor, which is ascribed to the fact that large quantities of roots are left in the soil, and only stalks and foliage removed.

The fodder should be cut as soon as the edges of the leaves begin to wither. Several different modes of harvesting and stacking have been adopted, with various different degrees of failure or success. When the quantity is small, it may be cut with a common hand corn cutter, bound into bundles, and placed in large, erect shocks, to stand until winter, till wanted—or if there is room in the loft of a shed, or on a bay of hay, it will dry perfectly. As this kind of fodder is finer and softer than common cornstalks from which ears have been husked, it lies more compactly, and admits but little air among its mass—and hence is more likely to heat or ferment. For this reason it must never be placed in large masses or in large stacks, else it will soon spoil, no matter how well it may have been dried in the field. We have seen fodder that had been exposed several weeks in shocks in favorable weather, and apparently quite dry, become strongly heated in three days after being stacked, sending up hot steam rapidly from the top, so that the whole had to be taken down immediately to prevent ruin.

Fields that contain several acres cannot be profitably cut by a hand-cutter. A common hay scythe may be employed, and by a little practice the operator will learn to strike the rows in such a mauner as to throw the tops uniformly in one direction, like cradling. It may then be raked up and bound by hand, and put in shocks, as already stated. A good vertical-toothed horse rake may be employed to gather it, if carefully run exactly parallel with the rows. Binding by hand is, however, so slow an operation that of late years we have dispensed with it, and made rapid work of the harvesting. If the ground is rather rough, or the fodder "lodged," it is cut with common scythes; if the surface of the field is smooth, the work is rapidly done with a reaping machine. It is suffered to lie two or three days, till partly dried, and then raked into winrows by means of a horse-rake, (one of the new vertical-teeth wooden rakes we find best,) and pitched into cocks or small heaps. From these it is pitched on the wagon and drawn to the stacking ground, adjoining the winter cattle-yard. If no ventilators are provided for the stacks, it must be put into quite small ones, not much exceeding in size a common hay-cock, or it will heat. But the best way is to provide ventilators as follows: Place three rails or poles upright,

Fig. 40.—*Poles for Stack.* Fig. 41.—*Section of Stack.* Fig. 42.—*Stack Complete.*

(fig. 40,) within a foot or less of each other, touching at the top, or witha small block placed between them. They should be set in the ground far enough to stand well, with a band around the upper end to hold them together—or if they are merely poles, the lower ends may be sharpened and more easily set into crowbar holes. The stalks are now stacked around the poles, which form a chimney for the escape of moisture and heated air, (figs. 41 and 42.)

The stacks should be made narrow and tall, and each one should not contain more than about a ton.

Another way, when the stalks are bound into bundles, is to set a single sharpened stake or pole into a crowbar-hole in the field, where each shock is to be made, (fig. 43,) and around this to place first a few bundles, tie the tops together, which will give the outside an inclined position, (fig. 44,) and on this inclined face place one, two or three successive layers of stalks, according to size, and the size of the pole, and then to bind them closely

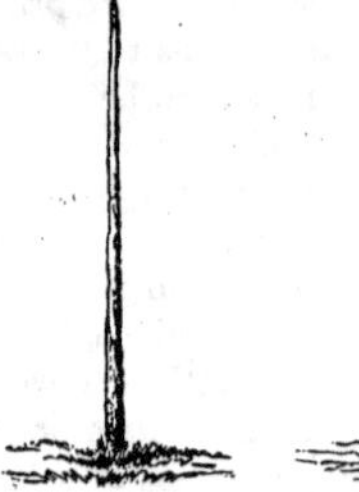

Fig. 43.
Stake for Shock.

Fig. 44.
Partly Formed Shock.

Fig. 45.
Large Shock of Fodder in Field.

around the top of the stake, (fig. 46.) An inverted bundle, as a cap, will throw the water off, as the leaves will point downwards from the stalks—while, without such a cap, the leaves will be apt to retain water up the axils. A cap of rye straw will be still better.

The farmer who has much barn and shed room, may spread his fine corn fodder over the surface of the lofts, on top the hay, and if not more than two or three feet thick, and placed sloping, or as nearly vertical as may be, will dry perfectly, even if put in quite green. The better the ventilation, and the freer the sweep of air through these lofts, the better will the drying process be facilitated, and the doors and windows should therefore be left open whenever conveniently practicable.

Corn sown as fodder, if rightly managed, forms one of the best and cheapest kinds of food for animals in winter. It is liked by cattle, which eat it up clean, both stalks and leaves, and in this respect it is greatly superior to common husked cornstalks. A portion of these fine stalks for horses promotes their healthy condition, and is much relished by them. But we must not be misunderstood—we mean good, sweet, well-cured stalks, and not the blackened, mouldy, half decayed matter resulting from careless and neglected harvesting, and half erect or prostrate shocks, sometimes erroneously called "corn-fodder." A larger amount may be obtained from a given area of land than from the best meadow. On good rich, moist soil, thickly sown in drills, and kept well cultivated while small, we have obtained at the rate of six tons of dry fodder per acre—costing, at present prices, not over three dollars per ton, including all expenses.

HINTS ON MANAGING VARIOUS CROPS.

DRAWING IN GRAIN.—The following excellent arrangement for harvesting and drawing in barley is given in the COUNTRY GENTLEMAN, as adopted by O. S. Lewis of Knowlesville, N. Y.: Mr. Lewis cuts barley with a mower before it is dead ripe; rakes with a sulky rake before it is dry enough to shell; puts it in good sized cocks, where he says it keeps well, and can be drawn without further drying when convenient. It is drawn in very fast, using three teams, one loading, one passing to and from the barn, and one unloading. The barley is unloaded with a horse-fork very fast; a smart boy drives the teams each way; another tends the horse on the fork, so there is no loss of time to the hands, either in the field or in the barn, while the work moves along very fast. A similar course is taken in drawing hay; also in drawing other grain, except that bundles are not pitched with the horse-fork.

CUTTING HAY IN THE EVENING.—This is done by some farmers, and with most excellent effect. They set the mower going at five or six o'clock, or earlier, according to the amount of hay it is desired to cut. If wet weather follows for a day or two, no harm, or but little, results, the hay being green. If, however, the weather is fair, as it is in most cases—no hay being cut when the sky threatens—it will get the first rays of the sun in the morning, and by noon will be pretty well dried. Stirring it freely—during the whole day—by three o'clock it will be fit to go in; this where the grass is rather light or the stem not heavy. Here is an advantage of several hours, which the morning cutting would not get. With the use of the tedder, hay, however green, can thus be cured perfectly in a day, and come out in the winter as fresh and green as it is possible to get hay.

PEAS THE CLOVER OF THE SOUTH.—The Reconstructed Farmer of North Carolina, says: "We trust that we shall be pardoned for again alluding to broadcasting peas as a sure, rapid, and in our judgment the most effective way of bringing up the lands in this region of country. It is the season of the year for making the experiment, and we wish to urge upon the farming community to try it. Try it by all means, on a large or small scale, as you please. You are treading on sure ground. The way is easy, cheap, and the experiments heretofore have been successful. Clover is the great fertilizer north, east and west of us. It may do on some of our heaviest soils, but rely upon it the field pea is the clover of this section. Clover burns out badly on most of our lands in July and August. The field pea is in its glory at that season. Clover wants three years to perfect its work. The field pea wants only *four* months. Who has not noticed the fine effect of pea-vines and hulls when the peas have been beaten out? It can be marked as far as the eye can see. Try it on land lying out, after oats, wheat or rye, or at the last plowing of corn."

WHEAT AFTER CORN.—Samuel Williams, of Waterloo, N. Y., says that

Joseph Wright of that village, whose excellent farming we have noticed on former occasions, pursues the practice of sowing wheat after corn with much success. The high condition of his corn ground doubtless contributes in two ways to this success—first, by causing early ripening and admitting of early cutting up and removal; and secondly, by the stimulus which a rich soil gives to the succeeding crop of wheat. Mr. Williams says that forty-five bushels per acre were obtained from the wheat crop thus grown last year. This year his best wheat was on ground sown the 10th of October last autumn, but owing to the injury caused by deep snow in many parts of the field, the average was only twenty-five bushels per acre.

Harvesting Beans.—Perhaps my plan, which works admirably, will be useful to others. Plant a stake six or eight feet long firmly in the ground; around this drive four pegs one foot from the stake, letting them stick up six inches; on the top of these pegs lay four bits of board thirty inches long, two to four inches wide, forming a square; and on these, with roots to the stake, place the beans as fast as they ripen; as beans seldom ripen uniformly, you may have to take two or three times at pulling and staking. Make the stack tall and slim, and be a little particular about bringing the top up sharp. A little string around the roots of the top layer, fastening close around the stake or pole, makes all safe. I have twenty or thirty stacks made as above, which have stood all the rains since the last of August, and are now in prime order inside, and but slightly browned outside. I have not seen a sprouted bean among them, although I have repeatedly looked for such. With a good crop of beans, it will require eighteen to twenty stacks to the acre. W. B. S.

To Prevent Birds from Pulling up Corn.—As soon as the germinating corn in the planted field makes its appearance, sow corn all around the borders of the field. After a few days walk around the lot; if you find the corn pretty nearly all picked up, sow again. Two applications are sufficient, and will last the birds until the growing corn is too strong for them. About a peck at a time will answer for a field of eight to ten acres, unless the birds should be unusually numerous. Timely observation in the field will indicate whether more is needed. I have practiced this method for many years, and have found it effectual. R. M. C.

Weeds after Harvest.—Weeds in pastures are apt to be neglected at this season, and not only present a disagreeable appearance to the eye of the neat farmer, but ripen and scatter seeds to the detriment of many future crops. If they grow in thick patches, cut them down with a mowing machine; but scattered mulleins, bull thistles, &c., may be cut up singly with a sharp grub hoe.

Weeds in corn, potatoes, and other hoed crops, frequently spring up in considerable numbers, as a sort of second growth, after the fields have been kept clean up to the present time. It is very important, and will save much future weeding, to destroy these weeds while they are yet green and soft, and have not formed seeds. Many sorts will bear a

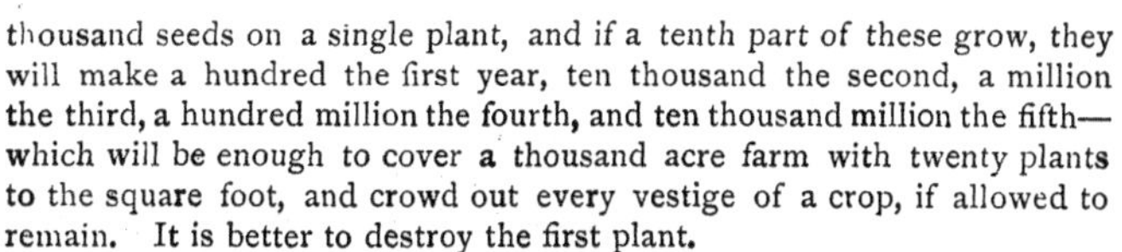

thousand seeds on a single plant, and if a tenth part of these grow, they will make a hundred the first year, ten thousand the second, a million the third, a hundred million the fourth, and ten thousand million the fifth—which will be enough to cover a thousand acre farm with twenty plants to the square foot, and crowd out every vestige of a crop, if allowed to remain. It is better to destroy the first plant.

Stubble ground may be cleared of most of its weeds by running over it with furrows, about two feet or more apart, and then harrowing the whole. The weeds will spring up in profusion, and the first plowing will turn the whole crop under. On soft, mellow ground, harrowing alone will answer.

CLOSING UP AUTUMN WORK.—Every snug farmer has a great satisfaction in putting his premises in complete order for the winter before the severe cold arrives. He don't leave anything in a slipshod style. All is neat and finished. Why cannot every one do so? We will give a little good, handy, practical advice to those who have hitherto been negligent. Take some bright, cool morning when you feel fresh and vigorous, and with a pencil in your right hand and a memorandum book in your left—or if you have no memorandum book, the back of a letter even, will do—and then spend an hour in looking over your premises and about your buildings. Note down on the spot, very briefly, all that should be done. A single word or two will be enough for a single item. Here you will see a gate with a defective hinge, or swinging open for want of a fastening. Note it down. Six boards are loose on the barnyard fence—the stable floor needs two new pieces of plank—you have no good feeding boxes for your cattle—the cold wind sweeps under the cow-shed—a pane of glass is broken in the carriage-house—your pigs are wet and uncomfortable in their pen for want of sufficient roof—your corn-fodder is spoiling because the shocks or stacks are not set upright or capped—yonder is a pile of brush on your meadow, and there is a neglected surface drain on your wheat-field, and so on. Do not be afraid of writing down more items than you can attend to—make the list as large as possible, and then you can select the most urgent, if not able to do all. After this list is completed look it over carefully, and underscore all that need immediate attention, and do these first, or copy it off arranging under three heads; first, those items that must be attended to immediately; second, those which are indispensable, but not so hurrying; and thirdly, those which ought to be done, but may on a pinch be postponed to another year. You will then see at a glance just what is to be done; and if you have not found it out before, you will now, that a man will do at least twice as much in the same time, if he has all his work systematically arranged before his eye.

Let us suggest a little more in detail. Do you see that heap of manure yonder, that covers in part the stable sill and rests against the barn boards? It is decaying the timber, and the manure is wasting. You intend to draw it out next spring for your corn? It will be worth twice as much if you draw it now, and your yard will be left with a neat appearance and your buildings uninjured. Draw out the manure and spread it as evenly as possible on the grass land you intend to invert next spring for corn. The rains will dissolve the enriching parts and carry it down into the soil, ready for the young crop of next year. Every load will thus be worth double or tripple the value you will get from it if spread next spring, left in lumps, and but little mixed with the soil.

Look now at your barnyard. Do you see that large, fine stack of straw, with a hollow or depression near the top, where it is letting in large quantities of water at every rain? It is spoiling the whole stack. Throw off the top and build it over again better, and save all your valuable straw. It is very common to despise straw because it is cheap and refuse matter, but it has many important uses, and should be saved. How much more comfortable a horse will be if lying on clean, dry bedding, than on half-rotted wet litter. Nothing is better than straw to cover potatoes and other roots for winter, provided it is dry and clean. We have found a coat of straw a foot thick, when packed on a heap of 50 to 100 bushels of potatoes, and with only three inches of earth or turf outside, far better than a little straw and a foot of earth. The dry straw absorbs all fumes and odors, and the roots are kept dry and sound.

Culture of Barley.—In the Northern States it commonly follows corn in rotation, and precedes wheat—or it may be followed by clover, for meadow or pasture. One of the important requisites is a good fertile soil. If the corn has been well manured, that will be sufficient, but if not, the best way is to *spread* the manure on the land in autumn or winter, that the enriching parts may become diffused by soaking into the soil. A wet soil will not answer, and if rather light or inclining to sandy, it may be plowed in autumn, and only stirred with a two-horse cultivator in spring, as it is important to sow early. But it is better to defer the sowing a little rather than to sow on a hard or wet or badly pulverized piece of ground. No crop needs more a thorough mellowing of the soil, but it need not be worked deep, as the roots are rather shallow. If not possessing a natural underdrainage, it should be thoroughly underdrained, if practicable; but in the absence of this, it is important to provide now, in autumn, plowed and cleaned channels to carry off all the surface water. This will facilitate the early working of the soil in spring. As to varieties, the common two-rowed is most uniformly successful. About two and a half bushels per acre is the common amount of seed. It does best, and ripens most uniformly, if put in with a drill, about an inch and a half deep.

Two modes of harvesting are adopted—one, to cut like wheat, bind in bundles and place in shocks; and the other, and most common, to cut with a machine and rake like hay, and throw into cocks. It is important that it be not subjected to rains, as they spoil the whiteness of the grain and lessen its market value. The use of the wooden barley forks, sold at the agricultural warehouses, facilitates harvesting.

We have for many years employed barley, when ground into meal, for feeding horses to advantage—preferring to sell oats and to replace it with barley, rather than have the latter manufactured into intoxicating drinks. Corn may be cheaper for feeding hogs and cattle, but the value of barley in a rotation renders it desirable, even if used exclusively for domestic animals.

KEEPING SWEET POTATOES.—The accompanying view and plan was furnished by A. L. Wood of Scioto Co., Ohio. The same plan might also be adopted for a *fruit house*.

Fig. 43.—*House for Sweet Potatoes*—(*may be used for Fruit.*)

The building is 16 by 24 feet, and 10 feet story in the clear. Studs 2 by 8 inches, on heavy sills, give 8 inches space, to stuff with dry sawdust—six inches of the same overhead. The windows, W. W., twelve panes of 10 by 12 glass, are near the inside, giving 6 inches between these and the outside shutters. D'. D"., outside and inside doors; dotted lines, temporary partition, removed for putting in the crop. X. is trap-door to cellar, 7 feet deep. A ventilator at B. lets up the damp air. A., stove for keeping the potato room warm; a small stove under O. is used in coldest weather. The front room being always warm, is used as an office, reading room and sleeping apartment, for which it is very convenient. In spring it is used for counting and tying plants. Boxes 1 to 6 hold the potatoes, 2½ by 5 feet, and 20 inches deep, extending one above the other to the ceiling. Scantling, 3 by 4, are under the lower boxes, and 2 by 2 between the rest to ventilate. The room holds 300 bushels packed in sand—closely packed would hold 400 bushels. The sand is dry and always sifted, and placed in the empty corner boxes when not in use. The inside temperature should not go below 46°, and will be 3° to 5° higher in the sand.

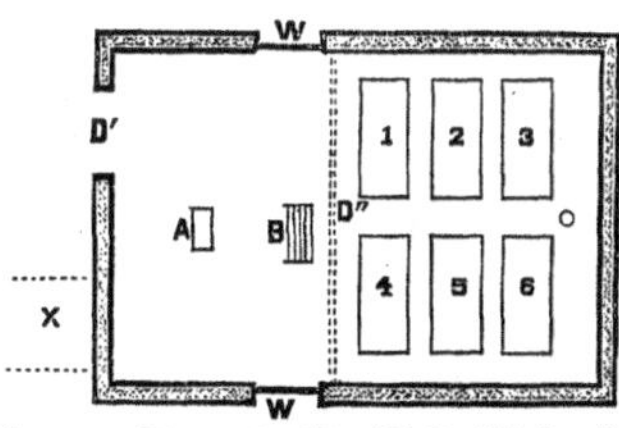

Fig. 44.—*Plan—Double Wall, filled with Sawdust.*

This house, substantially built, and painted white with green blinds, cost $750—rougher and without cellar, $350 or $400 would do.

CONTRIVANCES IN RURAL ECONOMY.

BRINGING WATER UP HILL.—A correspondent of the COUNTRY GENTLEMAN, has used a machine for two years to bring water eight rods up hill, without a cost of ten cents for repairs. A post at each end supports the wire, which is about the size of telegraph wire. At the spring it is fastened six or eight inches above the water. It is strained tight. A plank wheel with a groove is turned by a crank about eleven inches long. Two common door pulleys are attached to the ends of a piece of iron (weighing about two pounds,) bent like a semi-circle, the ends holding the wheels bending over one side, so that it will hang on the wire by the wheels. The pail or bucket is of tin, with parallel sides, the ears one-third down; one side slightly loaded, so that the bottom will not strike the water flat, or it will lift the pulley from the wire. A small copper wire, attached to the bucket, runs parallel with the other wire, and passes in the same direction through the upper post, where it winds on the wheel. If the large wire is stretched well, the posts may be fifty feet apart. The supports are turned up at the ends, and are flattened to about the size of the wire, with a notch to lay the wire in. The last post and brace should be 8 or 10 feet from the spring, and with a steeper inclination.

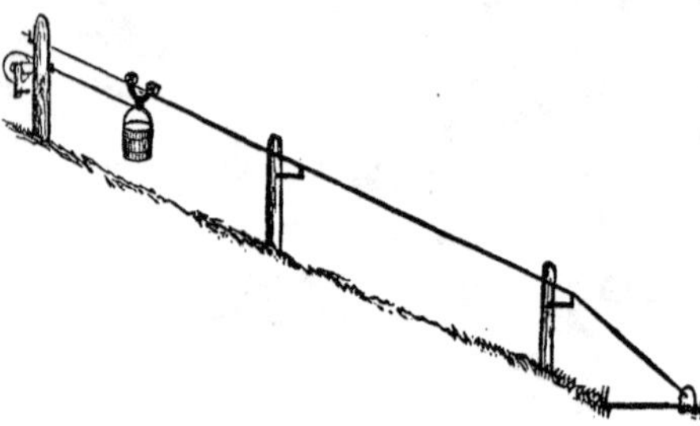

Fig. 45.—*Bringing Water up Hill.*

MEASURING IMPLEMENT.—Make the implement as shown in the cut, but lighter, with the points four feet and one and a half inches apart. It may be used for measuring rapidly, by walking it along a ditch or furrow. It is much more accurate than the common mode of pacing, and about as fast after a little practice. Four measures will make a rod, or if the points are made five and a half feet apart, three will make a rod.

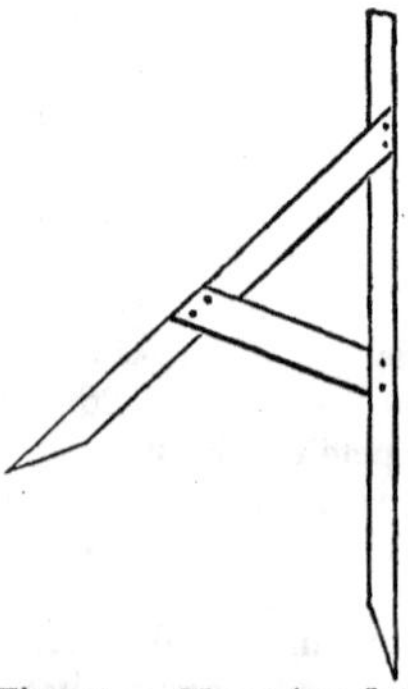

Fig. 46. — *Measuring Implement.*

CHEAP FILTERING CISTERN.—A correspondent furnishes in substance the following plan for a filtering cistern, recommended by its cheapness and simplicity, and which any mason can make without trouble, and at a cost of a dollar or two additional to any common cistern. The only drawback is a slight *taste of brick* for a few weeks at first. It consists simply of a brick partition built across, through which the water percolates slowly, but quite as fast over such a broad surface as it will be wanted for ordinary use. As there will be a great pressure of water against this wall, it must be firmly set in the cistern walls with cement, and all built up together; and it must be convex from the pump and towards the entrance spout—say a curve of six inches in five feet, or one foot in ten. Without this precaution, the wall will burst by the pressure of the water when it pours in rapidly in a hard rain.

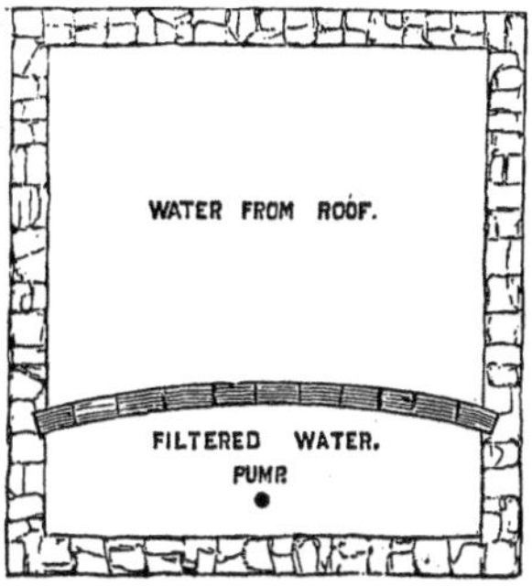

Fig. 47.—*Filtering Cistern.*

Convenient Wagon or Wheel Jack.—For greasing wagons and oiling carriages, a simple contrivance which one person can apply in a moment without difficulty, is needed. Many forms of construction have been adopted and recommended, but we have found nothing better than the one figured in the accompanying cut. It is made of a strip of plank, two and a half or three feet high, and five or six inches wide, set upright on a shorter and rather wider piece, morticed in and braced. A slit an inch and a half wide is sawed part way down, and a cross piece is spiked on at the top to make the two parts firm. The lever works in this slit on an iron pin or rod. A small chain is secured by a staple near the base of the upright, and a hook attached to the handle end of the lever. To use it, place the short arm of the lever under the axle near the wheel, then while bearing down with the right hand on the lever, hook the chain (held in the left hand) at any suitable link to the hook, and the wheel is then suspended from the ground. The whole is done in a moment, without moving a step. Several holes should be make in the upright, (not shown in the cut,) to adjust it to different heights.

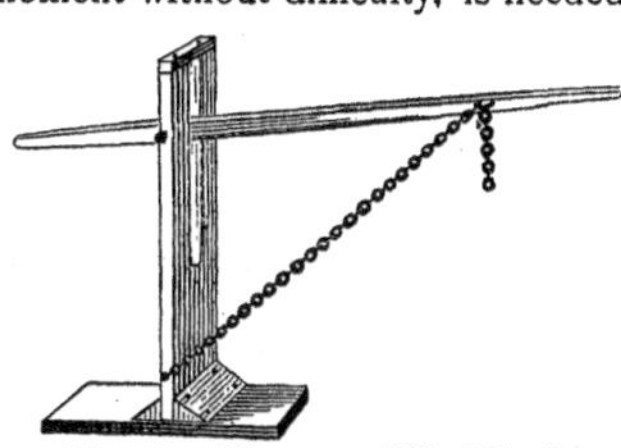

Fig. 48.—*Wagon or Wheel Jack.*

Movable Horse-Hitcher.—Whoever drives a horse and buggy has been annoyed frequently for want of a hitching post. The figures show the mode of making one which may be carried under the seat, and used whenever wanted, by screwing into the turf. It may be done in the open field, by the roadside, or on the pic-nic or crowded fair ground. The hitcher is best made of steel, the largest size steel rod used for spring-tooth horse rakes answering a good purpose. Iron rod will do, but it must be larger, (half an inch or five-eighths in diameter,) and it will not enter the sod so easily. Any blacksmith may easily make one. The coil should be about three inches in diameter, but if the soil is soft and light, four inches may be necessary, although it is better to have it longer so as to penetrate further into the earth. The ring must be large enough to give the screw a good purchase in inserting it. These hitchers, made of steel, might be made by the hundred for 25 cents each.

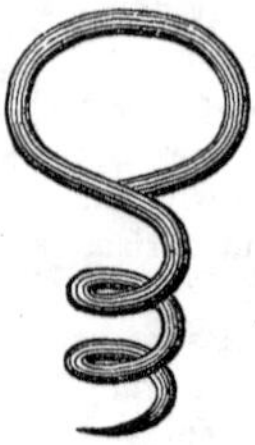

Fig. 49.—*Movable Horse-Hitcher.*

Fig. 50.—*The same Screwed into the Ground.*

This contrivance also answers an excellent purpose for tethering a horse or other animal, while nibbling grass in a back yard, fence corner or border if a cultivated field.

RACK FOR DRAWING WOOD.—The accompanying figures show a convenient, spacious and substantial form for constructing a frame or rack for drawing cord wood on a two-horse farm wagon. The side pieces are 12 feet long, 3 by 4 inches, and rest on the ends of six cross-pieces, which are two feet apart from centre to centre, and 1¼ by 5 inches in size. They are secured to the side pieces by the irons shown by A., (fig. 52,) made of three-eighths rod iron. At *a.*, *b.* and *c.*, the relative position of the pieces is shown. The stakes are 3 feet long, 1½ by 3 inches at the bottom, tapering upwards to 1½ inches square at the top. They are made of oak or other strong wood. The two at each end are secured with iron pins below the cross-pieces; the intermediate ones are merely set in and not pinned. The stakes incline about twenty degrees from the vertical, so as to give ample room between them.

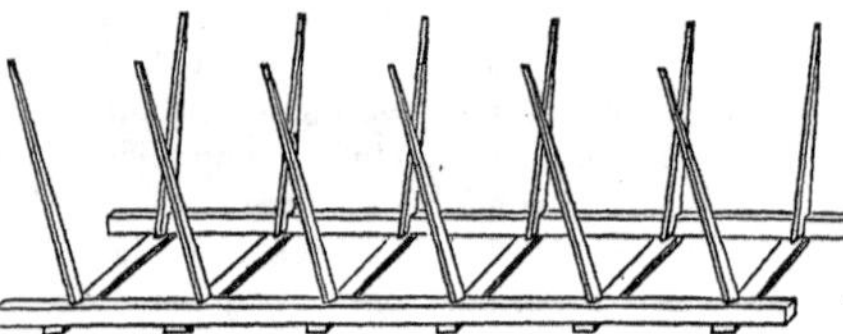

Fig. 51.—*Wood Rack.*

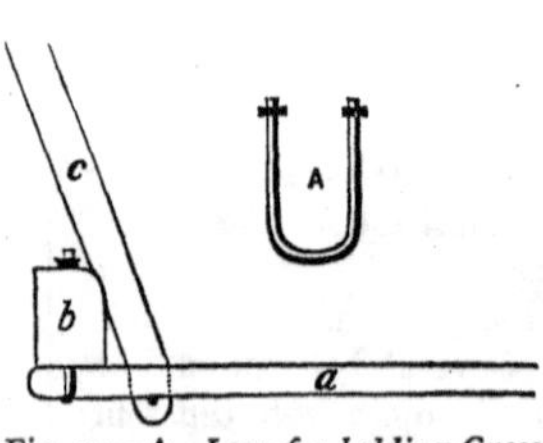

Fig. 52.—A., *Iron for holding Cross-pieces; b., cross-section of side timbers; c., lower portion of Stake; d., Cross-piece.*

TARRING GRINDSTONES.—I. Lamborn of Chester Co., Penn., makes the following good suggestion: When you have a grindstone in perfectly straight face across the stone, take a little good tar and make a ring around the stone in the centre, and it will cause it to ridge up in the middle, so as to be more convenient for grinding a perfect edge on a tool. You need not tell your careless neighbor to please to grind on the edge of the stone, for he *cannot* grind in the centre—the tar will prevent him from gouging out the middle and leaving the face irregular. I have tried this plan on my grindstone, and am pleased with it.

Out-Door Refrigerators.—The accompanying figure represents an inexpensive contrivance for keeping milk, butter, and other perishable articles in hot weather. Its size may vary with the wants of the owner, but the larger the ice-chamber is, the less frequently it would require filling, and if sufficiently large, it would keep ice all summer. The walls (fig. 53) are double, and filled with sawdust, as common in ice-houses. The door, *a.*, is double, with a space of three or four inches in it, filled with sawdust to exclude the heat. Two doors, one opening outward, and one opening inward, would be more perfect. Whether one or two, they should be made to fit very closely. These doors open into the refrigerator, which is kept cool by the ice above, and it may be lined with shelves. The joists, *b.*, must be stout, so as to hold several tons of ice above, provided a large sized building is erected. They are cut down towards the centre, so as to form a trough for the discharge of the water from the melting ice. On these joists galvanized sheet-iron is laid. On this the ice is deposited, and the iron being thus kept constantly cold, cools the air in the apartment below, by the natural descent of the cold air. By sprinkling sawdust over the iron floor, the thawing of the ice will be retarded, and thus its melting and duration may be entirely controlled, according to the depth of this layer of sawdust. The door, *c.*, receives the ice, and the window, *d.*, kept always open, is for ventilation. There should be one in each end. The freer this ventilation, the better the ice will keep—being covered with eight or ten inches of sawdust. It is better to line it with a few inches of sawdust at the sides, in addition to the sawdust walls. There is no use in a double roof.

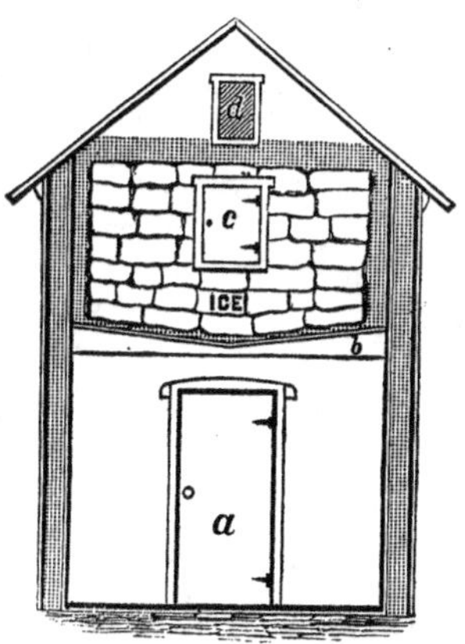

Fig. 53.—*Elevation and Section of Out-Door Refrigerator.*

Fig. 54.—*Section of Discharge Pipe.*

The water from the melting ice runs down into the trough, and thence into a lead pipe, which being bent, as shown enlarged in fig. 54, allows the water to escape freely, but excludes perfectly the warm air from without.

BOARD DRAINS.—The following mode of constructing board drains instead of tile, when the latter cannot be had, is given by a correspondent. It will be observed that a great advantage is derived from laying the tube *corner down*, as there is a stronger current of water, carrying off sediment, which would gradually settle in a flat bottom drain, and ultimately choke it.

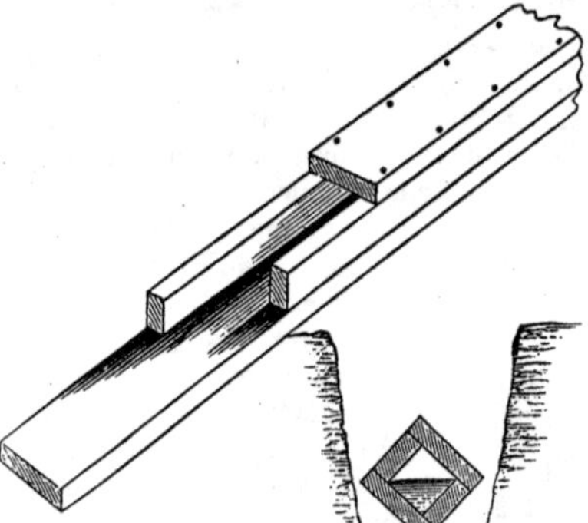

Fig. 55.—*Board Drains.*

For a Rod of Two-inch Drain.—Two boards, four inches by one inch, 16½ feet long—two boards two inches by one inch and 16½ feet long, nailed together with just nails enough to keep the boards in place. Now, by placing the ends of the boards three inches from each other, a very nice joint is made. In laying the drains, they can either be laid flat, or, what is better, on one of the corners, forming a diamond shaped drain. If there are not too many nails put in the boards, there will be plenty of room for the water to find its way into the drain.

BOARD FLOORS FOR HOGS.—Having had much experience in rearing and fattening swine, I can state confidently that hogs will thrive on a board floor, if properly taken care of. Their feet do sometimes become distorted, but seldom so much so as to interfere with their thrift. Such floors are preferably made of lath or rails laid at such a distance apart as to allow the urine, &c., to pass freely through. I know of no mode in which hogs can be kept so clean and comfortable as on such a floor in the summer season; in winter a tight floor is necessary for young pigs. J. P.

KEEPING OUT RATS AND MICE.—A. B. Allen says that sixteen years ago he perfectly excluded rats and mice from a brick stable, by laying the sleepers flat on the ground, and then filling between them with cement to their face, and before drying in the least, nailing hemlock plank firmly on them. No rat or mouse ever gnawed through that floor. The basement of his house had the floor laid in a similar manner, using matched inch pine instead of hemlock. To keep them from entering behind the lathing in the upper walls, three inches of cement was laid on the floor between the studs. He adds that barns may be kept tolerably clear by raising the floor a foot and half from the ground. Cats will then run under and catch any thing there. If laid close to the ground without cement, the floor becomes a perfect warren for all sorts of vermin.

CEMENT AROUND CHIMNEYS.—A correspondent of the COUNTRY GENTLEMAN made the roof perfectly tight around his chimneys, by making tar and dry road dust (sifted) into a thick paste, and applying with a trowel, extending four inches over the shingles. This formed a perfect collar till the roof required renewal.

CARE OF FARM IMPLEMENTS.—Carelessness and slovenly habits are not wholly confined to any particular locality, and hence we are not surprised to find that even among our enterprising western farmers, some indications of a want of care and economy occasionally occur. The Prairie Farmer gives a statement of a correspondent in which he says that during "a short day's ride" from home, he had the curiosity to count the instances where farm implements were left exposed to the weather, with the following result: Twenty-six wagons, nine buggies, two gigs, thirty-four sleds and three cutters were seen standing near houses, barns or sheds, fully exposed; four plows were standing in the furrows, and twenty-three by the roadside, turned up or laid flat; thirteen double-shovel plows, two of which lay upon an old sled, four hung upon fences, and the rest where last used; nineteen reapers were standing out, tilted up and lying in all positions; six were under shade trees and straw sheds, and three in fields where last used; seven mowers were backed up into the corners of fences or tilted around the hay stack, with rails piled on them to keep the cattle off; eight horse rakes were piled upon hog-pens or broken-down wagons; six seeders sharing the fate of other things. After the owners of these implements have paid all their taxes, they may buy a new set, as they evidently have more money than they know what to do with, judging from the freedom with which they throw it away.

TOOLS IN PLACE.—Farmers who have been in the practice of leaving their tools exposed to the weather the past season, should employ the first leisure evening in computing the relative advantages of exposure and shelter. Different implements are variously affected. A crowbar is only rusted by leaving it in the field; but that rust sticks to it all the next year, and makes rough and unpleasant handling. A spade and hoe are rusted at the blade, and hence do not run bright, easily and clean through the soil afterwards; and the handle at the socket becomes partly decayed, and finally breaks off. Figure up the cost of allowing a good hoe to become rusty, so that the man who uses it afterwards has to knock off the adhering soil a hundred and fifty times a day, or 1,500 times in ten days, besides doing only two-thirds of a day's work. Would it not be cheaper to carry it in at night? Do you remember how long you hunted for it in the morning? Implements with many joints or pieces, as plows, harrows and cultivators, are more injured and rotted; and those like mowing machines, which run by gearing, to a still greater degree, for all the difference between free and smooth running cog-work that has scarcely any friction, and a heavy, creaking, thumping motion, hard and wearing for team, results often from the rusty surfaces.

Cows Sucking Themselves.—The following contrivance, not entirely

Fig. 56.

new, is described by A. D. Newell of New-Brunswick: "I took an oak barrel head and cut it in this shape, (fig. 57,) five inches wide and eight inches long; I then sprung or bent her nose so as to get a point in each nostril; it then hung in front of her mouth. She can eat anything, but not suck. I have put small ones on calves, so that they can run in the fields with their dams, and not be able to suck them."

Fig. 57.

Probang for Choked Cattle.—The safest mode of removing potatoes or other obstructions in the throat of cattle, is described on page 205 of the 4th volume of Rural Affairs, and is particularly applicable to cases where the obstruction is in the upper part of the throat. When further down, a probang must be used. The great point is to have one with a *concave* end, so as to hold the potato in the middle of the throat. We have known a rounded or flat end of the piston to work itself between the potato and the side of the throat, and kill the animal. The following are suitable dimensions of a probang, made of hickory or tough white oak.

Fig. 58.—*Probang for Choked Cattle.*

It is three feet long, and an inch and a half thick—the concave at the piston end, and the handle at the other—worked down to five-eighths ot an inch between, so as to be flexible. Two men will hold the cow by the horns and nose, while a third with the instrument will very carefully and cautiously shove the potato down about half a foot, when it is carefully withdrawn. Cattle have been relieved and cured in less than a minute.

How to Repair a Chain Pump.—If the tube has got worn too large for the chain, so it will not raise the water properly, procure some light sole or heavy harness leather, cut into circular washers a trifle larger than the buckets; make a hole or slit in the centre; take the chain apart and slip on one of the washers next above the bucket, having it fit snugly. There should be only about four or five to any well, no matter what the depth is, as if more than two in the tube at once when drawing, the suction will be too great. Trial will show how large the washers should be left. A most efficient means of repairing a worn out establishment.

DESTROYING CROWS AND SKUNKS.—Take one dozen hen's eggs and break a small hole in either end, and with a small stick insert a small quantity of strychnine, and then place them about the cornfield, and you will have a dozen or more dead crows in the morning. The same remedy will answer for skunks by placing the egg in the hole. Be careful not to lay it where anything else will take it. A. J.

DURABLE FLAT ROOF.—Eighteen years ago I made a flat roof over the central portion of my dwelling, in the following manner: I first laid jointed flooring boards upon the joists, and covered this surface with roofing paper. Then I poured upon a section of the roof a small quantity of raw coal tar, spreading it evenly with a shingle, to the depth, perhaps, of a sixteenth of an inch. Upon this I then sifted common road dust, putting it on evenly to the depth of half or three-fourths of an inch—that is, as long as the dust continued to be wet through to the top by the tar. It took me but an hour or two to go over the whole roof—18 by 18 feet—in this manner. My first application was made in May, and about six weeks later I went over the whole surface again in like manner, finishing up with the fourth application in September. Since the application of the first coating of tar and dust, to the present time, the roof has not leaked a drop, and looks good for a century at least to come. Since the first year it has been like a firm sheet of stone, about half an inch thick, on which the family can sit, walk, run or dance, without injury to it. D. B. NEAL, *in Co. Gent.*

ECONOMY IN FARMING.

JOSEPH HARRIS of Rochester, says that a good grindstone, set true, and run by horse-power, for grinding tools, hoes, spades, and plow coulters, will pay for itself in a month. He makes his men grind their hoes every morning, and take a file into the field to sharpen them when they become dull. His men think it extravagant to grind away the hoes; but he can buy a dozen hoes for less than he pays one of them for a week's work. Tools cost nothing in comparison with labor. It does not pay to give a man a dollar and a half a day to load manure with a dung-fork with one or two teeth out. A dull, rusty hoe wlll cost more in a week than a dozen new ones. Good working horses are cheaper than poor ones. A man and team cost about $600 a year, and it is poor policy to save $100 in the original purchase, and lose $200 or $300 yearly in the amount of work done.

TWO OBSTINATE WEEDS.

OX-EYE DAISY.—This undesirable weed has been long known in the eastern portions of the United States, and in the early part of summer, fields that are thickly infested with it appear almost as dazzling white in the sunshine, as a sheet of newly fallen snow on a clear day in winter. We remember several years ago, when standing on a high piece of table-land in the south-eastern corner of Pennsylvania and looking at the faint and distant fields of New-Jersey, some fifteen miles off, of being puzzled to account for their white appearance, like snow ridges—then in a warm day of summer. We afterwards found that the whiteness came from the millions, or billions, of flowers of the ox-eye daisy, stretching for miles across the country.

Fig. 59.—*Ox-Eye Daisy.*

As the Colorado potato bug comes east, the ox-eye daisy goes west. Where they will meet, remains to be seen. In a recent ride through a part of the country where this weed was scarcely known some years ago, it was now observed in such abundance that on looking out the car windows as we shot swiftly past the banks where deep cuttings had been made, we were reminded of Bryant's description of the snow-shower—

> " As myriads by myriads, madly chased,
> Stream down the snows, till the air is white."

In most cases the ox-eye is an indication of too long a continuance of permanent meadows or pastures. It always is a proof of slipshod farming. A good, well-managed rotation will keep this weed under or out. Hoed crops, with the horse-hoe passing as often as once a week, and hand-hoeing as may be necessary, with densely sown and heavy masses of clover, and frequent alternations of the different crops which occupy the ground—will soon diminish the weed so much that hand-pulling may easily clear it out. Constant and clean cultivation is hard upon it, and clover crowds it severely—not thinly-sown clover, with bare patches, but with a peck and a half of seed to the acre, on a fine, rich, mellow surface, where it will germinate freely and grow well. Weeds, as well as insects, must be met by diligence and industry.

EXTERMINATING QUACK GRASS.—This is generally admitted to be one of the most obstinate of all weeds, and it has this characteristic, that it will never yield to half-way or slipshod management. One of the most successful experiments we have witnessed in its extermination was performed by Herendeen & Jones of Geneva, N. Y., on a wet, twelve acre lot, literally overrun with it, and all completed in a wet season. They took hold of it last season, and first tile-drained it thoroughly. They began then to destroy the quack grass, which in many cases formed an almost impenetrable mass. Although the whole season was very wet, they succeeded in wholly destroying it, and a finer piece of deep, mellow, dry, clean ground, is not easily to be found. For destroying the quack grass, they first plowed it under when about two feet high, by means of a plow furnished with a large chain to throw the vegetable growth in the furrow, drawn by a strong three-horse team. The surface was then harrowed many times, until a two-horse cultivator could be applied. It was then plowed again, harrowed and cultivated as before; and this process repeated about once a week until every vestige was dead. They find this quick work most economical.

Fig. 60.—*Quack Grass, (Triticum repens)—Right figure, before maturity; left figure, mature.*

LARGE AND SMALL FARMS.

THE ADVANTAGES of the smaller farms are:—1. They bring neighbors near together, provide good roads, lead to good schools, and to churches.

2. The owner or his sons closely supervise everything, and all may be kept neat and done in a perfect manner, and there is less eye service of hired men.

3. Less time is spent in going over a large surface, and in going to and from work.

4. Less labor is required in drawing manure to, and crops from distant fields.

The advantage of large farms are:—1. A more complete supply of farm implements and labor-saving machinery may be employed, the larger capital and larger use warranting the outlay.

2. There may be a more complete division of labor, and men skilled in each kind of work, kept at what they can do best.

3. Large and showy houses and grounds may be had.

4. The owner, if he makes a profit, may secure a larger income than from a small farm; in other words, the working of one head will bring in more money.

POULTRY HOUSES—HOW TO KEEP THEM PURE.

BUILT OF BRICK OR STONE, poultry houses are apt to be damp. Of wood they are not always warm enough. But the following is a cheap and excellent mode, the walls scarcely allowing the frost to pass. Nail common inch boards vertically; batten with two-inch strips on the joints, outside and inside. Then nail on the inside battens a complete coating of the felt commonly used on roofs under the slate—this felt is very cheap, and a few dollars will line a whole house. Then batten again, and lath and plaster on these inner battens. Or boarding may be used instead of the lathing and plastering if desired. Here will be two spaces of air in the wall, rendering it an uncommonly good non-conductor, at moderate cost. To exclude mice, fill the space at the bottom, six inches, with sifted ashes or with cement.

Fig. 61.—*Hollow Walls for a Poultry House—a., Outside Battens; b., Outside Boarding; c., Middle Battens; d., Felt; e., Inner Battens; f., Lath and Plastering.*

In a country residence, we find a good position for the hen-house on the division line between the house-yard and the barn-yard. The entrance door for the attendant is from the house-yard; but the opening for the fowls is into the barn-yard, where they can scratch and pick up scattered grain among the straw and fodder.

Every one knows that a brick or flagging floor is too cold in winter for such barefoot individuals as fowls. But a correspondent says he makes a good floor of soil, four inches thick, on a floor of flags. When the droppings fall on a smooth floor, it is impossible to keep the house sweet, as the odor taints the room before they can be swept out. But when they fall on a bed of fresh soil, most of the odor is absorbed. In one corner of the house, away from under the roosts, is a pile of good fresh garden soil. Every day a few shovelfuls are spread over the floor. The drier this soil, the better it will absorb.

CHICKEN COOP.

A DESIGN OF A CHICKEN COOP of rather neat appearance, has been furnished for the COUNTRY GENTLEMAN by J. W. Young of Pittstown, N. J., from the rough outlines of whose communication we have made the annexed sketch, by which farmers and others who have workshops may occupy the leisure hours of winter in making a few of these coops for another season. The board frame, *a.*, is of inch stuff; the cross slats of common strips of lath, one of which turns on a pin or screw at one end, and is represented raised, to afford entrance. When closed it drops like a latch into the place provided for it, and is fastened by a nail above it. The outer or roof boards are thin, and the whole is light and easily carried from place to place.

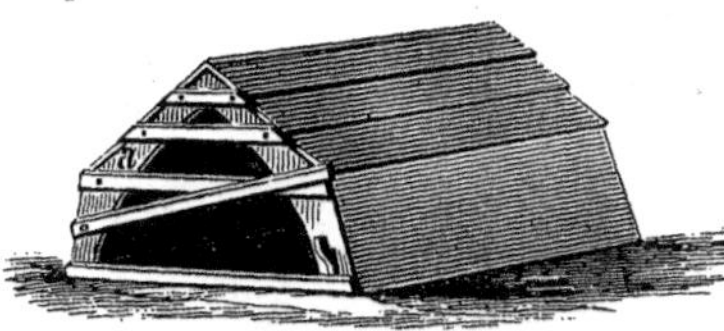

Fig. 62.—*Chicken Coop.*

Fig. 62 represents a platform on which the coop rests and keeps the chickens from the wet ground in time of rains. The dimensions are: Breadth, about 2 feet; length, 2 or 2½ feet; height, 18 inches; height of first or sloping sides, 9 inches; breadth of each sloping side of the roof, 13 or 14 inches. We suggest that when made, each of these coops receive a coat of crude petroleum, costing perhaps five to ten cents, and rendering them very durable.

Fig. 63.—*Platform for Coop to Stand on.*

AUTUMN, WINTER AND SPRING MANURING.

A YOUNG FARMER, whose rotation is corn, barley, wheat and clover, asks to which of these crops he should give his manure, and at what time of year.

Like nearly everything else in farming, the course to be pursued must vary with circumstances, and the farmer must exercise his judgment to some extent. But the following may be adopted as general rules: 1. The corn should have at least a portion of the manure, if practicable. It is scarcely possible to manure the land too much for this crop, provided it is properly applied, or so as to be well diffused through the soil. 2. The barley crop needs a good soil, but if the corn has been well manured, it will need nothing additional—the great additional points being thorough plowing and harrowing and early sowing. 3. The wheat requires more discretion in its treatment, and usually, on good land, will be sufficiently manured by the previous crops, with the exception of a top-dressing, after the last plowing, of five to ten two-horse loads of fine manure per acre. If oats are sown the second year instead of barley, a moderate manuring besides may prove useful, and sometimes necessary.

The usual accumulations of manure are in winter, but its fitness for application at different times of the year will be controlled by the materials employed in its manufacture. If composed largely of corn fodder, it will be unfit to apply till the following autumn, after rotting down in heaps. But if the corn-fodder is all cut with a machine before feeding out, it may be drawn out and spread as fast as produced. Nearly the same remarks will apply to straw, if used in large quantities as litter. In small quantities, it will not prevent winter application; or if cut up before being used for bedding, from 1 to 4 inches long.

Fig. 64.—*Badly Spread Manure.*

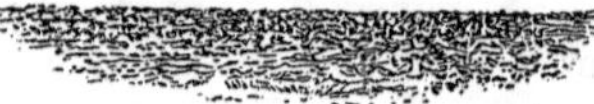

Fig. 65.—*Finely Spread Manure.*

Fig. 66.—*Badly Spread and Badly Plowed in.*

Farmers have little fresh manure in autumn. The cheapest application is in winter, drawing out and spreading over the fields requiring it, as fast as it is made. Several advantages result from this practice. It requires less handling over; it is soon out of the way; it is easily spread from the sled or wagon; it is drawn by men at a time when they may be otherwise idle; it removes the labor from the short and crowded period of spring; it allows the soluble manure to wash down into the earth and become intimately diffused; and it prevents the hardening and baking of the soil by the passage of the loaded wagons, when the ground is wet and soft, after the breaking up of winter. It should therefore be the aim to draw out, as it accumulates, all the manure which is short enough to spread well, to plow under in spring for corn or other spring crops, leaving the longest and coarsest to rot down in heaps for autumn sown wheat, or for spreading on sod which is intended for corn the next year.

We have already remarked that corn can scarcely be manured too much, if the work is properly done. If there is any danger of its running too much to leaf and stalk, which would be a rare occurrence, plant a smaller variety, and allow a larger number of stalks to grow. The succeeding barley, oats or peas, will receive a decided help from it—especially if the soil has a sufficient quantity of clay to hold the manure; and in good wheat districts, its effects will be sufficient to obviate anything further than a top-dressing. But if the soil is of moderate fertility—or if a heavy crop of oats precede the wheat,—(these two contingencies should never unite,)—an application before a shallow plowing, with thorough intermixture by the harrow, may prove advantageous in addition to the top-dressing at or near the time the wheat is sown.

We have not yet met the farmer who could make enough manure to obviate the necessity of using clover as a fertilizer, and a combination of the two generally gives excellent results. Manure spread on clover sod in autumn, as we have frequently had occasion to urge, is the best practicable or profitable preparation of ground for inverting the following spring for

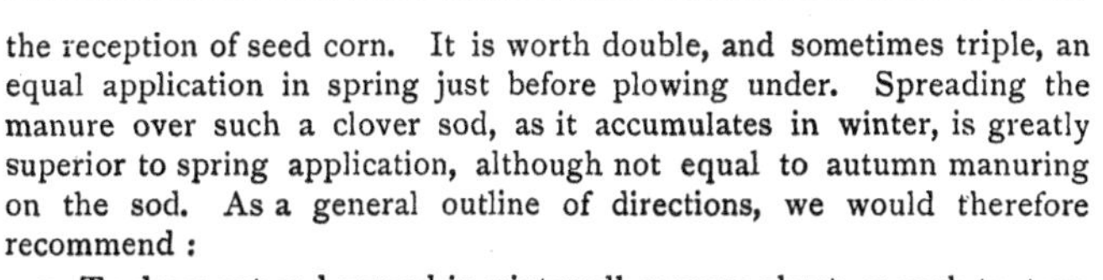

the reception of seed corn. It is worth double, and sometimes triple, an equal application in spring just before plowing under. Spreading the manure over such a clover sod, as it accumulates in winter, is greatly superior to spring application, although not equal to autumn manuring on the sod. As a general outline of directions, we would therefore recommend:

1. To draw out and spread in winter all manure short enough to turn under for corn.

2. To heap up for rotting down, all that is too coarse or long for spring.

3. To apply these heaps to sod intended for corn the next year, or to wheat fields after the last plowing, doing the work in portions at a time, as the last plowing progresses, so as not to tread the mellow soil with the teams or the wagons.

4. If applied in spring, break the manure and intermix it with the soil by harrowing and then plowing in. Ground intended for ruta bagas may be thus prepared well, as plenty of time is allowed for intermixture and preparation.

BONE MANURE.

IT OFTEN HAPPENS that bones may be had in large quantities with little cost, and when ground in a mill, and especially when made into superphosphate with sulphuric acid, they become a powerful manure on many soils. But it hardly pays for farmers to make their own superphosphate, as it requires considerable labor, some skill and experience, and the acid cannot be had so cheaply as at large manufactories. The next best way, if it can be ground, is to mix it in thin alternating layers with very wet soil, in heaps of a ton or more, and with about twice as much soil as ground bone. In a few days it will heat. Leave it undisturbed until nearly cool, and shovel it over, when it will heat again. Repeat the process until it becomes a fine powdered manure.

But if the bones cannot be ground, break them up with a sledge, by placing them on a thick flat stone, with a wide hoop around them. Then place this broken material in thin alternating layers with strong fresh stable manure, and layers of loam or earth about half as thick as the manure. The whole will heat, and the bones will work down and become soft, so as to mix well and form a rich compost when the whole is thrown over.

IRRIGATION.—Irrigate land one season, and the effect will be seen the next also—this on grass more particularly.

RASPBERRY CULTURE.

By A. M. Purdy, Palmyra, N. Y.

WHAT A CHANGE in this fruit from twenty years ago! Then the Black Cap was not to be found on the market stands, with perhaps a few exceptions of the little insignificant, seedy, wild sort. Well does the writer remember of rambling over fields in search of the raspberry, and perhaps gathering four to eight quarts after a hard day's work, with many a scratch and fall. How changed now! The market stands of our cities and villages seem to be loaded with the finest sorts, and yet the price, *the past season*, in most markets, was *higher* than five years ago, the fact being that the increase of population, and the taste and demand for this fruit, have increased so wonderfully, that it has forced prices up to paying rates; and too, how different with many families who generally intend to keep their gardens supplied with the choicest fruit. Now, by using discrimination in planting the proper kinds to keep up a rotation, their tables can be supplied for at least six weeks.

I propose, to the best of my ability, to show our method of culture, and how any family, at a very little cost and trouble, may supply their table with this choice fruit, and any party grow the same for marketing and make them very profitable. First we will give the method for field culture that we have found to succeed best and prove most profitable, and that too without the use of stakes.

We have found *any* soil that will grow a good crop of potatoes well adapted to raspberry culture, and have seen them growing on all kinds of soil with the best and most satisfactory results. In fact it is not so strictly necessary that the soil should be of that light character and easily worked as for strawberries, for the reason that the work can be mostly done with a horse and cultivator, while strawberries require a large amount of hand work. The land should be in good tillable condition—that is, having been occupied by some crop the year previous, so that the sod will be entirely rotted and subdued. *Never plant any small fruit on sod land*, if it can be helped. If it *must* be done, do the planting *in the fall*, so that the sod will get well rotted before dry weather comes on the next season, for if an unrotted sod lies under or next to the roots of any plant, in time of a drouth be assured the largest share will die out. One important thing must be observed. If the soil is of a sour, wettish nature, see to it that it is properly tilled and drained, for they will not yield good crops on such ground, neither will the plants that are layered to increase roots "take" well when water lies close to the surface.

We have practiced both the "hill" and "row" or "hedge" system, and, taking everything into consideration, prefer the latter—1st. Because we can plant from a half to two-thirds more plants on the same piece of ground, and as they yield but about one-fourth to one-third of a crop the first year after setting, we get twice to three times the amount of fruit the first year, thus making our ground pay us better the first bearing year. This is shown from the fact that in "hill" culture the plants are set 6 by 6 feet, or 1,210 plants to the acre, while if set by the row system they should be set 7 feet one way, and 2 to 3 feet the other—if 3 feet, then we have 2,076 plants. Now as a *one* year old raspberry may be allowed to bear at least a pint of fruit that season, if trimmed and cut back as we shall direct, we shall get the first bearing year, from the hill sets 605 quarts per acre, or 19 bushels; and from the last or row system, 1,038 quarts, or 38 bushels; and the second year, if they have received such attention as we shall describe, and such varieties are planted as we shall name, we name but a fair crop when we put the plants at two quarts to the bush, or from an acre of "hills," 2,420 quarts, or 75 bushels; while from the latter at least one-half larger crop can be obtained, (not as many again, for the bushes have now become so large that they meet together, and do not have that chance to fruit on all sides as when in hills;) still 100 to 125 bushels can be relied upon.

Another reason why the "row" system is profitable, is that the plants sustain each other, and are not liable to get twisted off by hard winds; and still another reason is that double the amount of plants can be obtained from the plants the first season, and at least half as many again the second season.

We usually lay off our lands so that the rows will be from 16 to 20 rods long. We mark out with a plow, being guided by a stake in the centre and at each opposite end from where we start.

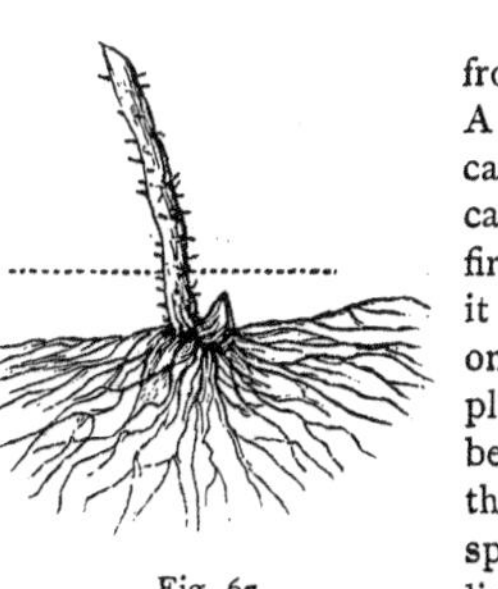

Fig. 67.

The plants are kept trenched in and taken from the trenches as they are needed to plant. A boy drops them along the furrow, and a careful person follows and sets them out, being careful to spread the roots and draw in a little firm dirt to come in contact with the roots. If it is the "sucker" variety, we usually leave on about 6 inches of wood to show where the plant is, so that it can be worked, if necessary, before the germ makes its appearance. If of the "layer" kind, such as the Doolittle, we spread out the roots carefully, (fig. 66,) the line indicating the top of the soil, with just a trifle of the cane that should be left on the plant above and the germ below the surface. Our object in leaving a little wood above is, as above stated, to indicate where the plant is; if there should be none of the cane on the root to show above ground, take pains and spat the top of the ground right *over* the plant with the back of the hoe, to show where it is planted, and then if dry weather follows, and the germ does not show itself, the crust can be broken and worked a little with the fork, (fig. 68,) an instrument that is admirable for working around them the first time, as it pulverizes the soil so nicely.

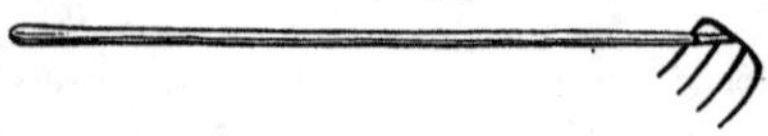

Fig. 68.

We usually plant potatoes or beans half way between them the first year, being careful, however, if it is the Black Cap family, or the "layer" sort, and we are desirous to get an increase of plants, to plant *early* kinds of potatoes, so that they will be ready to dig *before* it is time to layer the new growth; but if the "sucker" kinds, then later potatoes can be planted. As soon as the new growth gets about one foot high, we pass over the rows and nip off just enough to check its growth. If it is the *sucker* sort, we would allow it to get 3 feet high before nipping. It will usually throw up one or two sprouts, (fig. 69.) These should be nipped back as represented by the cross line in fig. 69. We cannot give the exact season for doing it, as they are so variable in different parts of the country; but lay it down as a rule to let bush get not to exceed 18 inches in height, and then nip it off just enough to check its growth. Soon these shoots will throw out laterals, and in two to three weeks will present an appearance as represented in fig. 70, the cross line showing where the bush was nipped as first described.

Fig. 69.

Now if very large, stocky plants are wanted early in the fall, we would not advise further nipping, but as soon as any or all of the above tip ends show a purplish, leafless, snakish looking appearance, bury them with a trowel at an angle of 45 degrees, and you will have, in four to six weeks, as fine roots as you would desire, and when these are taken up, 6 to 8 inches of cane can be left on them, and the bush cut back, as indicated by the dotted lines, (fig. 70,) for fruiting the next season—say not over 18 inches high and 18 inches across. If you desire to give your bushes a better shape for fruiting the next season, and want to increase plants from them to the *largest possible number*, then nip *all* of above stalks off back to where the dotted lines are—say just as they are forming the tip described, and in two or three weeks time you will have a bush having, instead of eight to twelve tips to layer as in the first case,

Fig. 70.

forty to sixty. In fact we have layered as high as *one hundred* very fine plants from one bush by nipping back the second time. The plants being layered late, and not having so much room to form fully developed roots, do not get as large or fine, but still we have had just about as good success with them. In both cases after taking up the roots we cut the bush back to where the dotted line is.

As soon as they are through fruiting the first year, cut out the fruit-bearing canes, having in the meantime however nipped back the *new* growth when it gets about two or three feet high. If a large *early* plant is desired this season, and *numbers* are not so much desired, don't nip the second time, but layer as soon as they have formed tips as described above. If, however, large quantities of plants are wanted, nip off these side branches when they get about two feet long, and they will give you as many tips as you can find space to put them in.

We prefer not to allow them to bear a large crop the first bearing, as it is apt to damage them for future planting. The less they are allowed to bear the *first* bearing year, the longer the plantation will last.

Remember, if it is desired to do away with stakes, this "nipping back" must be closely and systematically attended to, and judgment used in so doing. If the plant proves to be a splindling weak one, nip it back more, and if a strong stocky plant, it will not need such close pruning. It must be remembered, however, that when a bush has been layered to increase plants from, it will not stand up as erect as those that have been closely pruned and not bent over to layer.

We usually cut out the old wood with a pruning knife, wearing at the same time a pair of leather harvest mittens. This is thrown in piles together from two rows each side, and boys carry it out to the ends of the rows with forks, whence it is carted off to burn.

The raspberry crop is largely increased by a heavy mulching close around the crown, and a thorough and constant pulverizing of the soil with the cultivator, up to fruiting season. In the winter this mulching can be applied, and it not only acts as a mulch and protection against drouths, but enriches and supplies the plant with desired and necessary nourishment.

The red raspberries do not require that close nipping back that the blacks do, to make them grow stocky. If the new growth is nipped when they are about three feet high, it will cause them to thicken up and grow very strong. Especially is this the case with the Clark and Philadelphia.

We usually pick our fruit every other day—that is, by picking *half* the plantation every day and alternating.

We prefer the *square* splint baskets for shipping; and pints are far preferable to quarts, for the less fruit there is in a body the less apt it is to heat and spoil. These are packed in a slat crate, as represented in figures 71 and 72. They will receive four tiers of baskets of fifteen each, making sixty in all. The first tier is placed on the three horizontal strips that form the bottom; and before putting in the second, the frame or slat, (fig. 71,) is placed on the baskets already in position, so that each bar rests on the two edges of the contiguous ones holding them firmly in place. The frame is made of four bars, each two feet long, and half an inch square, across which six thin boards, two inches wide, and an eighth of an inch thick (slit from two inch plank) are fastened by small nails. The whole box is filled in the same way. When the top board is nailed on, the berries will carry safely long distances, if kept right side up. The elasticity of the slats assists materially towards this result.

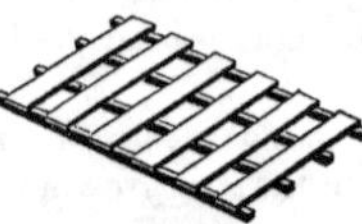

Fig. 71.

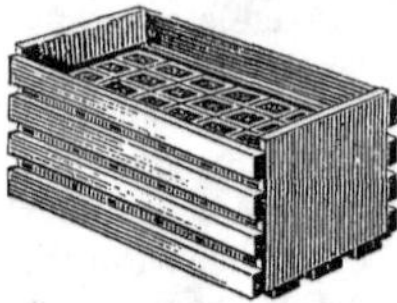

Fig. 72.

We usually make up plenty of these cases, and if express charges are so high as not to pay to have the case returned, we have our consignee "nest" the baskets together and send back by express, and the crates returned as freight or by canal boat. Still they are made up so cheap that if they are not returned it is no great loss.

Of the Red sorts, we can recommend for *general* cultivation and *profit*, the Kirtland, Clark and Philadelphia, (fig. 73.) The first is a medium sized fruit, but very prolific; bright scarlet, delicious, and yielding its *whole crop* at two or three pickings; early, making it one of the most profitable with us. The second is a magnificent fruit, large size, bright scarlet, productive and hardy; and the last the *most productive* raspberry we have ever grown of the red sorts; flavor No. 2, but hardy as a burr oak, and a *sure bearer every year;* sufficiently firm to market in pint baskets.

Fig. 73.—*Philadelphia.*

Fig. 74.—*Mammoth Cluster.*

Of the Blacks we would recommend the Davison Thornless and Doolittle for early, the Miami and Seneca for medium, and Mammoth Cluster (fig. 74)

for late, all very hardy and productive, and keeping up a supply for at least six weeks, although the main pickings for marketing last but about four weeks.

We usually pay two cents per quart for picking. The average price has been twelve to fifteen cents per quart. If planted with care, with oft repeated cultivation, 75 to 100 bushels per acre is a fair average crop from full grown bushes—that is, the second year after setting. If the old wood is kept out as described, plenty of enriching compost and mulch supplied, a plantation will last six to eight years. Two hundred dollars per acre is a fair average amount to realize from a well tended acre of raspberries—not counting in profits realized from plants. After the first year, there is no cost of plowing, or plants and setting, and but few weeds will grow if *thoroughly* worked the *first* year.

For family use, they can be set in a row next to a fence. An ordinary family, say of six persons, can rely on having their table *bountifully* supplied for six weeks, by setting six of each, Davison Thornless, Doolittle, Seneca and Mammoth Cluster, and six of each, Kirtland, Clark and Philadelphia. If it is in a section where peaches do not thrive, then set, by all means, twelve Lum's Fall-Bearing and Catawissa; cut out the old wood entirely each fall, and grow the new wood only each season, trimming and nipping them back until August; and they will yield a splendid crop of fruit *late in the fall*, especially if *heavily* mulched with coarse litter the early part of the season.

ORNAMENTAL PLANTING.

HARDY ORNAMENTAL SHRUBS.—At a meeting of the Horticultural Society of Western New-York, a vote was taken on the best, hardiest and most reliable shrubs, proved such by experience in the western part of the State. Only seven ballots were handed in, and the following was the result;

Deutzia gracilis,	6 votes.	Double Scarlet Thorn,	3 votes.
Wiegela rosea,	do.	Plum-leaved Spiræa,	do.
Purple Fringe,	do.	Double Flowering Almond,	do.
Japan Quince,	do.	Tartarian Honeysuckle,	do.
Spiræa lanceolata,	5 votes.	Prunus trilobata,	do.
Deutzia crenata, (double,)	do.	White Lilac,	2 votes.
Persian Lilac,	do.	Double Althea,	do.
White Fringe,	4 votes.	Forsythia viridissima,	do.
Double Syringa,	do.	Pink Flowering Currant,	do.
Snowball,	3 votes.		

The following had one vote each: Daphne mezereum, Josikea Lilac, Double White Almond, Purple Barberry, Siberian Lilac, Silver Bell, Magnolia obovata, Deutzia scabra, Rose acacia, Barberry, and a few others.

THE BEST RHODODENDRONS.—Tilton's Journal describes the great success in the culture of the Rhododendron at Mr. Hunnewell's place near Boston. The ground is prepared in the most thorough manner. They are sheltered, but not shaded; and not less than fifteen feet is allowed to each plant when fully grown; and thus sheltered and managed, they become clothed with verdure down to the ground. "Imagine," says the writer, "a Rhododendron fifteen feet in diameter, wholly covered with flowers and foliage! One such plant would be worth more than a hundred of the bare, lean, straggling stems too often shown." The following select list of sorts is given as combining variety of color and uniform hardiness, viz., *Purpureum*, *Grandiflorum*, *Archimedes*, (scarlet,) *Everestianum*, *Album grandiflorum*, *Roseum magnum* and *Chancellor*.

FORMING LAWNS.—There are three modes of forming lawns. The first—to mellow the surface and sow grass seed thickly, which coming up with the weeds, the seeds of which are in the soil, much labor is required afterwards to get all these out by hand. The second is to plow and re-plow, harrow and re-harrow, for a season, in order to work out all the foul seeds, allowing time between each operation for the seeds to germinate, and remembering that many seeds will not grow if buried over an inch deep; hence the necessity of repeating the stirring many times, in order to bring all parts up to the surface. Then sow fine grass seed, such as red-top, June grass, white clover, &c., mixed, and at the rate of at least one bushel per acre, rolling it in. This is to be done as early as possible in spring, and then, when it is a few inches high, mow it closely as often as once a week the season through. This will give a handsome green carpet-like velvet. The third mode, ususlly the most expensive, but the most speedy and certain, if well performed, is to *turf* the surface. First make the soil deep and mellow, and even at the surface; then pare from an old pasture the turf, cut very smooth, with perfectly parallel and straight sides, and of a perfectly uniform thickness of about two and a half inches; spread this turf over the mellow surface, as smooth as a floor, and roll evenly. If manure is applied to make the soil rich, it must be finely pulverized, and thoroughly and very evenly worked in.

CARTING SAND ON GARDENS.—We have covered a piece of clayey garden soil with a layer of sand two inches thick, which, when thoroughly worked through, greatly improved the character of the soil, and changed it from one of a heavy clammy character to a fine friable loam. This was done twelve years ago, and its improved character continues without any abatement. In fact the sand cannot work out—it must remain perpetually where it is put. It is here that the practice of drawing on sand has its great and peculiar advantage. Manures will dissolve and become abstracted by plants; but sand neither dissolves, evaporates, nor goes into plants.

NORWAY SCREENS.—A young planter inquires the cost of a good screen made of Norway spruce trees, intended to form both a screen and barrier

—not for the exclusion of rampant cattle, but to prevent the ordinary passage of men and animals.

For this purpose, trees set three feet apart, would probably answer the desired purpose completely in the course of six or seven years, and partially in four or five years, or even sooner—especially if the soil is kept clean, mellow and well cultivated for several feet on each side of the line. The trees, if they have been carefully taken up and set out well, will recover from the check of removal in a year or two, if two feet high—say one year is entirely lost by transplanting. If three or four feet high, they will not recover quite so soon; but much will depend on the care and skill of the operation, and on a moderate cutting back in the spring when set out. They will grow about three feet a year, (only one and a half if not cultivated,) and in six years will be fifteen feet high if desired, or if not shortened down any. Plants from one to two feet high may be had for about $15 per hundred, of the larger nurserymen; of some, cheaper; and a hundred will set out 300 feet, or 18 rods. The plants, freight, plowing, setting, &c., will cost, with the trees, about one dollar per rod, if of this size; but they may be selected of smaller size, so as not to come to more than 60 to 75 cents, set out. Cultivating the ground for a few years, till the trees are well under way, so as to keep them clean, and the necessary cutting back, need not cost more than 25 cents a rod, if repeated five times a year for three years, and twice a year for two additional seasons. By repeating it often, the work will be less expensive and be of much more use than if done at remote intervals. With interest, the screen, when seven years old, will not cost two dollars a rod, and will not exceed in expense a good post and board fence. The screen must not be sheared, but projecting and obtrusive limbs shortened back. The side limbs and smaller branches and shoots will form such a barrier that neither men nor animals will be disposed to go through, unless some great and special inducement invite a hard effort.

HONEY LOCUST HEDGE.—Of late years many miles of Honey Locust hedges have been planted in the northern States, the extreme hardiness of the tree giving it an important advantage in this respect over the more tender Osage Orange. Its straggling growth, however, renders extra care necessary in cutting back and training to the proper form, and preventing open spaces below. A fatal error on this point prevails extensively. Many of the young hedges are not cut down one-half as low as they should be during their early growth. We have taken the trouble to measure the successive heights at which these cuts are made, in a large portion of the hedges planted of late years, and which the owners obviously mean shall be well managed, so far as they understand management. One in particular, which has been carefully cultivated, and which has excited a good deal of attention, has the first cut the year after planting, five inches high; the next eight inches; the third sixteen inches, and the fourth eighteen inches—the height at the present time being nearly or about four feet—fig. 75 showing the hedge as seen endwise, with its thin and feeble appearance below. Fig. 76 represents a hedge cut as it should be—the first time very near the surface; the second, three inches higher, and subsequently four, five and six inches, or but little more. Fig. 77 shows the appearance of an attempted hedge, where

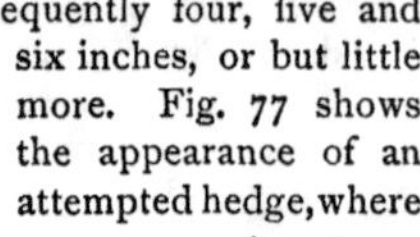

Fig. 75.—*Badly Cut Hedge.*

Fig. 76.—*Properly Cut.*

Fig. 77.—*Cut in Leaf.*

the cutting back is deferred till the leaves are partly or wholly expanded, producing a serious check, and preventing a vigorous growth of shoots. The cutting back just described, for the purpose of inducing a thick, dense mass near the bottom, is especially important for the Honey Locust, which, unlike some other hedge trees, tends to grow slender and with comparatively few branches—so much so that A. J. Downing once remarked to us, that it had "no hedginess about it," and he regarded it as of no value for this purpose.

Fences and Trees by the Roadside.—An effort to do away with roadside fences—a movement which, with suitable planting and care, would render our country thoroughfares much more beautiful and attractive than they are at present—is now going on in Western New-York, as we learn from a correspondent, who says:

"Along the road leading from Victor Depot to the East Bloomfield railroad station, (about six miles,) many of the farmers have taken away the street fences, and the change for the better along the road is very striking. The highway line is marked only by a row of apple or other fruit trees, and the grass is cut from time to time along the track. All the stones and rubbish are kept away, and the ground is becoming constantly smoother, and to travel over it is like travelling over a continuous farm, with occasional houses and out-buildings. I noticed that some had just piled their rails, and I could see that it had commenced on one of the cross roads.

"One cannot help regarding the people as more civilized than in regions where the practice has not yet begun, and will it not have an improving liberalizing influence on all who come in contact with it? I am glad to see that a course that is so evident an improvement is extending, and commends itself to all within its influence."

DESIGNS FOR GROUNDS AND GARDENS.

THE PLAN for a small village door-yard is shown in fig. 78. The house stands about thirty feet back from the street, allowing space for small trees and shrubs, and, if desired, some flower beds. The space in front of the house is planted with roses and other small shrubs, but these might give place to bulbs, annuals and herbaceous perennials of the larger kinds, in which case great care should be taken to keep them in the neatest and most perfect order, as there is no excuse whatever for allowing so small a piece of ground to present a slovenly appearance. The principal walk is up to the front of the dwelling, but a side passage to the right allows a more secluded access to the rear. The track on the left admits coal and other wagons to the kitchen and back grounds; but if the lot should happen to be too narrow for this passage, it may be entirely omitted, and the hand-cart or wheelbarrow employed.

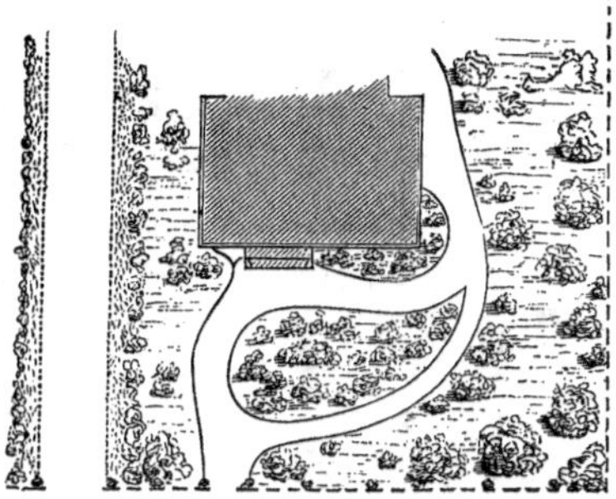

Fig. 78.—*Small Village Door-Yard.*

Fig. 79 shows the plan of a still smaller yard, the house standing within twelve or fifteen feet of the street. The passage on the left is for the wheelbarrow to the vegetable garden. The shrubbery in front must be of the most select kinds, and be symmetrical growers, so that the place may have a polished appearance.

Fig. 79.—*Smaller Door-Yard.*

Fig. 80 represents the plan of grounds where one acre is occupied with the dwelling and the surrounding ornamental planting and flower garden. It is supposed to face the west, and the arrangement of some of the details is made in accordance with this position. With a slight alteration it may be varied to suit any other aspect. The kitchen and fruit garden may be extended back to any desired distance, so as to embrace one or two acres, or more.

But little explanation is required. The carriage and wagon entrance, on the left, separates on approaching the house, carriages taking the right, and loaded wagons and carts the left, to the barn or rear grounds. Along the left boundary, and facing the south, are placed hot-beds, &c., the materials for which are easily obtained from the stable. The flower garden lies on the south side of the dwelling, the circular, elliptical and arabesque beds being cut in the smooth turf, and kept well cultivated with bulbs, annuals and herbaceous perennials. Similar beds may be occupied with the smaller shrubs; the larger shrubs and small trees, after they become will established, will grow in the turf. The nearer of the two seats or summer-houses, is placed where seclusion is desirable; and the rear one at a pleasant view of most of the grounds.

The walks need no explanation; the one beyond the rear summer-house passes along the fruit garden, and may or may not be flanked with ornamental shrubbery, as the owner may desire. An irregular screen of small evergreen trees and shrubs, nearly excludes the view of the fruit and kitchen gardens, and along the rear of this screen is the space for the horse to turn in cultivating the rows.

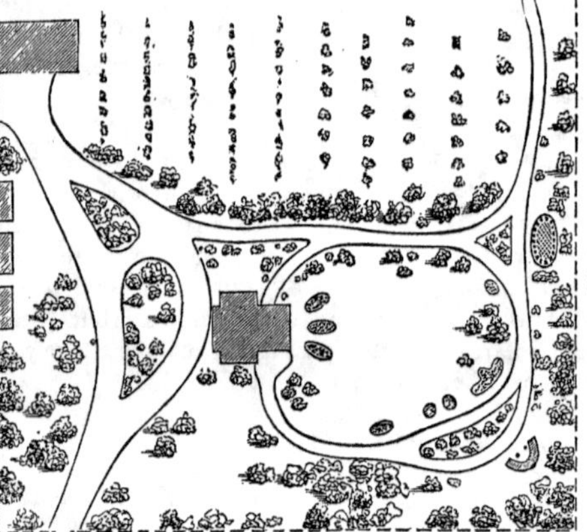

Fig. 80.—*One Acre.*

The portion of the garden nearest the barn is devoted to vegetables, planted in drills between rows of currants, gooseberries and raspberries, which thus obtain their share of good culture at little expense and trouble, and are too low to shade the annual crops.

On the right are the lines of dwarf apples, dwarf pears, grapes and other fruits requiring nearly exclusive possession of the ground, and like the smaller plants, kept cultivated by horses. A shallow plowing early in spring, before the buds swell, will mellow the soil and do no harm to the trees should a few roots be broken, while serious injury might result from tearing the roots after the leaves are out. The cultivation for the rest or the season is therefore done with the cultivator and harrow, the latter leaving the surface smooth to pass over in gathering the fruit in autumn.

One good gardener will keep such a place as this in finished order. The acre of ornamental ground should be mowed once a week in the early part of the season, requiring one day or more. Another day in the week will enable him to dress the walks and flower beds. The remainder may be devoted to the rear gardens, and to various other or miscellaneous jobs, in connection with the grounds. More labor might be expended in keeping the whole in a highly finished condition, bedding out flowers as they bloom from propagating houses, &c.

A village lot, about four or five rods wide, is represented by fig. 81. The principal object is to obtain a small flower garden and shrubbery on an area of about one-tenth of an acre, and allow space for a kitchen garden and a few of the smaller sized fruit trees on nearly twice as much ground in the rear. If desired there may be a small horse or cow barn in the rear corner on the left—the "cart-way" otherwise being intended only for conveying coal and other heavy articles to the kitchen cellar. A small screen of evergreen trees runs on the right of the cartway to separate it from the rest of the grounds, and the boundary on the left of this way may be planted with grapes, or with raspberries; the former being trained on the boundary fence, and the latter kept snugly within bounds by a slat running parallel with the fence, and about six or eight inches from it—the included space holding the canes spread out like a fan.

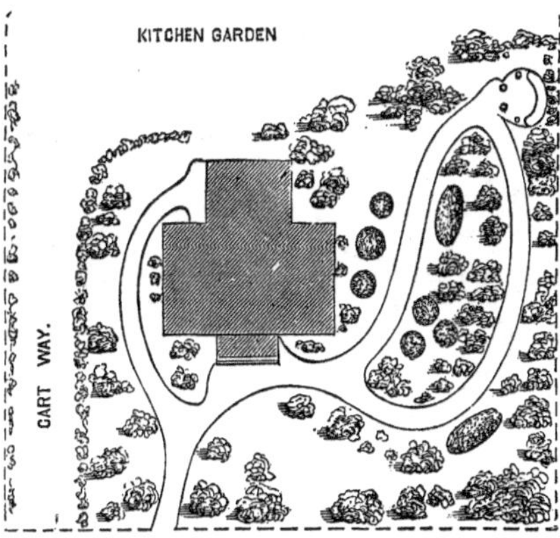

Fig. 81.—*Lot Four Rods Wide.*

No large trees are planted on these grounds, as they would after a while occupy too much space and shade the smaller shrubbery and flower beds. Small trees may occupy the most remote corners, large shrubs the more open space, and small shrubs only be placed near the flower beds, where it is important to preserve an open space for full sunlight. A few plants will flourish in the more shaded spots. The flower beds are mostly circular, with two elliptical ones. The circular beds are most easily laid out and kept in exact shape, according to the mode described on page 153 of the fifth volume of RURAL AFFAIRS. On the same page the mode of drawing elliptical figures is also distinctly pointed out.

For the smaller trees and large shrubs, the following are among the best, hardiest and most common :

Chinese Magnolia, Soulange's Magnolia, Tartarian Honeysuckle, (red, white and pink striped,) the large Philadelphus, Hawthorn, (white, pink and crimson,) common and Siberian Lilacs, Cornelian Cherry, Silver Bell, Virgilia, Snowball, Purple Fringe, Cercis, &c. Among the finest hardy evergreens of this size, are the Dwarf Norway Spruce, Dwarf Pine, the hardier Dwarf Arbor Vitæs, Cembrian Pine and Chinese Juniper.

Among the best of the medium and smaller shrubs may be named the Japan Quince, (pink and scarlet,) Dwarf Horse Chestnut, Sweet Scented Shrub, Japan Globe Flower, Wiegela, Dwarf Almond, Forsythia, Deutzia, the Spiræas, Tree Pœonies, and pre-eminently the Roses.

After such a place as this is well planted and under way, the ornamental part of the grounds may be kept in fair order by the owner if he has taste and zeal enough to get up at five o'clock and occupy an hour or two every morning before breakfast in dressing the grounds and conversing with the flowers, before he leaves for his workshop, office or counting room. The exercise and fresh air will give him health and appetite, and the beautiful forms with which he thus holds intercourse, will tend to increase the refinement of his mind. If he has children, they will like home better than other places, and become inspired with his taste.

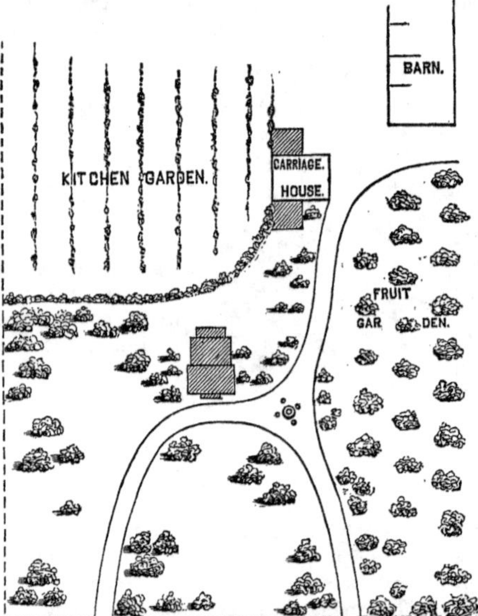

Fig. 82.—*Farm Grounds.*

Fig. 82 represents a comfortable farm residence, where the owner wishes to have everything neat and in good taste, but cannot expend much in ornamental gardening. The grounds are laid out in as simple a manner as practicable, so as to accord with good taste. As represented in the plan, they comprise from two to three acres, including the lawn, half of the fruit garden, and most of the kitchen garden. The dwelling is approached by a good and well made gravel road, and the surrounding grounds are planted with handsome shade trees; those towards the rear may be the hardier, more vigorous and symmetrical fruit trees, such as will flourish in grass—as for example, the Buffum, Boussock and Howell pears, and the Elton, Rockport and Black Tartarian cherries. The lawn should be mown at least three or four times early in summer, or it may be kept short by turning in, a part of the time, a flock of sheep, when they can be easily seen, and injury to the trees prevented. The fruit garden may be kept cultivated by a shallow plowing early in spring, and a few harrowings afterwards, and perhaps one or two rollings near the season of fruit, to keep it smooth to the pickers. The ice-house, hen-house, and other of the smaller buildings, may be placed near the carriage-house. An evergreen hedge or screen separates the kitchen garden from the front grounds. A water reservoir and hitching posts, are placed at the right of the house, at the intersection of roads.

By more expenditure of labor and attention, flower beds may be cut in a circular form near the dwelling, and the lawn may be kept in the best order by mowing every few days. The main object, however, is to present in this plan simple, neat and cheaply kept grounds for a farm residence, with little expense.

A WIND-EXPOSURE AND LAKE-PROSPECT.

A FRIEND who had purchased a building lot of several acres on an elevated ridge commanding an extensive and beautiful prospect on one of the lakes of the western part of New-York State, found himself in

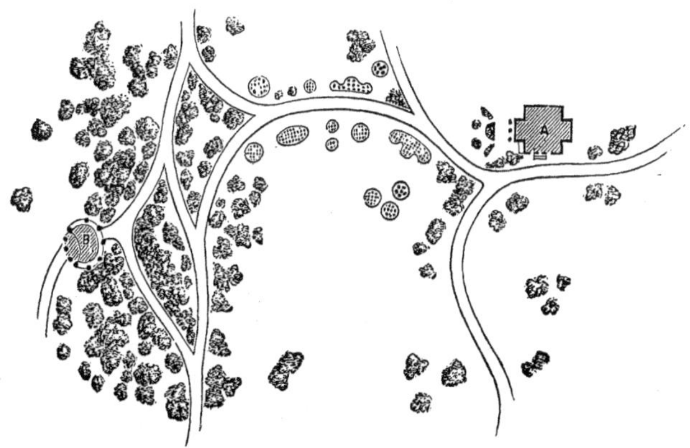

Fig. 83.—A., *Dwelling*—B., *Summer-House—Flower Garden between.*

the dilemma of either shutting out this fine view by trees, or of exposing his house to the sweep of the prevailing westerly winds in winter. He applied to us for relief from the difficulty, and we accordingly proposed the plan shown in the accompanying engraving, (fig. 83.)

As it was impossible to have the lake prospect in full view, and at the same time to shut off the sweep of the winds from that direction, it was proposed to relinguish the whole of the western landscape from the windows or verandahs of the dwelling, and to secure this view from other portions of the grounds. On this side of the house, and at a proper distance from it, and from the western boundary of the lot, irregular belts ot trees were to be planted, mostly evergreen, with some deciduous intermixed, or rather skirting their sides. The evergreens would serve as a screen, and the deciduous trees soften the hard outline which they would otherwise present. Such hardy and strong growing evergreens were to be employed as Norway Spruce, Menzies' Spruce, Scotch Pine, White or Weymouth Pine, common Hemlock, Black Spruce, Austrian Pine, &c. The exterior or outer portions of these belts should be of the lighter or more airy species, as the Hemlock and Weymouth Pine, more loosely scattered, so as to blend properly with the deciduous plantings. These, when well grown, would entirely shut off the west winds in winter, and exclude the lake prospect in summer from the dwelling.

But by placing the ornamental garden, (see engraving,) with its flower beds cut in the smoothly shaven turf, in the direction of these belts, the walks might properly be made to pass into them,—where a summer-house or two was to be placed, entered on the wooded, eastern or screen side, and on looking out towards the west, the whole of the view would be at once presented in this direction—including the village spires below, the lake bordered with woods, green fields, distant blue hills beyond the water, and the western horizon. A sufficient portion of the grounds west of these dense plantings were to be reserved, and more sparingly set with trees, so as not to cut off the view, but to give a pleasant aspect to the whole picture.

There are many places where a difficulty similar to that here alluded to is found to occur; and by resorting to a like expedient, necessary shelter may be secured, greater variety given to the landscape effect, and everything of interest secured to the grounds.

FRUIT CULTURE.

STRAIGHTENING UP TREES.—If newly set trees of moderate size have been well dug up, with plenty of roots, and the roots well spread on every side, they will maintain a stiff, upright position, and need no additional staking or stiffening. But they have not always received such careful attention; and in this case they may need straightening up. Nothing can be much worse for a tree than the bending about in the earth by strong winds. They will sometimes stand well till the leaves come out, after which the winds have more purchase on them, and staking may be required.

STRAWBERRY RUNNERS.—The formation of runners by the stools or

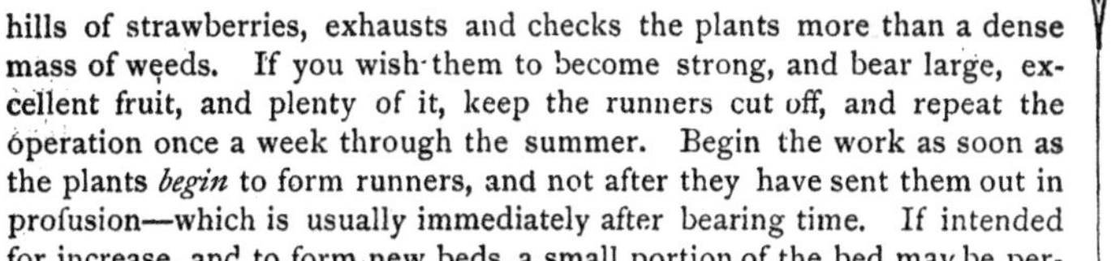

hills of strawberries, exhausts and checks the plants more than a dense mass of weeds. If you wish them to become strong, and bear large, excellent fruit, and plenty of it, keep the runners cut off, and repeat the operation once a week through the summer. Begin the work as soon as the plants *begin* to form runners, and not after they have sent them out in profusion—which is usually immediately after bearing time. If intended for increase, and to form new beds, a small portion of the bed may be permitted to run and root.

BLACK KNOT ON THE PLUM.—Remember the old and well tried remedy of cutting off as fast as the first indications appear. Do not wait till the tree is covered with these excrescences and destroyed. The complaint sometimes made, that cutting does not cure the malady, when the remedy is applied months too late, is not more reasonable than to denounce water for putting out fire were the engine and hose companies to come on the ground and commence playing on the heap of ashes the year after the conflagration. The labor and attention needed to keep the plum thus clear of the knot are not half as great as to keep a potato or cabbage patch clear of weeds, which every one so willingly performs.

THINNING APPLES.—Remember that it is not more than one-fourth the labor to thin out the small, defective, scabby or stung apples, when they are quarter grown than to hand-pick and assort them the next autumn—and the selected specimens which are left have a better chance to grow, and sell at a higher price in market.

GRAPES OVERBEARING.—Do not allow grapevines to overbear. Many a young vineyard has been injured, if not ruined, by carrying too much fruit. Novices often delight to show how many grapes they have on their young vines, and to tell how many tons per acre their new vineyards have yielded. They might as well boast of making a young horse sick by hard driving. Thin out the bunches as soon as they have set, and let the crop be small. It will be all the better in quality, and the vines will preserve their health and vigor.

GRAPE LAYERS.—These may be easily made on a small scale by the amateur, or by the farmer who wishes to plant out an additional vine or two, by laying down a good strong shoot of this summer's growth, and burying it a few inches deep at the middle. Roots will be thrown out at the joints, and by taking up before winter, and cutting in two, two good plants will be made from each layer. If the laying down is done too soon, the shoot will be soft, and may rot; if left too late, there will not be time enough for the roots to form. The proper season is just when it is beginning to assume a little of the hardness of wood. The formation of layers exhausts a vine in the same way that runners exhaust strawberries; they should therefore be sparingly made from bearing vines. If the summer happens to be very dry, the soil at the buried part must be kept moist by mulching.

CULTIVATING ORCHARDS.—We see a successful experiment stated in some of the papers, performed by H. Dayton of Alden, Erie Co., N. Y.

His orchard of two and a half acres, in grass ten years, which had produced wormy and scabby fruit, and in small quantity, was plowed very shallow in autumn, and harrowed and cultivated a few times early the following year. The result was it bore four hundred and fifty barrels of fine apples last year, worth about six hundred dollars.

Fruit Rooms.—The best fruit rooms are built above ground. They admit of more perfect control of the temperature, and may be kept cooler through autumn—a matter of much importance. For such rooms the walls must of course be double, and filled in with powdered charcoal, ashes, tan, or similar substance, which is rather better than to have the enclosed space occupied only with air. The ceiling or roof must also be double, and well secured from the intrusion of frost from above. Such rooms may be made to open on the north side, so as to be kept cool till the advent of the freezing weather of winter. Fruit, in an apartment thus managed, will keep much better than otherwise. We have experimented fully on this point, and observed the difference between the keeping of winter apples taken directly from the trees and placed on the shelves in October, and others placed, when gathered, on the floor of an out-building fronting north, and allowed to remain there nearly till the first day of December. The latter, at a careful estimate, *did not furnish more than one-quarter the number of decayed specimens through the winter* as those placed at once in the warm cellar when gathered.

The walls of fruit houses above ground may be built of brick, with a hollow space between the two portions or walls, or of double wood siding filled in. The windows should of course be double, or the sash at least, with double glass. The entrance door should also be double, or made of two parts, with a space between.

Fruit Ladder.—An improvement in the common tall step-ladder is made by attaching a basket support to the right-hand board. This support is made of rod-iron, about half an inch in diameter, and bent when red hot in the shape shown at *a*. This is inserted in the two holes shown in the ladder, (fig. 84,) so that the two ends pass closely below the step, which holds them firmly. On this the basket is set, and both hands are then at liberty to work, which is a great advantage when the limbs must be held for picking. Smaller rod may be used if pieces are welded across at one or both places shown by the dotted lines.

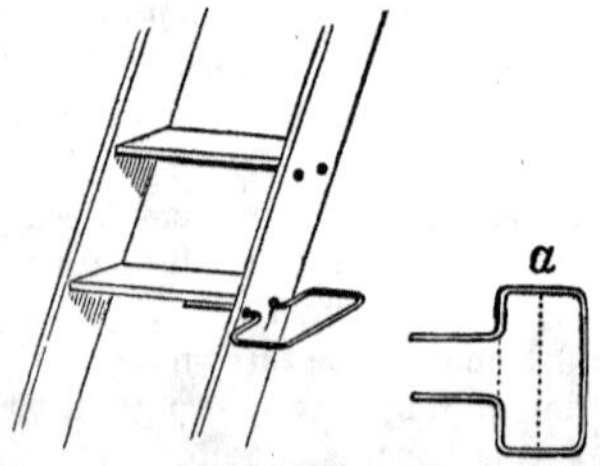

Fig. 84.—*Fruit Ladder.*

Drying Fruit.—A correspondent in Illinois informs us that he has tried one of the small fruit drying houses figured and described in a former volume of the Annual Register—that it heats quick, with little wood, and needs constant watching. For small quantities and home use it does well, but is not large enough for a market orchard, where wagon loads are in danger of rotting. He accordingly built four houses, each 6 by 12 feet, and 6 feet high. There were three tiers of drawers on each side, and six drawers to a tier. The studs were 2 feet apart, 2 by 3 inches; sleepers for drawers to run on, 2 inches square; stuff for drawer frames, 1 inch square; slats, half an inch by one inch—all cut to order at sawmill. Five days' labor built each house. A large stove is set under the drawers, and fresh air pours in at the bottom. A man and two children will gather, cut and dry nine bushels daily; the house may be refilled every day. Twenty-two thousand dollars of dried fruit were sold in 1867, at one station.

THE TWO MOST POPULAR CHERRIES.—At the Philadelphia meeting of the American Pomological Society, Coe's Transparent and Early Richmond had the largest votes—the former for the eastern and middle, and the latter for the western States.

HIGH PRICE FOR STRAWBERRIES.—The editor of the Gardeners' Monthly says that he made a journey of three thousand miles, and saw only one place where the hill system and constant clipping of runners was practiced in perfection—at Knox's, at Pittsburgh. The result was that while all through the region visited, strawberries sold at only about four dollars per bushel, Knox obtained as high as twelve dollars for his. The Jucundas were often large enough for twenty-five to fill a quart. These sold for one dollar a quart. The Fillmore and Agriculturist were nearly as fine.

PEAR BLIGHT.—Quin, in his "Pear Culture," says that out of twenty-nine cases of fire blight, (fourteen of which proved fatal,) there were 16 Glout Morceau, 4 Flemish Beauty, 4 Winkfield, 3 Belle Lucrative, and 2 Louise Bonne Jersey. This is additional proof of the liability of the Glout Morceau to the disease. He finds the old remedy best, of very promptly cutting off some distance below the disease, and burning the limbs—and re-grafting if cut very low down. Others have practiced successfully cutting down within a foot or two of the ground, and allowing a new low head to spring up from the stump.

FRUIT IN OHIO.—The Report of the Horticultural Society of Ohio says that the apple crop of the State in 1867, was 9,400,000 bushels—which, at 50 cents per bushel, would be more than four and a half million dollars. The peach crop was estimated at 1,400,000 bushels, worth more than three and a half millions.

SOUND ADVICE.—A. M. Purdy writes to the Rural New-Yorker, that no one must expect his strawberry plantation to pay on the "slip-shod" plan. "Don't wait till you *see* the weeds growing before you start the hoe or cultivator. Remember, you can go over a plantation four times, if done in season, where you can once if it gets weedy and surface-stiff."

BLACK KNOT ON CHERRY TREES.—A correspondent of the New-England Farmer says he has tested the old well-tried remedy of excision for thirteen years, and the trees continue handsome and vigorous, and bear abundantly every year, while all around him they are spoiled by the black knot. He cuts the branch off *as soon as it is affected*, and if the limb is too large, the knot is shaved off clean. We have pursued this course for twenty-five years on the plum, and have been equally successful. The remedy is perfectly simple, very easy, and requires but little labor. All that is necessary is prompt attention at all times.

Deep Planting.—The editor of the Gardener's Monthly says that "too deep" planting kills ninety-nine hundredths of all the strawberries that die in the year from transplanting.

Fruit and Drink.—P. J. Quin says in his new work on Pear Culture: "If the time should ever come that one-half the amount now spent for alcoholic decoctions should go for choice fruit, what a difference there would be in the homes of many of our poorer classes, now rendered almost desolate by the use of intoxicating drinks!"

Curculio.—An acquaintance asked how we obtained uniformly such heavy crops of plums. We informed him of the manner of stretching sheets, driving a spike into each tree to pound on, and jarring down the insects for thumb-and-finger killing. "Ah! but he had tried this way." "What! every morning, regularly?" "Yes—quite a number of times—*nearly* every morning." He, however, admitted that sometimes he omitted it for two or three days; and, on further questioning, it was found that he had performed the operation only "seven times—he was not quite sure,—perhaps it was six times." This is the way that the remedy proves a failure. Instead of trying it at least once a day, (twice a day when the insects are abundant,) without a single intermission, till the last insect is killed, it is done only occasionally or semi-occasionally. One might as well omit feeding his horse now and then for two or three days. "Neighbor Snoggs, my excellent old cow has dried up—what do you think is the matter?" "Do you milk her regularly?" "Oh, yes—or at least mostly so. I milked her yesterday,—and last Monday,—and, I believe, once the week before, but I'm not sure." This is the way in which many do their work on fruit trees, and wonder why they have such poor luck—and believe "these vaunted remedies are humbugs."

Propagating Blackberries.—Purdy & Johnston's Small-Fruit Recorder gives compactly the following directions from their extensive practice: "Cut the roots of blackberries up in the fall; pack away in sand, and sow in *shallow drills* in the spring, like peas—scattering over the surface a little coarse manure."

Capricious Markets.—A cultivator in Western New-York lately sent several half-barrels of selected Bartletts to his commission agent in New-York City, and although the market was overstocked, they sold for six, seven and nine dollars per half-barrel. Others contained very fine selected specimens of Doyenne Boussock, which many would regard as decidedly superior to Bartletts, and they could not be sold for more than three dollars and a half. So much for a fruit being popular and well known. Some of our readers may not have heard of the gentleman in the western part of this State, who, in the palmy days of the Virgalieu, marked his barrels, (all containing specimens of uniform quality,) a part with the well known name of "Virgalieu," and a part with the more approved name of "Doyenne." The former sold at nine dollars per barrel, the latter for only five.

SELECTING AND ASSORTING FRUIT.—The truth cannot be too strongly impressed on fruit-raisers, that on nothing does success in marketing more depend than on selecting good specimens only. The work should be commenced by thinning on the trees, while the young fruit is yet small, and its importance should not be lost sight of at any subsequent period. Cases have occurred where owners, in their eagerness to sell everything, have put a few poor specimens in a barrel of fine market pears, and these poor ones have spoiled the sale of the whole. Very fine pears are often sold at twenty-five or fifty cents apiece; on the other hand, a few bad ones will so reduce the market value of the rest as to cost the owner not less than twenty-five or fifty cents each. The editor of the Horticulturist says that a peach-grower, having discovered that his men brought in many poor peaches, had them assorted carefully before shipping, and forty baskets of cullings taken out of every hundred. The sixty good ones were then sold for $1.25 per basket, bringing $75. The unassorted ones would bring only sixty cents per basket, or $60 for the whole, while the freight on the forty baskets of poor ones, mixed with the good, would have greatly increased the freight and expenses.

CULTIVATING PEAR ORCHARDS.—Randolph Peters of Wilmington, Del., writes to Tilton's Journal of Horticulture that he has an orchard of sixteen thousand pear trees, half standards and half dwarfs, four, five and six years old. The Bartletts and Belle Lucratives are producing from half a peck up to a bushel per tree. A few were left without cultivation. These have not done half so well as the others, and the fruit averages only one-third size on these. He has scarcely lost a tree by fire blight in all.

PROTECTING THE TRUNKS OF PEAR TREES.—Orchardists are familiar with the disease that affects the bark on the bare trunks of standard pear trees, particularly at the south and west, where there is so much hot sun at mid-day. The Northwestern Farmer states that a fruit-grower has for several years protected his trees against the hot sun by adjusting a board to shield them from the two o'clock rays, with entire success—adding that "since he has tried it, he has lost no more trees, the bark on that side remaining as smooth and as soft as on any other part of the tree."

CANKER WORMS.—Robert Manning of Salem, Mass., gives the result of his experience and observations on this insect, in Tilton's Journal of Horticulture. Tar applied to a belt of coarse paper or cloth, placed around the tree, does well at first, but soon forms a crust, especially on cool nights, over which the worms pass. Mixed with oil, the tar is better, but the difficulty is not removed. Cheap printer's ink is still better. All need frequent renewal, all harden by cold, and the surface must be often stirred with a brush. A large string around the lower edge prevents the tar from running down on the bark. Troughs made of tin, lead, &c., to hold the tar or oil, are better but more costly. If the insects are already in the tree, they may be jarred down, or brought down by throwing sand among them, and when down they are eaten in great numbers by young ducks, who, after gorging themselves with the worms, lie down to digest them, and in a quarter of an hour are ready to begin again. This insect is making great progress in some parts of the country, and we have latterly observed many of the fine orchards in Western New-York, between Geneva and Canandaigua, stripped of their leaves. They spread slowly but surely, unless checked.

Registering Young Orchards.—Do not trust to memory or to perishable labels, to know how many trees or rows of Baldwins you set on this side, or how many of Greenings on that side of the orchard, or what kinds are placed in the different parts of the fruit garden. Register every tree carefully and accurately—first in a memorandum book, or on a slip of paper, to be copied afterwards into your account book or some other book which you will always readily have at hand; or, better, have a blank book expressly for a *Garden and Orchard Record,* where the place of every tree is noted, as well as your other plantings and experiments. Then, when the young trees begin to bear, you need not call in all your neighbors to ask them the name of this apple and that pear, with a very fair chance of half being named wrong, and endless confusion in sorts as a consequence.

Grafts.—Cut these in autumn if you have a good place to pack them. They have more vigor in spring than if exposed to all the cold of any severe winter—this is especially the case with plums, pears and cherries. Pack them in boxes of damp, not wet, moss; or in small boxes of damp, not wet, sawdust—large boxes of sawdust will heat. Mark every sort carefully and plainly. Another good way to keep scions through winter is to place them snugly in a box till it is more than half full; next nail in two or three cross-pieces to hold them, and then bury the box inverted with several inches of earth over it, on a dry spot or knoll. They will thus be kept from contact with the wet earth, and will receive enough moisture from below to keep them fresh and plump. Cuttings of currants, grapes, quinces, gooseberries, &c., are to be taken off soon, and they may be kept till early spring in the same way as grafts, or they may be set out at once, pressing the earth compactly against them, and covering well till spring with manure, litter, leaves or evergreen boughs.

Strawberries are injured in winter by severe winds, and by the successive heaving of freezing and thawing. They will always start earlier and fresher when covered. Sometimes snow will be an ample protection, but it must not be relied on. A thin coating of straw, evergreen boughs or even cornstalks, will shield and protect the surface of the ground, but it should not be applied till winter is close at hand, and after the ground is

frozen hard is not too late. Do not forget to loosen up this mulching very early the next spring, and stir and mellow the soil.

HARDY GRAPES—Which are pruned in autumn, should be laid down after the operation, as much cutting away always renders any plant more easily injured by frost. When it happens that the succeeding winter is quite severe, the vines will not only grow and bear better the following summer, but the crop will usually ripen several days earlier than if left fully exposed on the trellis to the winter's cold.

KEEPING FRUITS.—The following rules for keeping fruits are from the proceedings of the Royal Horticultural Society of England:

1. As the flavor of fruit is so easily affected by heterogeneous odors, it is highly desirable that the apple and pear rooms should be distinct.

2. The walls and the floor should be annually washed with a solution of quicklime.

3. The room should be perfectly dry, kept at as uniform a temperature as practicable, and be well ventilated, but there should not be a thorough draught.

4. The utmost care should be taken in gathering the fruit, which should be handled as little as possible.

5. For present use, the fruit should be well ripened; but if for long keeping, it is better, especially with pears, that it should not have arrived at complete maturity. This point, however, requires considerable judgment.

6. No imperfect fruit should be stored with that which is sound, and every more or less decayed specimen should be immediately removed.

7. If placed on shelves, the fruit should not lie more than two deep, and no straw should be used.

PLANTING BLACK-CAP RASPBERRIES.—A. M. Purdy of Palmyra, N. Y., whose experience is not excelled in raspberry culture, says he prefers to set raspberries, as well as gooseberries and currants, in autumn, and then throw a shovelful of coarse litter over or around each plant to prevent heaving by frost. The plants become settled in their places by spring, and receiving the early rains make a full growth the first season. In this way also the planter escapes the crowd of spring work which generally results from leaving all the planting till spring, and which is apt to make hurrying and superficial work. New raspberry plants, within a year old, are greatly preferred to two-year olds, which are of little value.

KILLING INSECTS.—The fruit-growers of Vineland have taken the business of killing destructive insects into their own hands, and among those who have competed for the premiums offered for killing the greatest number of curculios, we observe, from the reported results, that one man killed 4,400; another, 2,270; a third, 1,300; while no others came up to 1,000. The total was over 9,000. We think this a very moderate number for so extensive a fruit neighborhood, but they have made up on the rose-bugs, one owner having slain nearly 30,000, another 22,000, making in all over

100,000. This is the right way to take hold of these depredators. A few active, persevering *men* will do more than all the birds, repelling nostrums, &c., that the whole country can afford.

SELECT STRAWBERRIES.—Charles Downing, in the last edition of his work on Fruit, published in 1869, after describing 250 varieties of the strawberry, gives the following select list of sorts which have proved satisfactory within his own experience: Agriculturist, Charles Downing, Downer's Prolific, French, Green Prolific, Hovey's Seedling, Jucunda, Longworth's Prolific, Napoleon III, Royal Hautbois, Triomphe de Gand, Wilson's Albany.

MOLES.—The Small Fruit Recorder informs one of its readers, who has been troubled with moles working among his small fruits, that the use of strychnine, mixed with white sugar, and dropped into their burrows through small holes made with a quill, has been found by trial to be effectual in "fixing" them.

EARLY STRAWBERRIES.—The following method has been successfully tried in some places: Cover a good, well managed, clean bed of strawberries, the runners of which have been kept off, so as to form large vigorous stools, with dry forest leaves early in winter, three or four inches thick. Remove these leaves in February in the Middle States, and in March in the North, and place over the plants a frame with sash. Bank the sides with leaves, and cover the sash in severe weather. The plants will start early, and give ripe fruit at the usual blooming time. Airing and water must not be neglected.

MICE REPELLERS.—Generally a smooth, compact mound of mellow earth, free from grass, and made a foot high, late in autumn, is best. But sometimes a roll of sheet-iron or sheet tin is most convenient. Sheet tin is best, and will rust less than iron, unless the latter is well coated with gas tar. Roofing tin, fourteen by twenty inches, will make to each sheet four protectors, seven inches high and three inches in diameter, costing about five cents each. They last many years. They may be applied after some snow has fallen, with a little pressure and turning about. Fig. 85 represents one of these protectors, and fig. 86 several nested together.

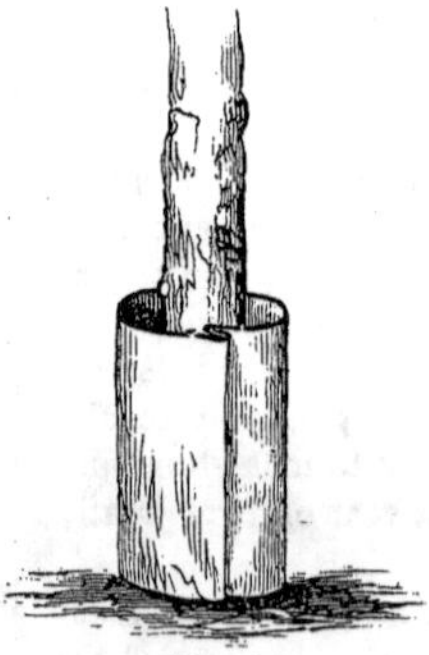

Fig. 85.

Fig. 86.

PEARS FOR KANSAS.—The Kansas State Horticultural Society, as reported in the Western Rural, recommends the following pears for cultivation in that State: Bartlett, as best; Seckel, Flemish Beauty, White

Doyenne and Howell, as next in value; and Doyenne d'Ete, Howell, Easter Beurre and Sheldon for farther trial. For dwarfs, they put Belle Lucrative and Louise Bonne of Jersey first; next, Tyson, Swan's Orange or Onondaga, Beurre Diel and Duchess d'Angouleme. Rostiezer and Glout Morceau are recommended for farther trial.

TOMATOES—TIME OF RIPENING.—Peter Henderson of Bergen, N. J., gives, in the American Agriculturist, a statement of his experiments with twenty-five of the varieties of the tomato found in cultivation. He treated them all alike, transplanting three times as they advanced in growth. His observations brought him to the conclusion that the extreme point of earliness was reached some years ago; that the only improvement now to be made must be in size, smoothness or solidity; that the difference in the time of maturity, of all the different kinds, does not exceed ten days at the farthest; that while the fruit may ripen in Georgia in May, in Virginia in June, in Delaware in July, and in New-Jersey in August, a certain amount of heat is required to do the work.

TIN LABELS FOR TREES.—After many years' trial, we find nothing so cheap, simple, convenient and durable as strips of sheet tin for permanent labels on bearing fruit trees. They may be seven or eight inches long, an inch or so wide at the larger end, and tapering nearly to a point at the other. Neither the breadth nor the length requires accuracy. They are cut out of scrap tin, and may be made at the rate of a dollar per thousand, or at a less cost. To write the name, lay the label on a table or board, and make the letters with the point of an awl or of a file ground to a sharp point, pressing firmly while writing. Each label is placed on the side limb of the tree, by bending the smaller end once or once and a half around it. The work is then done—in less time than the reader has occupied in reading these directions. Nothing further is necessary for many years. The point used for writing the letters scrapes away the tin coating, and admits the moisture of rain to the iron, rusting it and rendering the letters conspicuous. As the limb increases in size, the tin yields to the pressure, and never cuts the bark, and is at the same time stiff enough to hold on and prevent the label being moved by the wind, which so often defaces and wears out other labels. Out of some hundreds made and put on trees about eight years ago, all are now distinctly legible and in their places, and most of them appear likely to last at least eight or ten years more. The cut (fig. 87) is an accurate representation of one of these old labels.

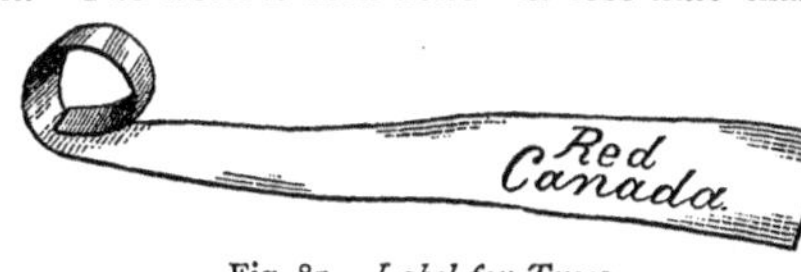

Fig. 87.—*Label for Trees.*

CORN FODDER FOR MULCHING.—In the proceedings of the Montgomery County Horticultural Society of Ohio, published in the Gardener's Monthly, J. H. W. Mumma gave a statement of his mode of protecting plants for winter, by means of cut fodder. After trying many substances, he prefers this. A machine both cuts the stalks and grinds them fine. Being used for his domestic animals, he easily obtains any desired quantity for winter mulching. He applies it after it is thrown out of the stalls. It is also used for flower beds in summer. One great advantage is that it is free from all foul seeds. A double handful is applied to each hill of strawberries, so as nearly to cover it. It is left on in spring, and the plants grow up through it, and the fruit stalks rest on the mulch.

Pruning Chisels.—A well managed orchard has the crooked or supernumerary branches removed before they become large, and an even, symmetrical head formed while the tree is comparatively small. Such branches may be easily and quickly cut off, while the operator is standing on the ground, by means of a suitable pruning chisel on a pole or rod. The form of the head may be seen to better advantage from the ground, and a cumbrous ladder is not required. A chisel is better than a pruning saw for such small limbs, working more rapidly, making a smoother cut, and preventing the annoying bending of the limb under the pressure of the saw.

There are some objections to the common square-edged chisel. A square cut is more difficult to make, and it often tears the bark. But if the edge is ground on the corner of a grindstone, as shown by fig. 88, or if the smith makes it in that shape, it cuts on both sides of the branch, pressing the opposite parts together, and makes a smooth, clean cut. It will also cut more easily by means of the sawing motion at the edge. With such a chisel, having a handle four or five feet long, and with a mallet to strike the lower end, a limb an inch or more in diameter is quickly taken off.

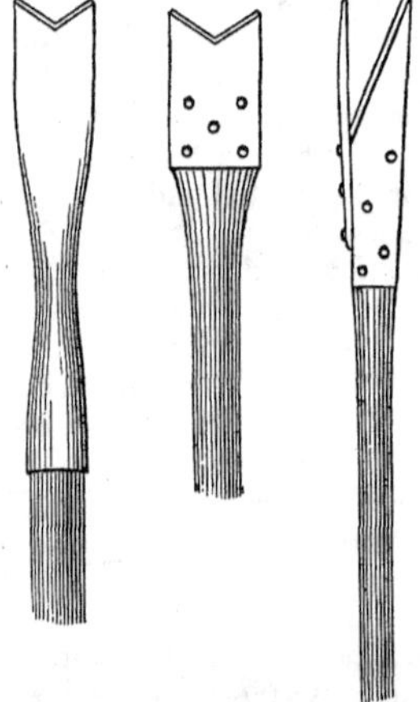

Fig. 88. Fig. 89. Fig. 90.
Pruning Chisels.

Another form is shown by fig. 89, where a piece of good saw plate or hoe blade is cut in the form represented, and screwed on the flat end of the long handle, which is used in the way just described.

A third mode of construction is shown in fig. 90, where the object is to give a more perfect saw motion or draw-cut. The steel blade is made with but one long cutting edge; and after this is secured to the handle, a smooth steel rod is screwed to the left side, so as to bring the blade up against the limb, as it is thrust upward.

By unscrewing the steel rod, the blade is easily ground on a common grindstone.

ANNUALS AND PERENNIALS—Their Cultivation.*

By Mrs. Sophia O. Johnson.

FLORICULTURE has taken rapid strides in the past ten years. The florists of Europe and the United States have devoted themselves to their trade with praiseworthy emulation; each country has striven with the other—Germans, French, English and Americans, have all been engaged in the strife, and to-day we profit by their exertions. Our old-time garden flowers are utterly eclipsed by the floral belles of the season! Compare the *Asters, Balsams, Stocks, Pinks, Petunias, etc.*, which amateur florists now cultivate, with those we raised in our childhood and the great improvement in each and all is seen at a glance.

The Camellia-Flowered Balsam, (fig. 91,) is a rival to its statuesque namesake. Many of its flowers are as purely red and white and as many petalled as the lovely Camellia. It is a gross feeder,

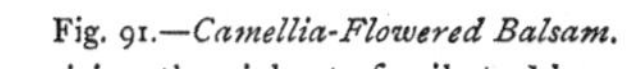

Fig. 91.—*Camellia-Flowered Balsam.*

Fig. 92.—*Balsam—Single-Stem.*

requiring the richest of soils to bloom in perfection. The plants appear to better advantage if transplanted nearly a foot and a half apart, and tied to stakes. The side branches should be nipped off to make a perfectly shaped plant. When the buds are formed, water twice a week with manure water;

* We are indebted to the Catalogues of B. K. Bliss & Son of New-York, and of James Vick of Rochester, for the cuts contained in this article.

horse manure is the best for this purpose, but failing to procure that, one tablespoonful of guano mixed into one gallon of tepid water will prove a good substitute. Ten different colored flowers are found among this species of Annuals. If desired for in-door plants, the seeds can be sown late in July, and at Christmas their gorgeous flowers will be in perfection. They are not ranked among green-house plants, but they are very fine for window gardening. They can be trained to bloom as a single stem, (fig. 92,) with three branches, or with five, or left to wander at their own sweet will.

Fig. 93.—*Delphinium Hybridum Novum.* Fig. 94.—*Lavatera Trimestris.*

Delphinium or Perennial Larkspur, is a plant well worthy of the attention of all amateur gardeners. It is a hardy species, will stand our coldest northern winters unprotected, and its perfect shades of blue, so rarely found among desirable flowers, render it an unrivaled addition to bouquets, vases and baskets of flowers.

Delphinium Hybridum Novum, (fig. 93,) is, as its name indicates, one of the recent results of the florist's care, and its flowers are rarely double and beautiful. The plants will often attain a growth of two feet in height, and are one mass of brilliant blue, purple or white tints. They will bloom well in any soil, but, like most other flowers, fully repay the care of the cultivator, and delight in a rich, loamy soil.

Lavatera Trimestris, (fig. 94,) is no novelty, but one of the flowers of our childhood, and though it may be considered an old fogy among plants, we cherish it tenderly for the sacred associations which hang around its rosy striped chalices. It is a hardy annual, once obtained, never lost, for it springs yearly from self-sown seeds, and its clusters of flowers are ever ready to your hand for an effective back-ground for a large vase. There is a white variety which is also a joy to us. These plants require no care, are constant bloomers from late in June until King Frost cuts them down, and a paper of seeds costs but five cents.

The *Double Portulaca*, or *Portulaca Rose* as it is sometimes called, from its close resemblance to the queen of flowers, is a "novelty" of some years, but its claims upon our notice increase yearly. It is a great addition to every *parterre*. Its blossoms equal the single varieties in brilliancy of coloring, but are very double, and the leaves are gathered closely together. It is seen in yellow, scarlet, white, orange, crimson, rose, and all these colors striped with white. Like all double flowers, it is not always true to its name; single flowers will creep in unbidden; these can be uprooted as soon as the first flower appears, yet it is well to leave a few plants for seed, as the double varieties produce but very few seeds, but they are more to be relied on for double flowers, as plants, like men, are apt to degenerate.

Fig. 95.—*Phacelia Congesta.*

Phacelia Congesta, (fig. 95,) is a curious plant, of a bright azure hue. It is a hardy annual, easily cultivated, and those who have a preference for blue flowers will be much pleased with it. It is not a profuse bloomer.

Petunias are classed in Bliss' Catalogue as half-hardy perennials, though most of our readers cultivate them as annuals, as they are quite tender. Thirty years or more ago the first small, single, deep purple Petunia was offered by Breck of Boston, as a great "novelty." A father who delighted in all rarities in the floral line, obtained a small packet of seeds, and though but a little child, I well remember his delight when the first bright consummate flower bloomed. How he would have gloried in a a bed of Petunias like the group we annex! But many years ago he was called up higher, where flowers never fade. Buchanan's Hybrids, (fig. 96,) are beautifully blotched and marbled with rose, crimson, and all shades of purple, and the velvety texture of the leaves is unequalled by any flower which grows. It is a plant of the easiest culture—any novice in gardening can grow it. It succeeds well as an indoor plant, and is desirable for window gardening, often blooming all winter with but little care. It is raised easily by cuttings — and the double varieties, which are highly esteemed, can be duplicated in this manner, as they do not seed readily.

Fig. 96.—*Group of Petunias.*

This flower is very effective as a lawn bedder, and shows well when planted in initials of very large size, or in crescents, stars, or circles. It sows itself—at least the single kinds—and perpetuates the bed from year to year.

German Ten Week Stocks, (fig. 97,) are great favorites with us, their spicy fragrance rendering them far superior to any flower we have mentioned. They are unequaled as a garden favorite. We are indebted to the German florists for vast improvements in this line of flowers. Some of the varieties are called Hyacinth Flowered, and they are decided rivals of that far-famed bulb. The flowers are more double than those of the rarest Hyacinth that ever bloomed, and their fragrance is as delightful. No garden is complete without them. They blossom in the brightest colors, and the purest white and creamiest yellow tints are frequent. They bloom much finer if weekly treated to a dose of liquid manure, and they delight

Fig. 97.—*Large Flowering German Ten Week Stock.*

in a rich loamy soil, in which their fibrous roots can easily expand. We love the Stocks in all their branches, and heartily recommend their cultivation.

There is a decided demand at the present time for climbing or running plants for vases and hanging baskets. The fashion of the times demands these tasteful adjuncts to every town or country house; so we will present a few varieties worthy of culture.

Fig. 98.—*Adlumia Cirrhosa.*

Adlumia Cirrhosa, (fig. 98,) is a desirable climber, either for baskets, trellis work or pyramidal frames, such as our cut presents. Its common name is Mountain Fringe; its foliage is very graceful, and it will often grow fifteen feet in a season. It is a biennial and is found in quaint country gardens, where it has propagated itself for years, and thrown its spiral arms lovingly around the dark green evergreens, making a charming contrast with its light traceries of bright emerald hue. It is a vine worthy of culture in every garden.

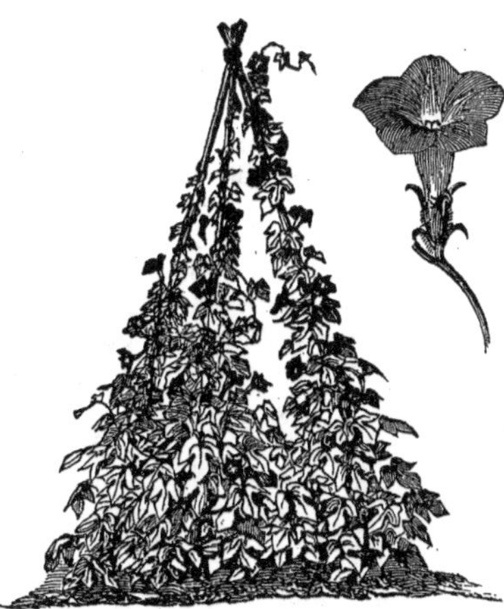
Fig. 99.—*Ipomea Hederacea Superba.*

Ipomea Hederacea Superba, (fig. 99,) is a recent "novelty," sometimes called Ivy-leaved Ipomea, from the form of its delicate leaves. It grows 8 feet high, is a tender annual, and its bright blue flowers margined with white bloom in great profusion. It is desirable to climb upon the supports of hanging baskets, and its flowers are welcomed everywhere for their exquisite color. Its habit is not equalled in grace or beauty by any climber which the florists' catalogues afford. It grows to perfection in a rich loamy soil, well watered with manure or guano. All these species of Convolvulus are desirable for in-door decorations.

Cardiospermum or *Balloon Vine*, is a very ornamental climber, beautiful for vases or hanging baskets, it grows four feet in height, and its seed vessels are inflated like a balloon—hence its name—when the seeds ripen the capsule bursts with a slight report. Its foliage is pretty, and though a tropical plant, it is desirable in every collection of flowers. It requires a sandy soil, and a sheltered situation from chilling winds, but delights in all the sunlight it can receive.

Last, but not least, upon our list come the *Tropæolums*, (fig. 100,) the most profuse flowered and easily cultivated of all climbers. Their colors are unsurpassed, and the flowers, amid the rich, dark foliage of the leaves, look like miniature tropical birds. There are many varieties of this species. B. K. Bliss offers no less than fourteen different colors, all vieing with each other for our favor. Many of them will bloom through the winter months, so that they are unequaled for window gardens. They grow readily from seed, but in July or August will root rapidly from cuttings. No plant is more easily propagated, and none more fully repays the cultivation bestowed upon it. For vases, verandahs or summer-houses, no vine can compete with it. It does not require a rich soil, but desires a gravelly loam. To make ours bloom in perfection we take the trodden soil of the highways—mixed with sand and gravel—and our plants are always the admiration of the passer by. As a bedding-out plant the Tropæolum is very effective—the shoots should be fastened down with hair pins and not left to straggle.

Fig. 100.—*Tropæolum Lobbianum.*

Momordica Balsamina or Balsam Apple, is of the gourd family ; the leaves are of a woolly texture, the flowers of golden hue ; the fruit when ripe bursts at the touch, with a slight report, and discloses a brilliant cherry pod. It is decidedly ornmental, and is much cultivated at the southwest, whence we porcured our seeds. It will climb ten feet in one season, but is a half-hardy annual. It will sow itself like Morning Glories, and spring up from year to year—is a native of the East Indies.

Lophospermum Scandens, (fig. 101,) is a very beautiful half-hardy perennial; the flowers are of a rich rosy purple, shaped like a Foxglove, and are especially effective for hanging baskets and vases. It will bloom the first season if the seeds are planted early in April, and it can be kept through the winter in a cool, frost-proof cellar, but it is desirable for a bay window. Any light, rich soil, will produce a good growth.

Fig 101.—*Lophospermum Scandens.*

The Lilium Auratum or Golden-Rayed Queen of Lilies, (fig. 102,) is perfectly hardy, enduring the coldest New-Hampshire winter with a slight protection of straw and leaves. It should rejoice the heart of every true lover of flowers. As a novelty its price has not brought it within the limits of all purses, but it is propagated so readily by offsets and seeds that soon the inhabitants of every little country village may behold its glories.

Fig. 102.—*Lilium Auratum.*

Six to twelve of these jewelled beauties are borne on one stem, and while it has no competition in beauty of coloring, it is unsurpassed in delicious fragrance.

Dr. Lindley of England, who introduced this rare plant into that country, says: "If ever a flower merited the title of glorious, it is this, whether we regard its size, fragrance or exquisite arrangement of color."

The blossoms are from eight to twelve inches in diameter; they possess six ivory-white petals thickly bedewed with chocolate crimson drops, and a golden ray extends through the centre of each petal. It blooms splendidly in pots, in a light sandy loam—stable manure is death to it, but fibrous loam and peat, with a good mixture of sand, is a desirable compost. The plants should be tied to stakes, or they are liable to be broken down by wind and drenching showers. If planted where they are protected from the noon-day sun, the flowers will remain perfect much longer, but if placed in too damp a soil, the bulbs will decay. They have been made, by proper culture, to grow eight feet high, and produce from twenty to thirty blossoms upon one stalk.

Fig. 103.—*Convolvulus Tricolor.*

Convolvulus Tricolor.—Few annuals are more desirable than the various species of Convolvulus. They are very beautiful either in beds cut into the turf, or in mixed borders. The *Convolvulus Tricolor*, (fig. 103,) is an especially fine variety, of a rich purple hue, with a pure white centre. It is a half-hardy annual and in some sections of the United States will sow itself as readily as the Petunia, and thus give a brilliant flower-bed with only the trouble of thinning out the plants. It is of a trailing habit, and a few plants will fill a bed six feet in diameter.

Fig. 104.—*Dwarf Nasturtium.*

Dwarf Nasturtium—(fig. 104.)—There are no annuals in all the florists' catalogues which have improved so much under hybrid cultivation as the above named variety. They rank now with the Verbena and Geranium; their rich colored and beautifully spotted flowers making them desirable in all gardens. They thrive in common garden soil, and blossom more freely if sand and fine gravel are mixed with the loam. Their colors are chiefly scarlet, spotted with yellow, deep maroon, rose tinted with maroon, crimson, yellow spotted with crimson, sulphur spotted with mauve, light yellow, nearly white. King Theodore is almost black, with very dark green foliage. No collection of annuals is complete without these charming varieties.

Sedum—(fig. 105.)—There are two species of this plant; one has a bright blue blossom; the other, a native of Kamtschatka, is of a brilliant orange hue. They grow freely on rustic or rock work, and are pretty for hanging baskets; being hardy perennials, they bloom without much extra care or labor.

Fig. 105.—*Sedum.*

Anagallis, (fig. 106,) are half-hardy annuals, chiefly blue in color. They are admirable for the edging of ribbon beds, as they grow only six inches in height, and are covered with a profusion of rich colored flowers. They are also desirable for window gardening, but for this purpose the seed should be sown in pots in June. Plants that have not blossomed can be potted in a rainy season, or after dark at night, and if shaded for a few days, will not lose their leaves, and will bloom all winter. Anagallis Marmora Dell' Etna is of a bright red; Napoleon III is a rich maroon.

Fig. 106.—*Anagallis.*

Fig. 107.—*Chrysanthemum.*

Chrysanthemum—(fig. 107.)—This fine flower is of brilliant hues, crimson, yellow and white, and makes a fine display in borders or ribboned beds. It is a hardy annual, grows one foot high, and is effective in large vases; is also often grown as a lawn plant, but requires much room to produce its best effect. Plants should be set eighteen inches apart. It thrives in common garden soil, but its flowers are finer if watered with liquid manure, and this applies to all annuals and perennials.

Fig. 108.—*Clarkia Integripetala.*

Clarkia Integripetala—(fig. 108.)—The Clarkia is an old favorite, but of late it has undergone some transformations under the skillful hands of the florists.

The species above named is of a rich crimson, and the petals of its flowers are not as deeply serrated as the older varieties. It is a hardy annual, and with the new double white variety, makes agreeable additions to our gardens. It will sow itself if the seed pods are left on the stems, and will flower early the ensuing season.

Eyrsimum Peroffshianum, (fig. 109,) is a very showy orange-flowered annual, brought from Palestine. There is no other flower possessed of the same color, and though it is not a graceful plant, if it is grown in thick masses it makes a fine display, and is a beautiful contrast to the Delphinium. In bouquets and vases its color is very effective. There is another variety, brought from Arkansas, of a pale sulphur yellow. These plants grow eighteen inches in height.

Fig. 109.—*Erysimum.* Fig. 110.—*Gilia Tricolor.*

Gilia Tricolor—(fig. 110.)—This is a hardy annual from California, and will bloom in any soil. It is white, purple and lilac, of delicate form, and especially pretty in small bouquets. It blossoms finely in pots, and will reward the amateur gardener if it is transplanted early in the season and cultivated as a pot plant. There is a pure white variety, which blooms profusely, and is very useful as a bedding-out plant. Its delicate foliage adds much to its beauty.

We wish that every dweller among flocks and herds could understand how much pure delight is to be obtained by an investment of a few dollars in seeds, plants and bulbs. Many more daughters would remain contentedly at home if allowed to beautify their surroundings, and many more sons would not forsake the ways their fathers have trod if their tastes for the beautiful were allowed full play. We know a young farmer, bred among New-Hampshire's granite hills, who yearly invests $5 or $6 in flower seeds and bulbs, and makes the little brown house of his parents a bower of beauty. His father sniffs and groans a little—"boys did not care for sich truck in his days"—but he throws no decided obstacle in his son's garden, so the flowers bloom sweetly, and nightly, after the cows are milked, the work done; the toil browned son weeds and hoes, stakes and trains, and in the autumn at cattle shows and camp meeting his jolly face is always seen aglow with pride and pleasure as the huge bouquets he has gathered and tastefully arranged, are praised and admired by the rustic belles of the adjoining villages.

The cultivation of plants from seed is a source of great pleasure to the amateur gardener—from the little tiny germ springs forth the two little leaves, then the stem and buds, and lastly the brilliant flower. The care bestowed upon such plants enhances their value to their possessor. A garden of bedding-out plants procured from a florist, cannot furnish half the delight which is to be obtained from a well cultivated bed of annuals. Try it ye sons and daughters throughout our land, and prove for yourselves the pure joys therein contained.

"There is a lesson in each flower,
A story in each stream and bower;
In every herb on which we tread,
Are written words which, rightly read,
Will lead us from earth's fragrant sod,
To hope and holiness in God."

HOUSE-KEEPING ECONOMY.

By A Young Learner.

MRS. NOVICE, a young housekeeper, appeared one pleasant afternoon at her neighbor Capable's kitchen door, and asked for a long promised lesson on domestic management.

Mrs. Capable met her with a bright smile. "I can give only odds and ends," she said, "gleanings from papers, cook-books, older housekeepers, and my own experience. Let us begin here first, and go from room to room, so that objects before our eyes may suggest the items to our memory."

The Kitchen.

"Our first topic shall be the cook-stove. We have wood to burn; we boil our kettles on the stove, and bake in a brick oven. To-day, Tuesday, we iron. In this deep sheet-iron pan, which is set into the top of the stove, we put our smoothing-irons to heat; they are hot much sooner when placed so close over the fire, less wood is needed, less space for it left, less warmth is requsite. With a coal stove you would not need it.

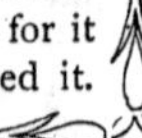

But a small charcoal furnace, used out of doors or in an open wood-house to avoid danger from the gas, is a great convenience for heating irons and boiling the tea-kettle on warm summer afternoons when you want no kitchen fire. A hot-water tank and a tin oven for keeping warm food, are valuable appendages to a cook-stove. Mine has not the latter, but we found a patent cast-iron rack, which is fastened by screws to the stovepipe, at a suitable distance above the fire; it is good for warming plates and keeping dinner hot.

"Lucy is making biscuits. Do not be surprised; it is no great attainment for a little girl; we sell more wheat, and buy occasionally the 'self-raising flour,' which needs only to be worked up with milk and butter, rolled, cut out and baked, to make a first-class biscuit; or prepared with milk, butter and eggs, and cooked in rings, for delicious muffins; no time is required for rising. The moulding-board on which Lucy works does not slip about. A cleat is nailed at one edge on the under side, so that it cannot be pushed back upon the table. Our large platters, which are too wide to lie upon the pantry shelves, are held nearly upright between them by cleats nailed a few inches from the wall; the platters are slipped in behind them."

"When I have unexpected company," said Mrs. Novice, her mind recurring to the quickly made biscuits, "I never know how to provide the table."

"A housekeeper ought always to keep on hand a supply of well prepared food for her own family. Private fasting and public feasting induce many difficulties. Guests should fare more plainly and the household more comfortably. There ought never to be any sour, heavy or doughy bread. Watch rising bread and it need not sour and require soda; knead it thoroughly and it will be fine and white; bake it instead of burning it, and it will not be raw and hollow hearted. Bread should be baked on the Tuesday and Friday of every week; weekly is too seldom; the old should be sufficient to last till the new has been one day baked. Previous calculation of the quantity is necessary to prevent borrowing and perplexity. In cold weather you can mould an extra supply of bread if you wish; after forming it into a loaf it can be kept a week in the cellar; when needed, you can make it into biscuits by adding a little shortening and working over; or you can mould it again, let it rise, and bake it for bread.

"Cream that has stood long and become bitter, makes poor butter; so does that which is skimmed from milk that has been heated to produce more cream. Some butter is spoiled by over or under salting. I always use an ounce of salt to a pound of butter. Laziness and bad butter are near relations. If I want a drink of buttermilk, I prefer to take it separate from the butter, and am warned of approaching rancidness whenever I see a roll of butter oozing milky drops.

"With good bread, good butter, and a supply of fruit, fresh, dried or canned, a woman need not tremble when she hears a carriage driving up.

In the spring, before apples are gone, I can them for table use ; after cutting up, I steam them to make them soft, pour over them hot syrup of sugar and water, and flavor with sliced lemon. Apple jelly is also good. I slice, very thin, thirteen apples, (without peeling them,) cover them with water, boil and strain them. (By the way, for squeezing jelly bags, we use something like a large nut-cracker, (fig. 111,) two wooden paddles, hinged together with a leather strip ; the bag is pressed between them and our hands are not burned.) To each pint of apple juice I add a quarter of a pound of loaf sugar ; boil to a jelly and add two sliced lemons. Let me here mention, lest I should forget it, though it is out of place, that *currant* jelly is much more easily made by boiling the strained juice about five minutes, pouring it boiling hot over the sugar rolled fine, and stirred till the sugar is dissolved, and then poured into moulds. It will be stiff when cold.

Fig. 111.—*Jelly Squeezer.*

"I must tell you how to keep ham through summer, and have it ready for sudden emergencies. Cut the whole ham in slices, fry it till nearly done, and put it in a jar in layers filled in with lard. A little more cooking makes it ready for the table at any time. Nice little ways, Mrs. Novice, of preparing common dishes, potatoes, hashes, toast, &c., are worth as much more than rich material as ingenuity is than abundance."

After some conversation, Mrs. Novice inquired if her friend always washed on Mondays.

"Always," answered Mrs. Capable. "If it rains on Monday it is quite as likely to rain on Tuesday or Wednesday. We use a wringer and a revolving clothes-dryer, and have the clothes ready to sprinkle and fold by evening. We do not crowd the business of the first half of the week into the last, which makes Saturday a nuisance in some families."

"What can be done with stains, mildews, &c.? There are some things that hot suds do not benefit."

"A fresh fruit stain is readily removed by pouring on boiling water, but when a stain is 'set' by the soap of many washings, it can be taken out by exposure to sun and air a few days, after rubbing the spot with yellow soap and wetting it with thick cold starch. If one application does not succeed, 'try, try again.' I have heard that mildew can be thus removed: Put one-quarter of a pound of chloride of lime in a gallon of cold water ; let the mixture settle an hour, then drain off the water and soak the injured articles in it two hours, wash them and hang them in the sunshine. When anything spotted with grease is put into the wash it is well to mark around the places with white thread, so that they may not be overlooked and neglected. I have lately learned an easy way to bleach cotton. I leave the unbleached garment in a pan of sour milk for a few days, occasionally shaking it well, then wash and boil it and it becomes white."

"Will anything take wheel-grease out of cloth? My only silk dress has a frightful spot of it."

"I will read you this prescription which lies among my scraps. I have no doubt of its efficacy: 'Lay the silk on a clean sheet, folded to eight thicknesses, rub the greased part with a soft cloth dipped in lard, moving the silk to a new spot frequently. After a time the wheel grease will all be through, leaving only clean lard. Clean this out in the same way by rubbing with nice soap and alcohol, using a clean cloth to rub with, and frequently changing to a new spot on the underlying sheet. Then lay the silk on a clean cloth and rub dry with a soft cloth. A friend cleaned a white Canton crape in this way, and you cannot find the spot where it was greased.' "

Mrs. Novice expressed her thanks for the receipt, and then began to look about the room. "You use a skirt-board and bosom-board for ironing, I see," she said, "but what is that black board for, hung on a nail by the fire?"

"For a very different purpose. We put it on the kitchen table when we cook, and when we take a kettle from the fire we set it on the black-board while mashing potatoes, draining vegetables, &c. It keeps the soot off from the table."

"And I observe another convenience, too, if I may speak of it. Those bars for drying small articles, (fig. 112,) which you can draw out one by one or push back against the wall, out of the way. There is a rod which runs through all of them, so that they can be turned on it. And your pumps are both in the kitchen; you have painted tables and an oiled floor that almost keeps clean of itself, and opposite windows that make the room so airy and light; fuel under cover, and a door that opens into the wood-house instead of directly into the open air. How comfortable and pleasant everything is!"

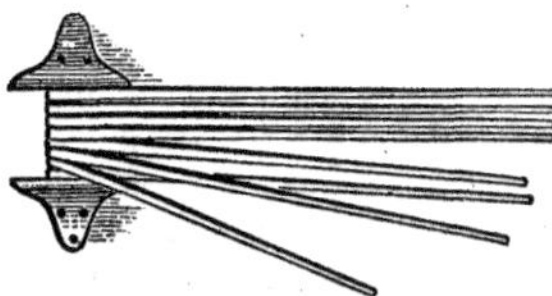

Fig. 112.—*Clothes Bars against the Wall.*

The Cellar.

It was large, dry, clean and well ventilated. When the ladies descended to it by an easy flight of stairs from the kitchen, they first observed the soap barrel.

"Soap is much softer and 'goes farther' after it has been made a year," remarked Mrs. Capable. "One gallon of old soap is worth a gallon and a half of new. Sometimes we make cold soap; we boil the grease in strong ley, put it hot in the soap barrel, fill it up with cold ley, and leave it all to turn to soap at its leisure."

"There is the vinegar cask," she continued. "I save many gallons a year by contributions to it of the odds and ends. The rinsings of molasses jugs when they are washed, the juice left from preserves, are instances. Vinegar is much sharper the second year after making."

Mrs. Novice much admired the Zero refrigerator, but said that she could not afford to have one of her own, though ice could be conveniently procured.

"But ice can be kept in the house a great while without a refrigerator," answered Mrs. Capable.

"In a blanket?"

"For some time it can; but I am told it will not melt in less than a week in a case made in this manner: Make two woolen bags nearly equal in size, no matter how coarse, old or faded the material; put one inside of the other, and fill in the space between them with feathers; hen's feathers are as good as any; sew the tops together all around, and tie the ice within this double receptacle. When I mentioned this plan to Mr. Capable, he said that wool would make a better interlining than feathers, and it is as good a non-conducter."

Mrs. Novice said she intended to have a safe made like her friend's—a plain wooden cupboard fitted with shelves, the door to which mainly consisted of fine wire netting, which admitted air and excluded flies. She has already the "swing shelf," fastened by posts to the ceiling instead of the floor, which is indispensable to every milk-room.

The Bed-Room.

After this Mrs. Capable took her friend to the spare bed-room, where she might get some kind of useful suggestion. The bed was neatly made over a sponge mattress. "Oh, I must tell you about that mattress," exclaimed Mrs. Capable; "it does very well for a bed that is seldom occupied; is soft and durable, but by constant use becomes damp and unhealthy. Those crocheted covers to the pillow cases keep them from dust and hide them when tumbled and creased. The spread is only a thick sheet, knotted with tufts of candle-wick. I have made several and arranged the knots in different patterns. They look well, cost little, and can be washed as easily as a common sheet. The girls made that pasteboard box for combs and brushes before we had a toilet bureau for the room, and it is so convenient and pretty that we have left it yet. It is, you see, flat on the back and rounded on the bottom and front, and covered with fancy paper, (fig. 113.) A strip of cloth firmly pasted to the back and lid, forms a hinge between the two. I have an old cigar box covered with pretty calico and nailed on the inside to the wall for clippings and all sorts of refuse in my room. It saves me many a tramp from the bedroom to the kitchen stove."

Fig. 113.—*Box for Combs and Brushes.*

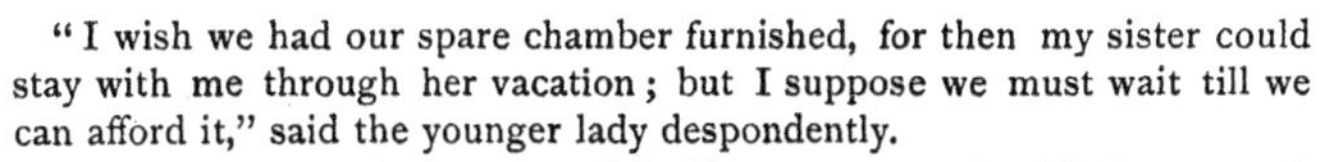

"I wish we had our spare chamber furnished, for then my sister could stay with me through her vacation; but I suppose we must wait till we can afford it," said the younger lady despondently.

"If you have a bedstead and bedding, you can furnish it at once," answered her friend.

"Oh yes, we have an old-fashioned bed in it now, but there is no bureau, no wash stand, no chairs, no wardrobe—nothing. Mother says she will give me one of her carpets for it when we are ready to fit it up, but I can't tell when that will be."

"Send for the carpet to-day," was the prompt reply, "that is if you are willing to offer your guests a cozy, though not a fashionable sleeping place. Oh, the bad nights I have spent in some stylish chambers! I think a comfortable bed and pure air eclipse all the upholstery in the world. And now I will inform you how to have necessaries without money. That little old rocking chair that Freddy broke, can have the broken rockers taken off, be covered and made into a serviceable seat. You can have a beautiful toilet table by setting a deep box on end, fitting it with shelves and draping it with a white cover and curtains all around. There should be an opening at one side in the curtains, so that they can be drawn back upon the tape that shirs them, and give access to the shelves. Such a pretty table as my mother had in the days of our poverty, when I was a child! Her walnut and mahogany sets do not look half so handsome to me now; and yet it was only an old-fashioned oval table of grandmother's, with a bit of white Marseilles over a rounding cushion on the top, and a full flounce of barred muslin, neatly starched and ironed, concealing the slim, awkward legs, and giving the whole room a clean and fresh look."

"I will try, my kind adviser, to fit up the room. But what shall I do for window curtains? You don't like green paper shades any better than I do."

"White holland costs but a few shillings a yard, and keeps clean indefinitely. It can be nailed to the window head, rolled up by hand and held in place by home-made cord and tassels if you cannot purchase 'fixings.' Perhaps two of Freddy's out-grown baby frocks would make a prettier curtain, shirred at the top, parting each way and looped back with ties of ribbon."

"There—we forgot the wash stand! Surely that article cannot be improvised."

"Oh, yes—ask Mr. Novice to fasten cleats to the wall, in one corner of your room, and nail securely upon them a triangular shelf of good plank, (fig. 114.) You can lay a towel over it, or make a tasteful cover purposely for it, and it will hold the washing apparatus. But now let us step in to see Cousin Dora."

Fig. 114—*Corner Wash Stand.*

An Invalid's Room.

Dora's disease was incurable. She occupied, in her cousin's house, a room which was the focus of the cheerfulness, tidiness and order which prevaded that dwelling.

"Neighbor Novice has come for *information*," said Mrs. Capable pleasantly; "can you describe to her, Dora, some of your little contrivances?"

"Oh, you must see my new easy chair. Arm-chairs are costly, you know. But mine is made of a cask; some of the upper part is cut away to form a back and arms, and a seat fitted in, (fig. 115.) The staves and hoops are very thoroughly nailed together, so that it will not drop to pieces. Cousin Edward made it the week that his ankle was sprained. Then the girls made stuffed cushions, and covered it all over with an old French calico I used to wear, (fig. 116.) But when they had got it done they had to take off the arm cushions—the staves were too sharp to rest elbows on—and nail on flat strips of boards over them to make the arms comfortable. I have a wide board, like the leaf of a table, with long prongs attached, which run through these sockets on my chair, (fig. 117.) This holds my work or writing desk when I am able to sit up. This chair-back is also a great convenience—an old chair-back with legs behind it instead of below it, which holds it firm in its place, as they cannot slide back farther than the headboard of the bedstead. It is more comfortable than a pile of supporting pillows when one wishes to sit up in bed; and I have lent it to several sick persons, who have all considered it a very useful article."

Fig. 115.—*Barrel Cut for Chair.*

Fig. 116.—*Barrel Chair Complete.*

Fig. 117.—*Work-Board.*

"Do you take no medicine, Miss Dora? I see none about the room."

"My bottles need not always be in sight." She opened the door of a small closet with her right hand. "I keep them in this square basket. Half way up in it there is a piece of pasteboard tightly fitted in; this has holes of the right size cut in it, and the vials stand up in it as if they were in stocks. Do you know how to wash a high and narrow bottle? Put in a good many carpet tacks, fill up with suds, shake well, and it will be clean. Do you see my hyacinths there? They are blossoming in certain old cast-off, wide-mouthed bottles of mine. The bulbs are set on the tops of the bottles, which are filled with water and a little charcoal. The roots soon strike down and look very pretty and white through the transparent glass. The water needs no changing, but occasionally a little more must be poured in to make up for what the plant absorbs."

"Dora is the more cheerful because she is generally occupied with planning or performing some useful work. Cousin, please to show your knitting work to Mrs. Novice."

The mittens were knit of Saxony yarn, and Dora had sold many of them at 75 cents a pair. Some were scarlet, with white lined cuffs, and some were drab, with blue, but all were alike in pattern, warm, durable and handsome, (fig. 118.) They were knit plain on both sides, which made them slightly elastic, the rows running lengthwise, the thumb-piece made separately and sewed in. The shape is shown in the figure. They were finished with cuffs crocheted in plain rows, beginning at the top and widening at the beginning of each row. The stitches of each row were taken through the back side only of the top stitches of the preceding row, which made a ribbing. When the cuff was deep enough, a zephyr of another color was fastened on, and a piece of the same shape crocheted, narrowing where the other half was widened. Then the whole strip was doubled together, sewed up at the side and set on to the mitten.

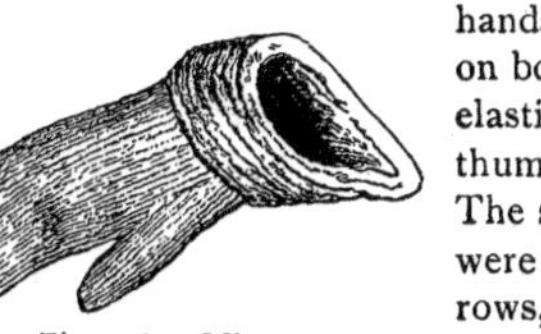

Fig. 118.—*Mitten.*

"The slippers which Dora makes for us reduce our shoe bill very much; they are so comfortable and convenient for house wear." Dora exhibited a pair; they were made of single zephyr—coarse yarn perhaps would have answered—crocheted in plain rows, beginning at the toe, and forming the shape of a canvas-worked slipper before it is made up. Each row rose from the back of the preceding row, like the mitten-cuffs. Widening was done from the toe to the top, along the instep, to fit the foot. A shell border surrounded the top of the slipper and a crocheted rosette was set on. The work was bound around the bottom, and sewed over-and-over on the wrong side to a cork sole, also bound. An elastic cord was run into the top. Mrs. Capable mentioned that she covered the soles with cloth on the outside when they began to get worn. She thought the soles of old gaiters might do instead of cork.

As the two friends were passing to the sitting room, Mrs. Novice observed, "You have oilcloth on your hall floor. Is that better than a carpet for halls?"

"I think so," was the reply. "It is much more easily kept clean and is more durable. We wash ours weekly with milk and warm water, which brightens the colors. If you ever put one down, let me beg you to lay several thicknesses of paper under it. When we had our first oilcloth, I did not know the importance of this, and when it wore out and we tried to take it from the floor, it refused to come. Where feet had trodden most, it stuck so to the floor that it could not be wholly removed."

The Sitting-Room.

"Now you shall see my arrangements for sewing. A large work-basket gets over-laden and disordered. I keep the implements, and nothing else, in a small basket; clothes to be made or mended are laid in this drawer.

When I go from home I take this bronze morocco case with me, (fig. 119;) it was a Christmas gift, and has proved most useful. It is lined with blue silk; has a needlebook, pincushion, and a place for scissors; a pocket for work; a receptacle for combs and brushes lined with oiled silk; and here you slip in a tooth brush—I always carry mine in a little oiled silk bag, so that it dampens nothing. This cutting board (fig. 120) is a great favorite; it goes to all the sewing circles, for the ladies find it very convenient to sit when they cut out work the whole afternoon. I can sew on dress facings without bending over a table when I have the board in my lap. It is hollowed in on one side so as to fit close up against one; and as it is marked with feet and inches on the edge, I can measure without going after a tape line. I have been making skirts lately. You can wear a white skirt twice as long by making it a little short, and putting on a full border with buttons and button holes; the border should come down a little longer than the skirt, and will be soiled first; then it can be unbuttoned and taken off, to leave the fresh edge of the skirt. We do not wear many white cotton skirts; it is better to change the underclothing twice a week than to fill up the clothes basket with skirts; white serge skirts can be worn under any but the thinnest dresses, and seldom require cleansing. A very good balmoral and hoop skirt, all in one, for winter wear, is easily made of red flannel, closely gored, with half a dozen tape shirrs at intervals on the under side, into which the springs of an old 'skeleton' are stitched."

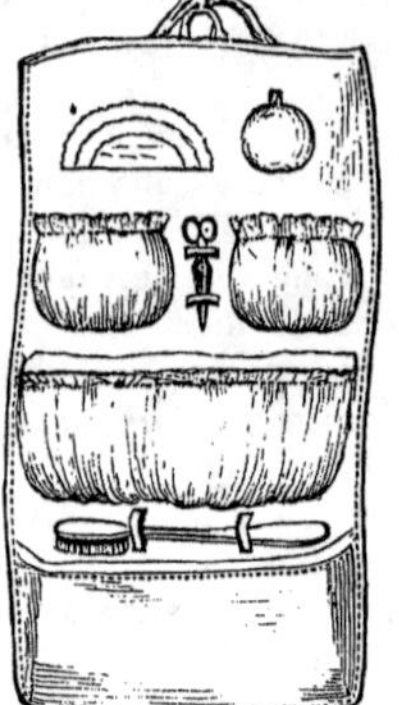

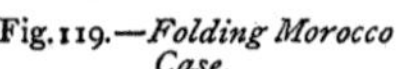

Fig. 119.—*Folding Morocco Case.*

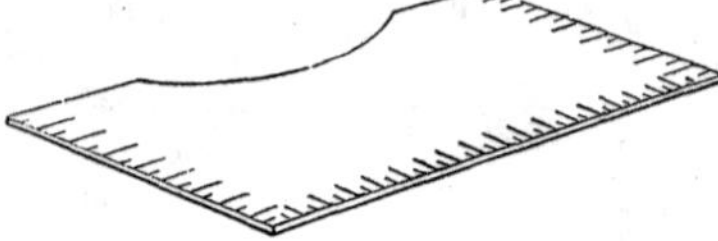

Fig. 120.—*Gradulated Cutting Board.*

"I am very grateful for your instructions in economy to-day, and hope I shall remember half of them. Until our farm is all paid for, I expect to be poor, and I must learn to save."

"My husband owes no debts; the last cent is paid up on the old place, and he lays up a little every year. But, my young friend, when you have lived as long as I, you will see that there is never a time when economy ceases to be a duty. I am glad to find you a believer in small and well-applied expenditure. At the same time I hope that Mr. Novice and you will see that cash is not the only kind of capital; that the domestic resources are of four kinds: Money, time, strength, and thought; that time and strength are to be husbanded, because, once lost, they cannot be regained; and thought, being the cheapest, must be made to save the other three.

"If you think yourself worth the least of all your husband's possessions, you can start out a drudge and end your career in broken health or early death. You can be the one to jump when anything is wanted. You can wait on the 'men folks' so obsequiously that they will grow very particular as to the time and quality of dinner, and will never be able to hang up their hats or find a candle. You can train your little boy to selfishness by attending yourself to his personal wants and excusing him from every disagreeable duty—as many fond mothers have done, with one invariable result.

"Do you see my two youngest boys out there in the back kitchen?"

"Knitting! knitting there by the fire, like a couple of old grandmothers."

"I never talk about them in their presence—it would harm them; but they are too far off to hear now. My sons all know how to knit, and have made many stockings, and mittens, and scarfs. They enjoy it as an amusement in long winter evenings. If a button comes off from their clothing, they do what it would be unpardonable laziness for their mother and sisters not to do for themselves, however busy—they sew it on again. While they are small they do my errands, wipe the dishes, help churn, and such like jobs; and when they are older and have men's work to do, they do not ridicule my young helpers, for they know something of the magnitude of household tasks, and still assist me when hardiness and strength are requisite. I never milk; never whitewash; never put down carpets. I did not spend my honeymoon in teaching my husband to depend on me for what it was better for him that he should do. And he has been more than self-reliant; he has been unselfish and helpful; many times hanging out the washing for me when I was not well, and helping me at night to care for wakeful babies. So now, after twenty years, he has a healthy wife, as good as new, and healthy children, full of life and good humor. And the best of all is, that while the woman and the family are in some instances degraded into mere dollar-making machines, the direct aim of all our economy has been to promote and leave room for the social, mental, moral and religious."

The sun was getting low, and Mrs. Novice put on her pink sun-bonnet, and hastened home to meet the time for supper.

How to Peel Peaches.—As the time for putting up peaches is at hand, we have procured from a lady friend the following recipe for peeling peaches, which we confidently recommend to our lady readers: Take a kettle of very strong lye, and heat to boiling; take a wire cage—similar to a corn-popper—fill it with peaches and dip into the lye for a moment; then into cold water. With a coarse towel wipe each peach, and the rind will peel off smoothly; then drop into fresh cold water, and the operation is complete. You need have no fear of injuring the flavor of the peach.

DOMESTIC AND RURAL ECONOMY.

PRESERVING HARNESS AND BOOTS.—A correspondent of the COUNTRY GENTLEMAN thinks the following preservative will make boots last twice as long as without it, keep dry feet and preserve health. Take two pounds beef tallow, half a pound of rosin, quarter of a pound of beeswax, and four ounces of castor oil. Melt slowly, adding and mixing well a heaped tablespoon of lampblack. Do not let it boil. When thoroughly melted and mixed, it is set aside for future use; when wanted for use, put a small quantity in a tin basin and melt it, but do not let it boil. Apply to the leather quite warm, with a stout piece of cotton cloth, folded tightly and sewed strongly together, so as to be about two and a half inches long. Hold this in the fingers, and rub the composition well into the leather, holding the latter frequently over a hot stove. Apply both to soles and upper leather. In applying to harness, add about as much neat's foot oil, to be applied warm near a stove.

OILING HARNESS.—Well kept harness will last many years; neglected, allowed to be dirty, dry and unoiled, it often wears out in two years. It should be washed once a month with castile soap, using a good sponge, and oiled at least twice a year. Take the harness apart, soak it in water only blood warm for a longer or shorter period, according to its dryness and stiffness, from one to three hours. Wash with castile soap, work till soft and pliant, and hang up to dry moderately. When half dry, oil the pieces, and let them dry a day longer; then rub with a coarse cloth.

The following mode of making the best neat's foot oil is given by a correspondent:

Take the feet of a beef, crush the bones well with a sledge or axe, and boil them in a large pot of water for twelve hours. Make two quarts of tallow from fresh beef or mutton suet, and pour it into a four-quart can, (which should have a lid to keep out mice,) and place it on a stove. Add a lump of pure yellow wax as large as a hen's egg, stirring as it melts. Then fill up the can with neat's foot oil, and, removing it from the fire, continue to stir until the intermixture is thoroughly complete. This, when cold, will be of about the same consistence as hog's lard. Keep in the can a bit of sponge always ready for use. It ought to be damp when it first goes into the grease, as it will remain more flexible always afterwards than if greased when dry.

F. P. LeFevre, Lancaster Co., Pa., adopts the following mode, which he thinks has great advantages, being quickly done, without waiting to dry, making the leather soft, and preventing the penetration of water: Apply one or two coats of lampblack and castor oil, warmed enough to penetrate freely; then wash with soapsuds, with a sponge. The previous oiling will keep the water out and loosen the dirt. When dry rub with a mixture of oil and tallow, equal parts, with lampblack or Prussian blue, to give it color.

TRAVELING—PACKING THE SATCHEL.—Nearly every person takes frequent journeys, and it often happens that something is forgotten in the hurry of packing the traveling-bag. A business acquaintance in New-York keeps a satchel constantly packed for a journey, and he is always ready at a minute's notice. We have adopted another mode, which, on the whole, we like better. It is this:—Take a card and write on it, at leisure, in a clear, distinct column, the name of every article which you will want while from home. This may be done by taking a list of everything in your bag, and adding to it as any omission is discovered. Then, whenever you are about starting on a journey, glance along down this list, and see what you want this time. You may thus pack a bag and get ready at any time in five minutes, and never forget an article. As a sample of such a list, we give the following, from which anything wanted is at once selected:

Watch,	R. R. Guide or Map,	Shaving Tools,
Match Box,	Lunch,	Money,
Pocket Compass,	Collars,	Tracts, Cards, &c.,
Spy Glass,	Cravats,	Paper,
Door Fastener,	Shirts,	Ink,
Thread and Needle,	Stockings,	Pencils,
Hair-Brush,	Gloves,	Envelopes,
Drinking Cup,	Overshoes,	Postage Stamps, &c.

Then, in starting or in changing cars, remember the three words, "*Over-coat, Satchel, Umbrella.*"

THE VIRTUES OF OIL.—A lady came to me with a new pair of scissors —they were screwed too tightly together, as she thought, but she could not unscrew them. I suggested a slight touch of oil, which was obtained from the sewing-machine feeder. The tenth of a drop was placed on the finger tip, and drawn along the edge or side of each blade. "Oh," she exclaimed, "they work like a perfect charm! There is nothing like having a doctor!"

A neighbor was sick of a fever, and at every turn of the door, the sharp creak of the hinge annoyed him severely. No one thought of the simple remedy—which was applied by first touching a feather to the kerosene wick, and then drawing it between the joints. The creak ceased instantly, and the patient fell into a refreshing sleep.

The front door had to be slammed hard in order to make the latch shut. A single drop of oil made it catch without the least noise, and with perfect ease.

The hired man was conveying manure to the garden in a wheelbarrow. At every revolution the wheel went "skreech-y-shriek, skreech-y-creak!" I examined the axle, and found it fast wearing out. A little oil stopped the harsh music at once.

"What makes this sewing machine run so hard?" asked Mrs. —— anxiously. "Don't you hear that soft creak?" "Why yes, but I didn't think that would make any difference." "Put a drop of oil on that pivot, and see." She did so, and exclaimed, "I declare! how nice it runs now."

A farmer was sowing grass seed with a hand Cahoon machine. He said it did the work well, but went "plaguey hard." "Put a little oil," said I, "on those pivots." He oiled them, and then said, on trying it, "Why this beats all! Why it goes as easy as open and shut—I would not believe it's the same machine!" X. X.

CUTTING CORN-FODDER.—Cutting corn-fodder an inch or more long, is not good for cattle. A farmer, who had some thirty or forty head of cattle, cut enough stalks in half a day to last them a week, by means of an eight horse power attached to a cutting machine, adjusted to cut scarcely an eighth of an inch in length. The stalks were thus reduced nearly as fine as chaff, and were eaten without difficulty. A large part of the sweet portions of the stalks, that are usually rejected and wasted, were hus consumed, and about double the amount of food thus obtained from them. Sweet, well-dried stalks, gave a great deal of the best feed. Water-soaked, half-fermented stalks, were not good for much, in whatever shape they might be fed. This mode of cutting admitted the mixture of meal without difficulty. Cutting by hand is laborious, and does not pay.

CUTTING FEED FOR HORSES.—An accurate farmer has furnished the COUNTRY GENTLEMAN a statement of his experiments with feeding cut feed and meal to his horses, accompanied with weighing and measuring. He cuts oat straw about an inch long, with a rawhide cylinder machine, and this chopped straw is then treated with corn meal and bran, mixed in about equal quantities as to weight, so that each horse has about a bushel of cut feed, and three quarts of the meal and bran, twice in each day. Sometimes hay is cut instead of oat straw, or both are mixed. It is found that two hundred pounds per week of this mixture of corn meal and bran, added to the cut feed, will keep a pair of working horses in the best condition. This he is satisfied from experiment is *less than two-thirds the cost of keeping them on uncut dry hay and whole grain.* The corn meal alone is not so good for horses, as when diluted with bran. An excellent meal is made of ground oats. The fodder is cut by horse power on stormy or spare days, and stored in large bins, so as to furnish always a surplus on hand.

SHYING HORSES.—L. A. D., in the Scientific American, says that a horseman should never "shy" himself when the horse shies, or show the least nervousness, nor notice it in the horse, and far less punish him for it, and adds: Allow me, having had a great deal of experience in managing horses, to add another bit of advice to nervous horsemen. Whenever they notice their horse directing his ears to any point whatever, or indicating the slightest disposition to become afraid, let them, instead or pulling the rein to bring the horse towards the object causing its nervousness, pull it on the other side. This will instantly divert the attention of the horse from the object exciting his suspicion, and in ninety-nine cases out of a hundred the horse will pay no more attention to the object, from which he will fly away if forcibly driven to it by pulling the wrong rein.

PRICE OF CORN AND PORK.—The best pork raisers whom we have met with, make a pound of pork from five pounds of corn meal. When pork is five cents a pound they consequently get one dollar a bushel for their corn. Everything was done in the best manner—good breed, cleanliness, regularity, corn ground and scalded with three pails hot water to one pail of meal, standing twelve to eighteen hours. With what is termed ordinary good management, it requires about twice as much corn, or 10 lbs., to make a pound of pork—and the common rule is, corn at 50 cents, makes pork at $5; corn at one dollar makes pork at $10; and so on. Scalded meal is worth double corn in ear.

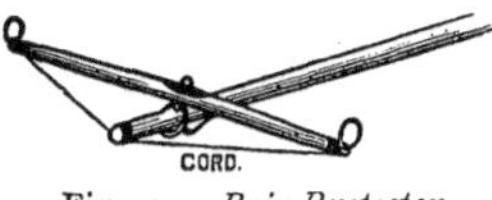

Fig. 121.—*Rein Protector.*

REIN PROTECTOR.—Annexed is a contrivance, (fig. 121,) to keep the reins from getting under the end of the wagon tongue while teaming. The cord passes through the ring in the end of the tongue.

ECONOMICAL PAINT.—We have tried many kinds of cheap paint, some nearly worthless, and others of considerable value. But the best, by far, yet employed, is first a coat of crude petroleum, and then one or two coats of the Averill or some equally good paint, if there is such to be found. We have tried this mode of protecting wood, on several tenant houses. These houses were made of rough pine plank outside, neatly battened, studded or battened on the joints inside, and lathed and plastered on the studding. The roofs were covered with slate laid on felt. When the painting was completed, the exterior had a handsome appearance—better than if painted on a planed surface. We would recommend painting on the rough surface for any battened or vertically boarded house, even if a large and handsome dwelling.

These houses are 16 by 24 feet, with a kitchen wing 12 feet square. The eaves are 14 feet high, making good chamber rooms. A common laborer applied the petroleum with a whitewash brush—half a barrel, costing $6, being required for each house. The cost of applying was two dollars and a half—eight dollars and a half for the petroleum coat. Twelve dollars' worth of the Averill paint, (which is furnished by the company in New-York, in kegs ready for the brush, and tinted, and sent in this shape by railroad,) was sufficient for each house, three dollars more for painting—making twenty-three and a half dollars for the oiling and painting complete. The rough surface takes the crude oil better, and retains more paint. Hence one coat of the Averill mixture on the oiled surface will answer well for a long time, when another may be added at less expense. This paint is particularly recommended for *adhering*—a few weeks only being required for the petroleum to dry in before the paint is applied.

One of the houses was sided with good hemlock, making it much cheaper than the pine houses; but more time was required in painting the rough surface. It makes, however, as neat an appearance as any, and forms a handsome cottage when surrounded by flowering shrubs and embroidered with climbers.

RULES FOR LABORERS.—Employers commonly give *verbal* rules for the government of laborers, which are soon forgotten, or irregularly observed. A much better way is to fix on certain leading directions, have them printed in large letters on stiff pasteboard, and nail them up in the barn, workshop, shed or stable. If any infraction occurs, it is much easier and more effective to point silently to the rules than to be compelled to go into a lecture.

Every agricultural warehouse or store should have such rules suitably printed and for sale for the use of farmers. Fifty cents paid for such a card would be worth as much as ten dollars paid for a plow.

The following is a specimen of such rules as may be adopted, which may doubtless be improved in some particulars.

1. Be regular and uniform in hours of labor.
2. Do every operation in the best manner.
3. Finish one job before beginning another.
4. Clean every tool at night or sooner when done with.
5. Bring in all tools and machines at night.
6. Treat all animals kindly and gently.
7. Never talk loudly to oxen or horses.
8. Study neatness in everything you do.
9. Never enter the house with muddy boots.
10. Never use profane language or get in a passion.
11. Take a general interest in the success of the farm.
12. Study to improve constantly in knowledge and skill in farming.

PORTABLE REVOLVING LEVER.—A correspondent of the Prairie Farmer gives the accompanying figure of a lever, (fig. 122,) answering the purpose of a crane, to save labor in scalding hogs, and for other purposes. It needs but little explanation. Its size may be obviously varied to suit its intended purposes, but the dimension given are—pole 22 feet long, (of *light* wood,) foundation, two pieces, 6 by 6 inches, and 6 feet long; hard wood post 6 feet long, and braces 2 feet long of ½ inch iron. The post has a tenon on the bottom fitting a two-inch auger hole; a clevis on the top with a pivot entering a hole, and washer at top of post, and a clevis, hook and bolt on the end of the pole.

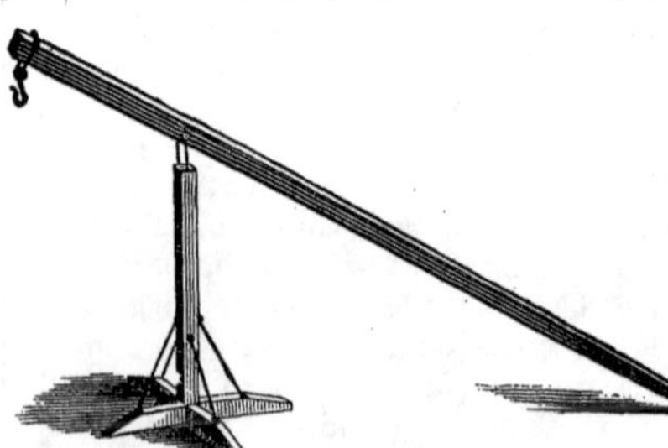

Fig. 122.—*Portable Revolving Lever.*

A small rope may be attached to the small end of the pole. It will be obvious that the pole should be much the strongest at the pivot, and taper moderately to the shorter end, and much more towards the longer end.

SPOUTS FOR MAPLE TREES.—I would say to all who are using wooden sap spouts, to throw them away, and get some of sheet-iron—galvanized sheet-iron is the best. They should be 1¼ inches at the small end, and 2 inches at the large end, and from 6 to 10 inches long; but if hooks are used to hang up the buckets, they need not be more than 5 inches long. My spouts cost 12½ cents per pound. I bent them myself, by cutting a groove in a block with a gouge, and then taking a round iron and pressing the spout into the groove. Tin will do for spouts, but they have to be bent smaller. I have some made out of old tin pails, which have been in use 18 years. The spouts will pay for themselves in one year, to anybody that uses spouts driven after a gouge; all you have to do is to hew off the outer bark of the tree, and drive the spout into the inside bark with a mallet or axe. I use hooks for hanging up about one-third of my tubes, made of nail rods, and costing 1¼ cents each.—L. H. K., in COUNTRY GENTLEMAN.

HOW TO KEEP BACON.—In answer to an inquiry, several correspondents have given the following different modes for keeping hams, either of which may be employed according to circumstances, or the facilities possessed or at hand:

Mix equal parts of slacked lime and wood ashes; spread three inches of the mixture on the bottom of a box, then a layer of bacon; cover with lime and ashes, lay a few laths on, then a layer of bacon, and continue until the boxes are full. Set in a dry, cool place. All ashes will answer, if no lime near by. For a few pieces for a family, cover each piece of bacon or ham with paper, and pack in a salt barrel, with ashes between each piece, and fill the barrel up with ashes. The meat will be as good at the end of the year as when put in. I have tried it thirty years and never failed. D.

Do not pack it down in anything, but take each piece and hang it in a loose bag; stuff the bag tight with cut hay, and your hams will keep sound and fresh for an indefinite time. I have hams two and three years old, perfectly sound, and retaining their juices, and they improve in quality like old wine. J. S.

If he will pack his hams, shoulders and dried beef in barrels, and cover them with powdered charcoal, his meat will keep sweet, and will not be touched by flies, mice or rats. E. B.

Malt screenings will keep bacon better than bran. G. G.

I have always been opposed to "packing," but if a man will pack, ashes are better than bran, and sticks laid in between it to let it have air, are better than either; then cover it with what you please, to keep the flies off. If you want good, fine flavored bacon, *give it air and keep it dry.* A dry atmosphere is a *sine qua non* to good bacon. The prime necessity of fire under meat is the drying process, and *not* the smoke. Bacon hung up in a dark, dry place, is not likely to be disturbed by flies. Whenever the atmosphere is sufficiently damp to settle upon the meat and, (as some say,) *cause it to sweat,* drive out the dampness with fire, and burn a little brimstone in it. Thus you will drive out the flies and, "save your bacon." AMATEUR.

Nice sweet timothy hay, cut fine, is the best thing extant for packing bacon, as it imparts a pleasant aromatic taste to the meat. Bran will sour and spoil the bacon. J. M. E.

A WASHING FLUID.—We have lately tried a washing fluid which has proved very successful in extracting the dirt from cotton and linen, and the proportions are such that they cannot injure the clothes :

Five pounds of sal-soda.
One pound of borax.
Half a pound of fresh unslacked lime.
Four ounces salts of tartar.
Three ounces liquid ammonia.

Dissolve the soda and borax in one gallon of boiling water ; when well mixed, pour in the liquid ammonia and salts of tartar. Boil the half pound of lime for five minutes in one gallon of water ; set it aside to settle, and when clear pour it off carefully, not allowing any sediment to mingle with it. Pour the two gallons of solution together, and turn upon them eight gallons of cold water. Put into a cask or jugs.

The night before washing, take six tablespoonfuls to a tub filled with clothes, mixing it with four pails of warm water. Soak them over night ; next morning add hot water enough to wash the clothes with good soap-suds. Boil the clothes. Wash out another tub full of clothes in the same water used for the first boiler.

One trial of this fluid will show its good effects.

An excellent soft soap can be manufactured from this compound. Take one quart of the fluid ; slice into it three pounds of yellow bar soap, and add to it two pounds of sal-soda. Boil it in three gallons of water for ten minutes, and it will make four gallons of soft soap which will prove unequaled for all purposes wherein soap is needed.

In using these receipts for washing, the clothes do not need to boil more than half an hour, and in many cases persons prefer to pour boiling water upon them, and let them stand until it is cool enough to wring out. By thus doing it is thought the clothes are whiter.

These receipts have been sold through the country at high prices, and a good deal of money has been made from their manufacture.

WATER-PROOF BLACKING.—A correspondent of the COUNTRY GENTLEMAN gives the following mode of making a thoroughly water-proof blacking. We can vouch for its efficacy, having seen and tried substantially the same :

Take an old pair of India rubber shoes (boots or any old India rubber ;)

cut them up and pull off the cloth lining; put the rubber in about a pint of neat's foot oil, and set it on the stove until the rubber is entirely melted, stirring it once in a while, and don't let it boil or burn. It will take about two days to melt the rubber. As soon as the rubber is melted, stir in one-half pound of beef or mutton tallow, and one-half pound of beeswax. If it is not black enough, you may add a little lamp-black, but I don't see any use in it.

Now to apply to the boots: Wash them clean of mud and blacking; when they are nearly dry apply the water-proof all over them—if the weather is cold, work near the stove. The best thing to use in applying this blacking, is one's hands and considerable elbow grease, to rub it well into the leather.

Rules for Making Butter.—1. For making good butter, the first thing is to have good sweet pasture, free from weeds or any growth that will give a bad taste to the milk. Good upland grass is better than coarse grass growing on wet places. Some dairymen think that limed is better than unlimed land, but this is a matter of minor importance. Others regard the practice of sowing plaster in spring, and repeating it early in autumn, as tending to sweeten grass.

2. Good, well selected cows, are the next requisite.

3. Perfect cleanliness, from beginning to end, is indispensable—the most so, perhaps, of any one thing. No dirt or dust must drop into the milk, for which reason the animals should have a clean place to lie on, and never be allowed to stand in mud or manure; vessels all thoroughly washed—scalded whenever necessary to preserve perfect sweetness—including pails, pans, pots, churns, workers, and tubs or firkins. They must be first washed clean with *cold* water; for if hot water is used first, it will curdle the milk in the cracks or corners, and prevent its washing out.

4. A perfectly pure *air* is of great importance. Bad odors will taint butter. The dairy house should therefore be far away from manure yards and everything else of the kind. Keep tobacco smoke off the premises.

5. Let the butter be well worked, so as to press out all the buttermilk. It is impossible to have a good article if this is not done. Perhaps this is the most common cause of failure. If much milk is left in, it soon ferments and makes rancid and worthless butter.

6. In laying down for winter, use *new* firkins—never use them a second time; and pots or jars must not be used, if they have ever had bad butter in them, or pickles, or anything else that will taint them—the taint can never be wholly removed.

7. The best dairy salt is important. Butter in hot weather must be covered and excluded from the air with saturated brine.

Sagging of Posts.—Every farmer in the country has witnessed the inconvenience of a sagging gate. New ones are well constructed and well secured to firmly set posts, by stout hinges. The owner takes special pains to make a good self-fastening latch, and the whole contrivance being in perfect order, he promises himself much satisfaction from these convenient and permanent entrances to his yards and fields. All goes well the first summer. But he finds after the next spring that the latch strikes too low, and will not catch the socket. The soft earth has given a little, and the constant pressure of the heavy gate has caused the post to yield, a hair's breadth at a time, till it has varied a little from the true perpendicular. Being now often left unfastened, it beats against the post, and the latch is broken. The subsequent hard usage it receives makes the post settle away still more. Subsequently the gate rests on the ground, over which it is laboriously dragged day after day and year after year, until the hinges are broken.

Some of our readers can figure with approximate accuracy what the damages are likely to be from cattle breaking through this unshut gate into the wheat and corn fields. It is obvious that a firm, erect post to every gate contributes greatly to successful and satisfactory farming.

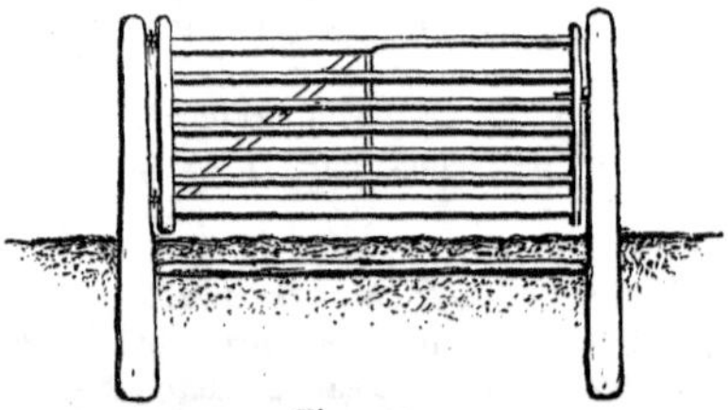

Fig. 123.

A commonly recommended mode of obviating the difficulty is shown in fig. 123, a trench being cut between the posts directly under the gate, and a piece of durable timber, (such as a white oak rail, a locust pole, or a cedar scantling,) laid in and compactly covered with earth. The nearer it is to the surface, the better will be the bracing which it will afford; the deeper it is buried, the longer it will last. It is therefore best to procure the most durable wood and place it near the surface. This mode of bracing accomplishes the desired purpose partially. Unless the earth is very hard and firm at the lower part of the post, it will slowly yield to the outward pressure, and in process of time the bottom will be thrown out, as shown in fig. 124, and the gate will settle on the ground. A more thorough mode therefore is to dig a deep trench and lay in two connecting pieces, as shown in fig. 125. The bottom one must be dove-tailed into the bottom of the posts, to prevent spreading, and a pin or a few spikes in addition will render the connection firmer. This lower timber, being excluded from air and changes of moisture, will last an age, and the upper one only being likely to decay, will need replacing. By this mode the gate willl be sure to keep its place. But it is attended with considerable

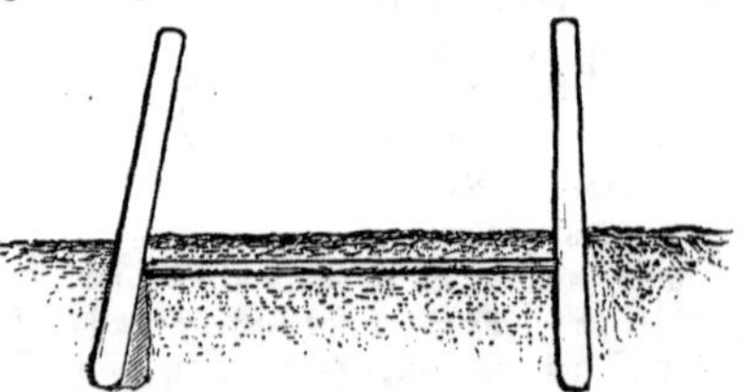

Fig. 124.

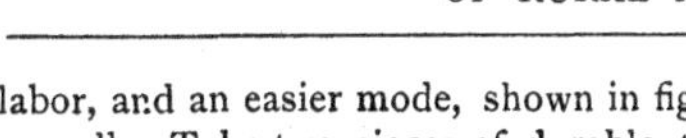

labor, and an easier mode, shown in fig. 126, will answer nearly or quite as well. Take two pieces of durable timber, (short posts, or an oak or

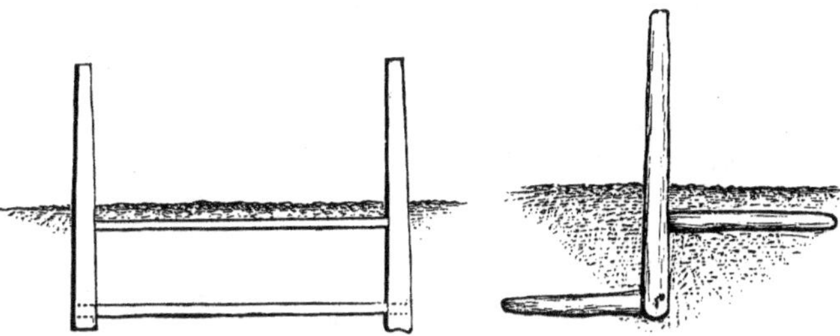

Fig. 125. Fig. 126.

chestnut rail sawed in two,) and place the *ends* against the post in the manner represented, beating the earth firmly about them. If only three or four feet long, they can never be moved a hair's breadth by sliding endwise in the soil when firmly placed. Some would prefer to place them *across* the post, but this is quite a mistake, as earth will thus yield by hard and continued pressure; while no practicable force could move them in the slightest degree endwise. It often happens that such pieces of durable timber are found on every farm, and a moderate degree of labor will cut a short, deep trench on the outer side, and a shallow one on the inner, and firmly place them in position.

LAMBREQUINS.—In answer to inquiries concerning lambrequins, a correspondent of the COUNTRY GENTLEMAN says that material, quantity and pattern must depend entirely upon one's purse, taste and size of the window. They are made to drape over under curtains, either shades or drapery. If desired cheaply, they can be cut out of green, blue or crimson empress cloth, into deep points, scollops or battlemented squares; they must be lined with stiff cambric, and trimmed either with worsted fringe or wide gimp. They are nailed to fixtures of black walnut, gilding, or plain pine wood covered with the same material as the lambrequins. They are very tasteful when made of chintz, edged with bright blue, green or turkey red for bordering; also of plain cotton, starched stiffly and trimmed with home-made crochet or netted fringe. The depth of them must be regulated by the height of the windows. If they are twelve feet high, the lambrequins can be made

Fig. 127.—*Lambrequin.*

a yard in length at the sides of the window, and cut up in a graceful half scallop, with a very large scallop in the centre, which is laid in a deep fold in the middle, at the top of the lambrequin, and hangs gracefully from the fixture. Any woman possessing taste and ingenuity can produce a fine effect at a small expenditure, and add greatly to the adornment of parlor or chamber. The material, if not of sufficient width, can be easily pieced out so as not to mar its effect. Merino or thibet cloth will look nearly as well as empress cloth or reps.

We add an illustration, (fig. 127,) showing a simple pattern often used.

How to Destroy Red Ants.—Take a *white china plate*, and spread a *thin covering* of common *lard* over it, and place it on the floor or shelf infested by the troublesome insects, and you will be pleased with the result. Stirring them up every morning is all that is necessary to set the trap again. A young bachelor sends this receipt, having used it with perfect success in his aunt's closet.

Cheap Bird Houses.—One of my neighbors uses tin cans which have been emptied of fruit, for houses for the birds. The smaller insectivorous birds can easily build in them. He fastens them by a band of tin, which is placed around the can and nailed through the ends. Better do this than allow the cans to litter up the yard, and besides it *pays* to protect the birds.

Butter-Workers.—Two principal forms are adopted for butter-workers, variously preferred by different manufacturers. One consists of a brake or lever, fastened at one end by a swivel joint, so that the face of the brake may be brought down with force on the butter, which rests on a trough-table, which is best if with a marble top.

Fig. 128.

Fig. 129.

Another form has a grooved roller, similarly attached, which is used by rolling and pressing the butter at one operation. The former is represented by fig. 128, and the latter by fig. 129, which also shows the pail into which the buttermilk flows.

Farm Accounts.—H. H. Walter states in the Boston Cultivator, that since he has kept farm accounts he has cleared double the money which he had done before, as it enabled him to see just where he could obtain the largest profits.

THE

ILLUSTRATED ANNUAL REGISTER

OF

RURAL AFFAIRS.

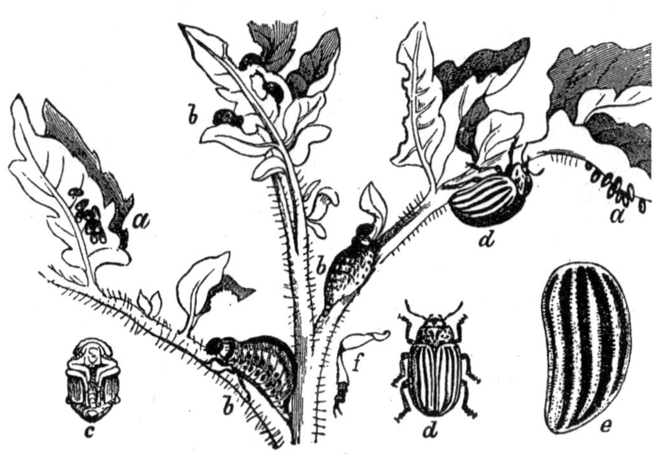

COLORADO POTATO BEETLE.

DESTRUCTIVE INSECTS.

INSECTS AFFECTING FARM CROPS.

COLORADO POTATO BUG, (*Doryphora* 10-*lineata*.)—Dr. Fitch figured and described this beetle in vol. 5, p. 207, of this work, in the year 1868. Since that time it has been steadily moving eastward, and has already reached Ohio and Michigan. As it is important that every one should know it well, so as to be ready to destroy the first comers, and to keep them in check, the accompanying figures of this insect are given, showing it in various stages, and enabling the cultivator to recognize it readily, both in the larva and beetle state. The deep orange eggs, freshly deposited on the under side of the leaf, are seen in the above engraving at *a a*; *b b b*, larvæ in different stages, of a reddish yellow color; *d d*, perfect insect, yellow with black lines; *e*, magnified wing-case. All of these changes are passed through in less than a month; and there are about three broods a year. The first two, when about to change to the pupa, go into the ground, and come out in the beetle state in ten or twelve days. The last one remains under ground all winter, and makes its appearance in time to lay its eggs on the leaves of the young potato plant.

The American Entomologist describes another nearly allied species of Doryphora, (*D. juncta*,) that a casual observer would pronounce identical, but which has some small but well marked points of distinction, and which feeds on some other species of Solanum, but which pertinaciously refuses to touch the potato, and specimens have actually died from starvation in cages where there were plenty of fresh potato leaves.

There is only one way to deal with the potato bugs when they take possession of a potato field—and that is to kill them. One mode is by poison, and the other is by catching. Paris green is used as the poison, and is applied by mixing one pound with many pounds of flour or plaster, and sifting it through a coarse muslin cloth on the potato plants, while the dew is on. The bugs eat, drop and die. Paris green, being a deadly poison, must be used with caution.

S. B. Johnson of Illinois, gives the following particulars for the use of Paris green, in his lecture on the potato, delivered before the Madison Co. Farmers' Convention:

"How are we to save ourselves from this scourge? The little corner patch of an acre or so can be managed on the tin-pan and fire plan. But here are 20, 50 or 100 acres in a plantation. We know of only one way in which it has been done effectually on a large scale; and that is by the use of *Paris green*. This is a most virulent poison, and must be used with the utmost care. Secure the best. It can be purchased by the canister (14 lbs.) for 45 cents per pound. Puncture the bottom of a quart tin bucket with holes about the size of bird shot; solder midway on the side a handle with a socket three or four inches deep, into which thrust a stick four feet long. Having muffled nose and mouth, mix thoroughly one part of Paris green with eight parts of gypsum.* With this long handled bucket, and by keeping on the windward side, the muffler can be removed, and you may march with safety into the battle-field. Commence as soon as the plants appear and dust every hill thoroughly. Go over the field twice a week if anything in the shape of a bug is to be seen. It may be disguised by countless myriads in the yellow eggs lain on the under side of the leaf, and some day, when least expected, the naked, defoliated stalks are reeking with the filthy larvæ. By the use of plaster instead of flour, a stimulant is employed of great value to the crop. After the bug is vanquished it would be of advantage to continue the application of the plaster until the crop is ripened."

* Some mix it with twenty or thirty parts.

Various appliances are resorted to for capturing them or killing them by mechanical means in large quantities. Should they continue destructive, there is no doubt that some trap or machine, or mode of crushing them, will be contrived, that may be driven rapidly along the rows by horse-power, for killing or cleaning them out, but as yet no rapid and efficient mode has been devised.

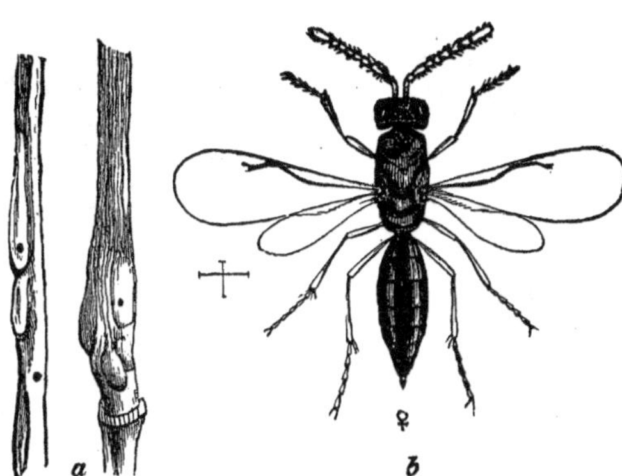

Fig. 2.—*Joint-Worm.*

THE JOINT WORM—(*Isosoma hordei.*)—The wheat crop in Virginia, and the barley crop in New-York, have been extensively damaged by this insect. The nature of the injury is shown in the above cut at *a*, (fig. 2,) where the straw is swollen from its pressure. Like the Hessian fly, it occupies the straw just above the joint, but it differs from that insect in penetrating the substance of the straw, while the Hessian fly reposes at the bottom of the pocket formed by the sheath and straw. The sheath is represented as removed from the straw in the above figure, to show the character of the injury, and the small round holes through which the insect escapes in the fly state. At *b* the fly is represented largely magnified, the cross lines showing the natural size. This fly makes its appearance early in summer, and lays its eggs in the stems of the growing grain. It soon hatches and does its work of mischief. When full grown the larvæ are about an eighth of an inch long. They mostly remain here till the following spring, although some escape the same autumn. The proposed remedy is to burn the stubble containing them, but we are not aware that it has been practiced to any satisfactory extent.

THE COLORADO GRASSHOPPER.—This destructive insect, which may be compared to the Eastern Locust (another grasshopper) for its devastations and vast multitudes, so nearly resembles our common grasshopper of the east that many have not observed the difference. In color and general appearance they are very nearly the same, but the western insect has wings much longer than the body, while in our common grasshopper the wings scarcely extend beyond the body, as the annexed figures of the two indicate, (fig. 3,) where *a* is the western grasshopper and *b* the common sort. It is by means of these long wings that the Colorado insect is enabled to sweep through the air for miles at a time, while the eastern insect can fly only a few yards. It is fortunate for the States east of the Mississippi, that this great destroyer cannot pass many hundred miles from its native canons among the Rocky Mountains, without losing its vigor and vitality.

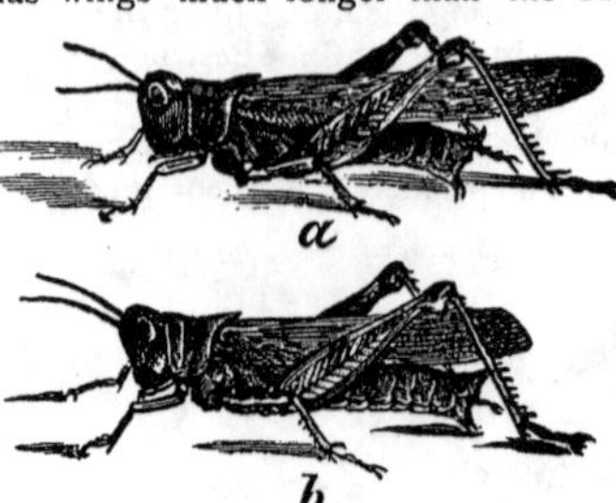

Fig. 3.—*Colorado and Common Grasshopper.*

Insects which Affect Fruit Trees.

In former volumes of this work, Dr. Fitch has given descriptions of a large number of destructive insects, their habits, and the modes recommended or adopted for destroying them. In the present article we figure and describe several additional species, with further information relative to a number formerly described.

It has been estimated that the aggregate amount of the damages to the fruit crop annually committed by two insects alone—the codling moth and the curculio—throughout the Union, is not less than twelve million dollars. The New-York State Agricultural Society ascertained that the midge had caused the yearly loss of fifteen million dollars to the wheat crop ; and competent persons have estimated that the entire amount of depredations in the United States from the different species of insects cannot be less than three hundred million dollars annually. While such enormous depredations are committed, and while so little is known of their habits by cultivators at large, every additional contribution to the knowledge which shall enable us to attack them understandingly and effectually, cannot fail to be valuable.

There is one department of insect study, which we can only briefly allude to, that possesses great importance, and which is very little understood except by scientific men. This is the knowledge of *useful insects*—those which confer a great favor on the cultivator by thinning the ranks of his foes. The work of birds has been indiscriminately recommended as "destroying the insects," without knowing whether those insects are useful or noxious. In one instance birds were seen devouring, as was supposed, a destructive caterpillar ; but it was found, on a scientific examination, that they were only picking a parasitic insect from the caterpillars. The parasitic insect, by destroying these caterpillars, was assisting the cultivator, and the birds were feeding upon his best friends. One of the most useful class of insects of this kind is the lady-bug—of which we represent a single species — the Convergent Lady-Bug in the accompanying cut, (fig. 4,) and which with many others, is very useful in destroying plant lice—*a* representing the larva; *b* the pupa, and *c* the perfect insect. We present this figure in order the better to explain an amusing occurrence, showing the blunders of ignorance, related by Dr. Fitch. The rose bushes of one of his neighbors was grievously infested by plant lice. He complained to Dr. Fitch, that although he took the greatest pains to go over his bushes every morning and destroy all the "old ones," yet his bushes were ten times as badly injured as those of his neighbors, who took no pains with them. On examination it turned out that he had been killing off the lady-bugs, supposing them to be the "old ones," which were doing all they could to rid his bushes of the pest.

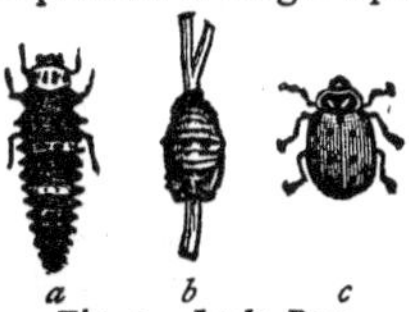

a *b* *c*

Fig. 4.—*Lady-Bug.*

Insects which Affect Large Fruits.

The Apple Worm—(*Carpocapsa pomonella.*)—This is the most formidable enemy of the apple in the United States. By eating the core and filling the interior of the fruit with cast-off matter, it renders it unfit for market or for the table. It also does much damage to the pear, but does not attack stone fruit. It was introduced from Europe near the beginning of the present century, and has been steadily spreading and increasing up to the present time. In many orchards in the eastern States it ruins nearly the whole crop, and it is now penetrating into the States beyond the Mississippi.

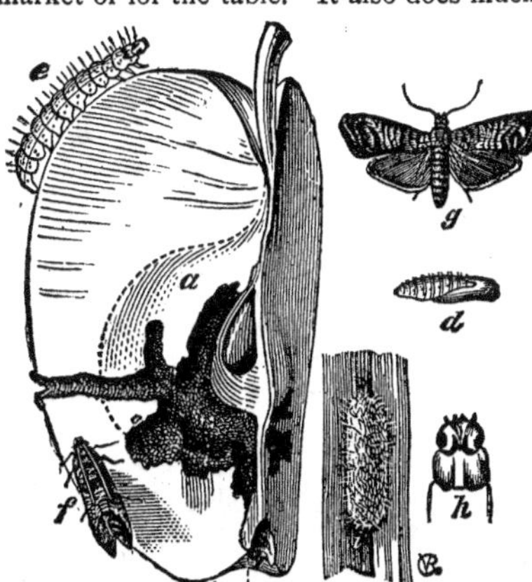

Fig. 5.—*Apple Worm or Codling Moth.*

The accompanying figure (fig. 5) represents this insect, (which in the perfect state is known as the Codling moth,) in the different stages of its existence, and shows the manner of its devouring the interior of the apple. The burrowings are shown at *a*; the place of entrance is at *b*; *e* is the worm or larva; *h*, its head magnified; *d*, the pupa; *i*, the cocoon; *f* and *g*, the moth, which is distinguished from all other moths, (says the American Entomologist,) by a patch of burnished coppery scales at the tip of its front wings. The moth appears first early in summer, and lays its eggs in the blossom end of the young apples, a single egg in each. The young larva soon hatches and burrows towards the core, eating as it goes. In three or four weeks, or more, it is full grown, and the young apples fall to the ground nearly at the same time. The larva passes out through a round hole which it makes, and crawls for some place to spin its cocoon, usually to the rough trunk of the tree. The moth or miller comes out in a few weeks for a second brood, but the apples have now grown so large that fewer fall to the ground from the injury, but they are more or less spoiled for use and market. The insects are often found in them after the crop is gathered for winter, and hiding in various places, spin their cocoons, and come out in spring to perpetuate their mischief. C. V. Riley says that in a barrel of wormy apples, which he broke up early in the spring, he found about two hundred such cocoons; and estimating that one barrel would furnish a hundred winged females, each of which would lay two hundred eggs and spoil as many apples, and allowing a hundred apples to the bushel, he arrived at the result that two hundred bushels of apples may be ruined by the insects from one apple barrel, if allowed to escape.

The remedies for the prevention of the work of this formidable insect are of two kinds, and are founded on the destruction of the larva while in the fruit, and of the cocoons before the miller comes out. Animals which would pick up and devour the young and infested fruit as soon as it falls, would perform the first named service. Swine, if sufficiently numerous, answer the purpose well; but as few owners of large orchards have herds large enough, it is proposed to employ sheep, which are known to eat the young apples readily, and which may commonly be had in large flocks. The bark of the trees may possibly need protection from them. In the few instances where they have been thoroughly tried, year after year, they have given smooth and fair crops. Oliver Chapin of East Bloomfield, N. Y., recommends the practice of employing boys to pick the infested apples from the trees, stating that one boy would pick several bushels in a day. The second remedy—destroying the cocoons—may be effected in part by passing hay ropes, or strips of old carpet, around the trunks of the trees early in summer, and afterwards crushing the cocoons which form under these ropes; and also by placing pieces of old carpets, &c., in the forks, and then crushing those which adhere to them. This may be done rapidly by means or a common clothes wringer, and the operator will then have the satisfaction of knowing that he is "killing them by machinery." But the best and easiest mode of destruction is doubtless the employment of sheep.

THE TENT CATERPILLAR—(*Clisiocampa Americana.*)—This insect, called also the American Lackey-moth, is generally known throughout the country by the owners of apple orchards. It sometimes becomes numerous and destructive, devouring the foliage on large portions of the trees, and then, for several seasons, it will nearly disappear, till favorable influences cause another return in large numbers. The eggs which furnish the caterpillars are deposited by a brown moth or miller, (fig. 6,) about midsummer, in masses or cylinders, which encircle the young shoots, each mass containing about three hundred eggs. The position and appearance of these eggs is shown at *c*, (fig. 4.) The eggs, when laid, are then covered with a vesicular water-proof varnish, which protects them both from cold and from rain (fig. 8.) They remain in this condition till the following spring; and as soon as the apple buds begin to open, almost to a day, the young caterpillars hatch and make their appearance, ready for this new, fresh, tender food. The hot weather of the previous August and early September made no impression on them; but the mild spring weather, just at the right time for their food, brings them out. When they first appear, they are not so large as a cambric needle, nor more than the tenth of an inch long. If cold or stormy weather occurs, arresting the growth of the young leaves after the worms are hatched, they can live without food for ten or

Fig. 6.—*Moth of Tent Caterpillar.*

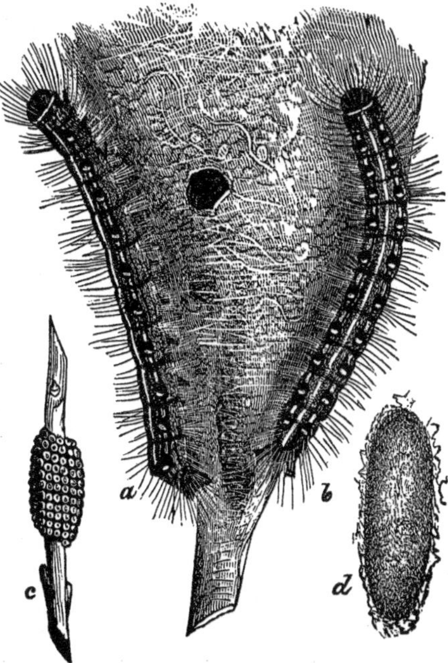

Fig. 7.—*Tent or Orchard Caterpillar.*

Fig. 8.

twelve days. They immediately commence stretching their web across a fork in the branches, and thus manufacture a tent for shelter. This increases in size with the growth of the caterpillars until it sometimes becomes nearly a foot in breadth. Like other larvæ, they moult or shed their skins four times. They are represented as full grown and of natural size in fig. 7, *a* and *b*. The uniform white line along the back distinguishes them conspicuously from the Forest Tent Caterpillar, sometimes miscalled the "army worm."

Although mostly infesting apple orchards, the Tent caterpillars are occasionally seen on the pear, plum, peach and cherry, and the wild cherry often throngs with them.

In five or six weeks they scatter in various directions to undergo their change to the pupa state, when each spins a cocoon, and then remains in this state some three weeks—*d*, fig. 7.

The perfect insect or miller, represented in fig. 6, measures about an inch and a half from tip to tip of the wings; it has no sucker to take food, eats nothing, and lives only to lay its eggs. It has but one brood in a season.

C. V. Riley says that the only bird known to devour these caterpillars greedily is the American Cuckoo. But this bird is too few in numbers to make much impression on them. An active man, with a quick eye, will collect hundreds of the rings of eggs in a day, in autumn and winter, and every such cylinder of eggs destroyed at this time prevents the formation of a nest of larvæ. Some years ago when they promised to be very abundant, we employed a man three days, and in that time he destroyed, on old and young orchard trees, and in a nursery, three thousand nests of eggs and of the newly hatched insects, nearly a million in all. A sharp blade, set at an acute angle on the end of a light pole, will enable the operator to cut off the eggs by means of a quick jerk, when they are otherwise beyond reach. If recently hatched, the same tool may be employed; but when they become larger, and spread over the tree, they may be destroyed early in the morning, when mostly in the nests, by a swab on a pole dipped in lime wash; or even by winding them on the end of the pole only, and crushing them under the foot. All that is necessary in order to keep an orchard cleared of them is a moderate amount of timely labor and attention. It is important, for economy of labor as well as for thorough work, to secure them before they hatch.

The Tent caterpillar is sometimes confounded by superficial observers with another insect, known as the Fall Web-worm, which hatches out, not early in the spring, but after mid-summer, and which was briefly noticed and described by Dr. Fitch in vol. 3, p. 303 of this work. Both make a web or tent; but the Fall Web-worm has a wider range of trees for its food. It spins a cocoon late in summer, and does not come out till the following summer. The moth or miller is white, and it deposits its eggs in an irregular mass on a leaf, where they soon hatch and the larvæ begin their work.

FOREST TENT CATERPILLAR.—This insect (*Clisiocampa sylvatica*) resembles, in some particulars, the Tent caterpillar of the orchard, (*C. americana*,) but differs in being less confined to nests, and in the markings of the larva and moth. It appears only occasionally in large numbers. In the year 1867 it was quite destructive in Western New-York, and was given the erroneous name of "army worm," the true army worm being a southern insect, which destroys sometimes hundreds of acres of grass in a few days.

About forty-five years ago the Forest caterpillar was so abundant in Western New-York that it nearly stripped the foliage from large forests in the early part of summer; and although the leaves were replaced in a few weeks, the check given to the growth was a serious injury, and many branches died, partly from the effects of the severe winter following.

Like the common Orchard caterpillar, the miller deposits its eggs in the form of a ring or cylinder, on the young twigs; but instead of the rounded form given to the mass of eggs of the orchard caterpillar, the eggs of the forest caterpillar form a distinct even-sized cylinder, with square ends, as at *a*, fig. 9. Each mass contains about 300 or 400 eggs. The eggs are small, about the twenty-fifth of an inch long, and the fiftieth part of an inch in diameter, and are represented magnified at *d*, *c* showing the appearance of the end, with its sunken centre. These eggs are deposited about midsummer, and the larvæ hatched early the following

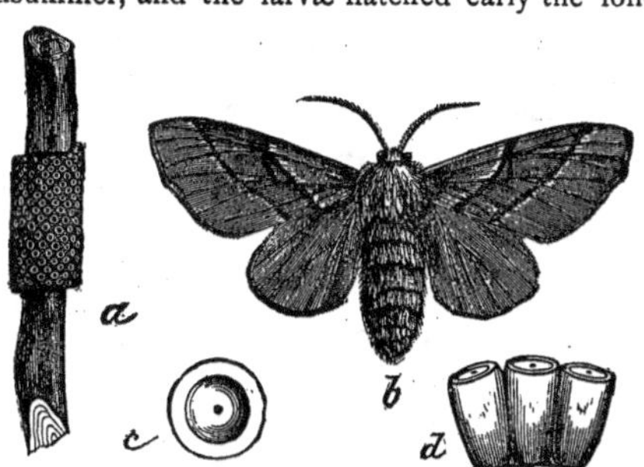

Fig. 9.—*Tent Caterpillar of the Forest.*

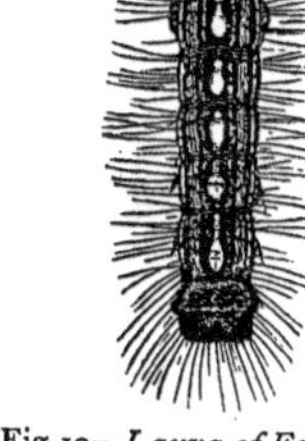

Fig. 10—*Larva of Forest Caterpillar.*

spring. They are very hardy, and endure any cold snap that follows. They commence spinning a web wherever they go. They moult four times, somewhat changing color each time, and when full grown are accurately represented by fig. 10. It will be seen that the middle of the back is marked by a row of spatula shaped white spots, which most readily distinguish this from the common Orchard caterpillar, with its continuous white line. The perfect insects of each, shown in fig. 9 at *b*, and in fig. 6, may be readily known from each other by the markings of the wings, the Orchard caterpillar being lighter and the Forest caterpillar being darker between the bars.

The Forest caterpillar spins a web close to the tree, but as it grows larger it wanders far away, and hence is generally supposed to have no web. In its travels it generally selects smooth surfaces, and seems to have a special liking to the cap-boards of board fences. It often swings down on a web from trees, and when numerous in forests proves quite annoying to persons traversing the woods. It devours the leaves of different kinds of trees, but seems to prefer the basswood, of which large trees have been stripped entirely bare. In the orchard it is particularly destructive to the foliage of the apple. On account of its wandering character it is more difficult to attack and destroy in masses, and for this reason more care should be taken to cut off and destroy the rings of eggs before they hatch, from the orchard trees when they are found.

The American Entomologist describes several insects which destroy this caterpillar, and commonly keep it in subjection, except during those occasional years when it appears in the greatest numbers. But generally "these cannibals and parasites do their work so effectually that it is seldom exceedingly numerous for more than two successive years in one locality."

THE CANKER WORM—(*Anisopteryx vernata.*)—The young larvæ hatch early in summer, and pierce small holes in the leaves, and as they grow larger they consume all the leaves except the larger veins. The male (fig. 11) has wings; the female (below) is nearly destitute of them. The larva is a measuring worm, nearly an inch long, ten-footed, black, dull yellow or greenish, very variable in color, commonly with an ash grey back, and a pale yellowish stripe along each side.

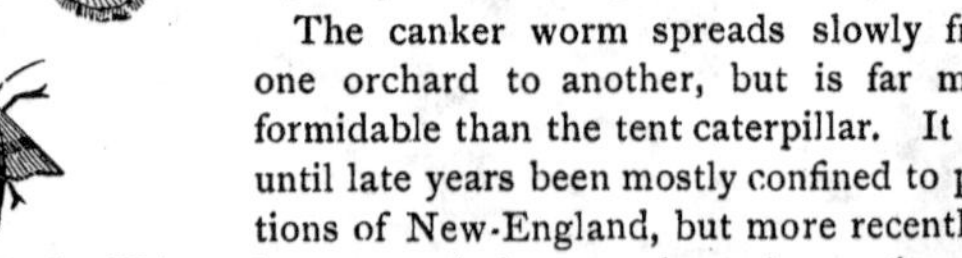

Fig. 11.—*The Canker Worm.*

The canker worm spreads slowly from one orchard to another, but is far more formidable than the tent caterpillar. It has until late years been mostly confined to portions of New-England, but more recently it has spread largely through portions of Western New-York, and will doubtless find its way elsewhere. It should be well known to cultivators that they may destroy it when it first appears. It attacks both leaves and fruit; and when numerous the webs and the denuded branches together give the trees at a distance the appearance of having been scorched. As the female cannot fly, various expedients for preventing it from ascending the tree in winter or early spring have been devised. Belts of canvas or coarse paper extending around the trunk of the tree, have been covered with tar and train oil mixed together—the application requiring frequent renewing. Dennis' lead troughs, filled with fish oil, were used to some extent, and proved effectual, although somewhat expensive, the troughs being held by wooden wedges, and grass rammed in between the troughs and the bark, preventing the insects from passing. A more recent and cheaper expedient, represented by fig. 12, consists of belts of sheet zinc, about four inches wide, passing round the tree the bottom standing outwards like an inverted funnel. The lower edge should be as smooth as possible. Sheet iron will not answer, the insects clinging to the rusty edge. The shape into which the sheet zinc should be cut is shown in fig. 13, the lines being marked with a pair of compasses for the shears to follow. When applied, the ends are lapped past each other, so as to fit the tree, the pressure holding it to its place, after being secured by means of small copper wire thrust through punched holes. The longer the arc, the more nearly horizontal will be the rim ; a shorter arc will give it more inclination. It is well, before cutting the zinc, to try the form on the tree, by means of a piece of pasteboard or stiff wrapping paper cut with shears, from which a convenient pattern may be made.

Fig. 12.

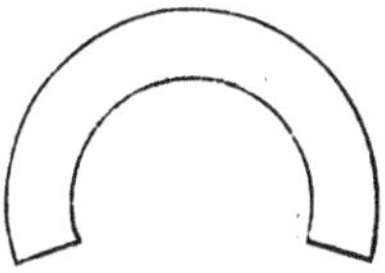

Fig. 13.

Another mode, described about thirty years ago by Dr. Harris, consists of an open box placed around the base of the tree, the outer sides holding a trough of oil, with a projecting edge nailed on the upper margin to shed the rain. This keeps the oil away from the tree, and prevents injury from it. We lately observed in Tilton's Journal of Horticulture, a communication from J. G. Barker of Cambridge, in which he gives an account of his success with this contrivance. He applied the boxes around fourteen trees in the autumn of 1867. The remainder of the orchard, fifty-six in number, he protected by the old tarring process. The following season, of those which had the boxes, "scarcely a leaf was touched, and hardly a worm was to be seen, except what blew from other trees," and they bore finely, the crop being sufficient to pay for the boxes, which cost about $2.50 per tree. The remaining fifty-six were stripped of their foliage as bad as ever. The next year, 1868, the rest of the trees were boxed, at a cost of $2 for each tree, and they proved a perfect protection—the apples more than paying for the boxes, which may be used for many years to come.

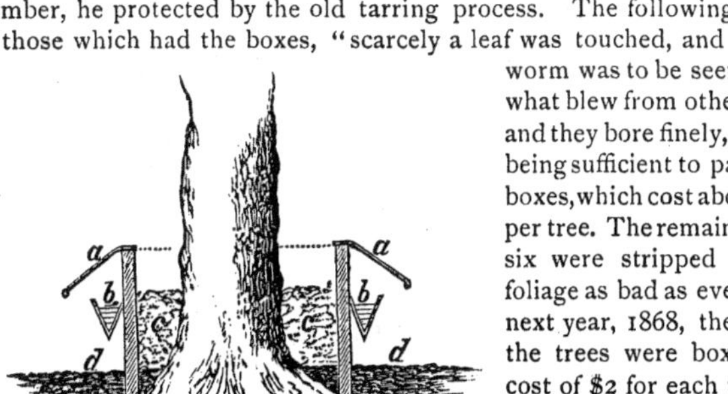

Fig. 14.

At our request, Mr. Barker has kindly furnished a detailed description of these boxes, with sketches, from which we are enabled to make the accompanying drawings. Fig. 14 is a section of the whole contrivance—*a a* being the zinc roof over the oil troughs, *b b; d d*, the surface of the earth; *c c*, the tar or lime which is used to fill the box around the tree.

Fig. 15 is a smaller view of the same. The box is square—large enough to leave about four inches of space around the tree; is sunk some four inches in the ground, and rises about ten inches above the surface. The trough is in shape like the letter V, two inches deep, and is made by a tinman before nailing on the box; it is tacked on two inches below the upper edge of the box, and then the roof is placed in position and secured by a single screw into the upper edge of each side or board. It must, of course, be placed in a level position, to hold the oil. This is done by means of a spade used in setting the box in the earth. The box and roof are nearly completed in the tin-shop, but the corner of both must be left open till placed around the tree, when the parts are soldered together. The roof is about four and a half inches wide, with the under side turned under about the fourth of an inch, to keep it stiff and in shape. In order to examine the oil, and to see that all is right, it is necessary to loosen one of the screws. The box will vary somewhat in size with the magnitude of the tree; with a trunk six inches in diameter, the box should be about fourteen inches square and fourteen inches high; for a trunk a foot in diameter, it should be about twenty inches square; but a variation of two or three inches would not be of great importance. A few inches of tanbark or lime placed within, is for the purpose of preventing the moths from ascending inside. One pint of crude petroleum (costing 3 cents per tree, at 24 cents per gallon,) is enough for each tree. The boxes are commonly placed around the trees the latter part of September, so as to prevent the autumn ascent of the wingless female moths, and are kept there as long as there is danger. Mr. B. remarks to us, at the conclusion of his description, "I assure you it has been a great pleasure to us, after years of labor and no fruit, to be enabled, with so simple an arrangement, to protect our trees perfectly, and to have an abundance of fruit."

Fig. 15.

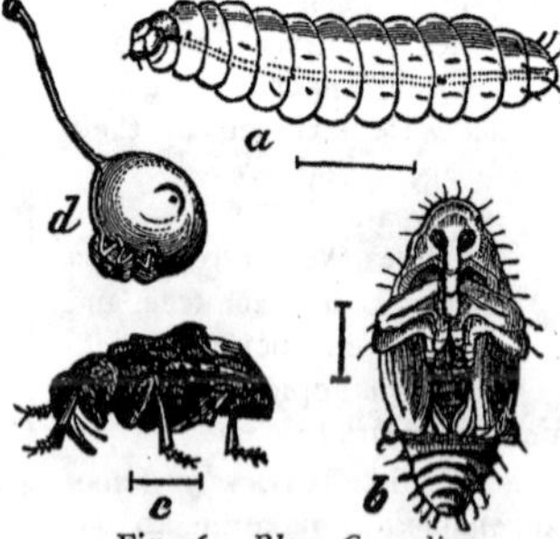

Fig. 16.—*Plum Curculio.*

THE CURCULIO.—An extended account of the habits of this formidable depredator was given by Dr. Fitch in the third volume of RURAL AFFAIRS, p. 298, with cuts showing the beetle the natural size. We give the accompanying engravings, (fig. 16,) representing enlarged views of the larva at *a*, the pupa at *b*, and the perfect insect at *c*. At *d* the natural size is shown, with one of the insects in the act of making an incision, and with the crescent mark on the upper side. The American Entomologist, to which we are indebted for this engraving, states that it attacks the following different fruits by way of preference in the order named: Nectarine, plum, apricot, peach, cherry, apple, pear, quince. With us it always attacks the apricot in preference to the plum.

The following appliances for destroying this insect, including the stiffening of the sheets, striking on iron spikes set in the tree, and the other details, we have thoroughly tested by several years trial, with entire success, and since the first notice of the efficiency of striking on iron spikes was published in the COUNTRY GENTLEMAN, four years since, this mode of jarring is becoming widely adopted throughout the country.

We must say, in the first place, that there is no royal road to freedom from the attacks of this insect. Like everything else valuable, it is only reached by labor. Some years ago one of our best horticultural journals suggested the offer of $50,000 as a premium for a satisfactory and easy mode of destroying the curculio. Now we might as well offer a $50,000 premium for getting rid of weeds without labor. The thing cannot be done. The mode we adopt for destroying this insect, has been, with more imperfect appliances, practiced under our eyes for more than forty years; and when applied with a tenth part of the care and perseverance which every good gardener and farmer is willing to adopt for extirpating weeds, it has afforded us profuse and delicious crops, while without its application the trees bore few or none. In a plum orchard of seventy trees or more, the annual cost of securing abundant fruit for the past few years, *has not exceeded five cents per tree.* But half-way work will be of little or no use—the remedy must be perseveringly and thoroughly applied.

Many remedies, as every one knows, have been used to accomplish the desired purpose by avoiding the simple, straight-forward, efficient mode by direct attack and killing. Repellants amount to nothing except to consume time. It is now generally admitted that jarring down on sheets is better than anything else. After trying different modifications of this remedy, we find the following most cheaply made and easily used:

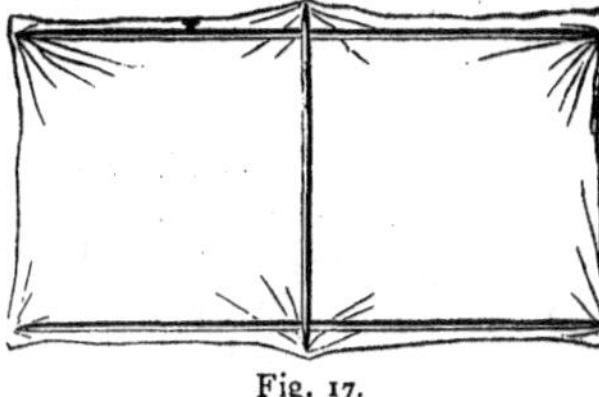
Fig. 17.

Procure the widest sheeting that can be found in market, except it be for quite small trees. Cut off two pieces about three yards long; for small trees they may be shorter. Stiffen them by means of light rods, in the manner shown in the cut, (fig. 17.) The ends of the rods are first sharpened, and a small notch made about two inches from the point, to prevent the sheeting from slipping on too far. Thrust the sharp ends into the four corners, so as to produce tight stretching. The middle cross-rod keeps the whole extended, and serves as a handle—the operator, alone, taking one in each hand, or both in one hand. Two of these stretched or stiffened sheets are all that are needed, one being placed under each side of the tree. The trunk or branches are then struck, and the insects, opossum-like, fold themselves up and drop. A quick eye detects them in a moment, and one pinch of the thumb and finger despatches them.

If the trees are quite large, the sheets should be of corresponding size, and they may be stiffened more thoroughly, as shown in fig. 18. Suitable sticks may be made by cutting green rods or poles, where they can be had, peeling off the bark, and allowing them to dry a few days. One-half or three-fourths of an inch in diameter will render them stiff enough.

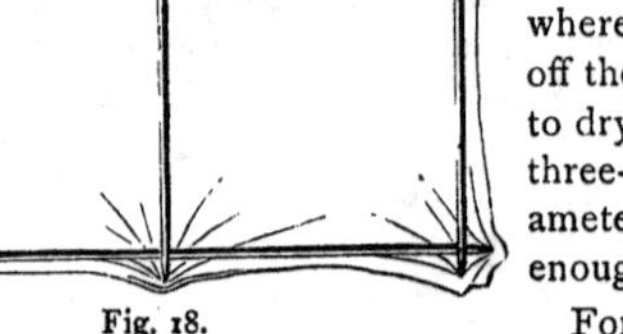

Fig. 18.

For jarring down, a *sharp blow* is important. Merely *shaking* will be of little use. The following statement was made nearly forty years ago by David Thomas, in one of the early volumes of the Genesee Farmer: "Under a tree in a remote part of the fruit garden, having spread the sheets, I made the following experiment: On *shaking* the tree well I caught *five* curculios; on jarring it with the hand, I caught *twelve* more; on striking the tree with a stone, *eight* more drop on the sheets. I was now convinced that I had been in error; and calling in assistance, and using a hammer to jar the tree violently, we caught in less than an hour more than two hundred and sixty of these insects."

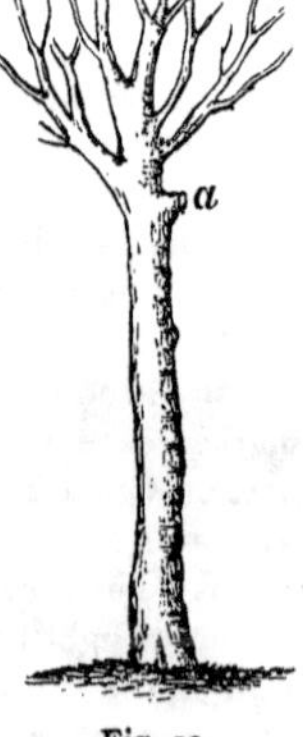

Fig. 19.

Downing recommended, and many have followed, the practice of binding several layers of cloth over the mallet, to soften the blow and prevent the bruising of the bark. Dr. Hull has adopted this mode in his famous curculio catcher. But it defeats, partially, the very object the operator has in view; it fails to bring down all the insects, and a part escape. *This is one reason why so many fail in the use of the jarring mode.* Formerly, in order to make a sharp sudden jar, without bruising the tree, we adopted the practice of sawing off a small limb, leaving a stump an inch or two long, (shown at *a*, fig. 19,) on the end of which a large hammer or axe could be struck with safety. This was more efficient than a cushioned mallet. But more recently we have employed a still better mode. Procure a sufficient number of large sized cut-spikes, bore holes into the trunk or large limbs to receive them, and drive a spike firmly part way into each hole, fig. 20. If the spikes are too long, break off the points in a vise. On the heads of these, a blow of a large hammer will bring down every curculio. Its efficiency contrasts strongly with the old modes. When the trees are small, one spike in each tree is sufficient; when they become quite large, it will be best to insert one in each of the larger limbs, as shown at *b b*, fig. 21. Instead of spikes, short pieces of rod-iron will answer a good purpose.

Fig. 21.

When the insects are very abundant, it may be more expeditious to kill them in hot water. In this case make the frames double, or with joints at the middle, (using two sticks in place of one,) so that the sheets may be folded together like the covers of a book, forming a trough down which the insects may be shot into the vessel of hot water.

Fig. 20.

The work of catching the insects should be commenced as soon as the young plums are as large as small peas, and continued several weeks, till no more can be found. At first they may increase for a few days, as they continue to make their appearance; but they will soon be found to diminish as their ranks are thinned by the thumb-and-finger warfare. In one of the most abundant seasons, about two hundred were killed every morning in a quarter-acre orchard for the first week. By the end of the second week the number had diminished to about sixty daily. They continued to grow fewer, until at the end of the fourth week only five were found at the last examination, when the work was discontinued. A magnificent crop of delicious fruit was the result. It is well to persevere long, as new crops of the insects often continue to come, after the earlier ones are all destroyed. The best time is early in the morning, when they are more torpid than at mid-day. Once a day will commonly answer, unless in seasons of extraordinary abundance, when a second examination should be made at sundown. The work should not be intermitted a single day. It is such intermissions that often cause failure.

Dr. Hull's curculio catcher consists of a large hopper-shaped frame covered with muslin, and attached to a heavy wheelbarrow, the front frame of which is driven against the tree, jarring it, and bringing down the insects into the hopper. It requires that the stem of the tree be trimmed up three or four feet high, like fig. 19. The mode we have described allows the limbs to branch within a foot of the ground, as shown in fig. 21. By this mode of training we can pick half the plums while standing on the ground, and the remainder with a small step-ladder.

In using the jarring process for destroying the curculio, it must not be forgotten that the practice of turning in pigs and poultry for a month or so early in summer, is a useful auxillary. If they pick up and eat every larva in the fallen plums, they destroy a vast number which might otherwise make havoc another year. Sweeping up the fallen fruit daily accomplishes the same purpose. In some cases swine have thinned the insects so much that uniformly heavy crops have been obtained year after year—the animals being sufficiently numerous to make thorough work, and it must be yearly without intermission. In connection with jarring down, these depredators may be so effectually thinned out, that the crop will be saved in such places, and in such seasons, as are most abundantly infested with them. The application of these two remedies is both easier and more effectual than many others which have been strongly recommended, such as covering the trees with lime wash or tobacco water, smoking trees daily, placing putrid substances under them, spading in the rising curculios, cutting canals under the trees to fill with water, laying brick pavements, making mortar floors, and other modes hard to apply, and of little or no efficiency.

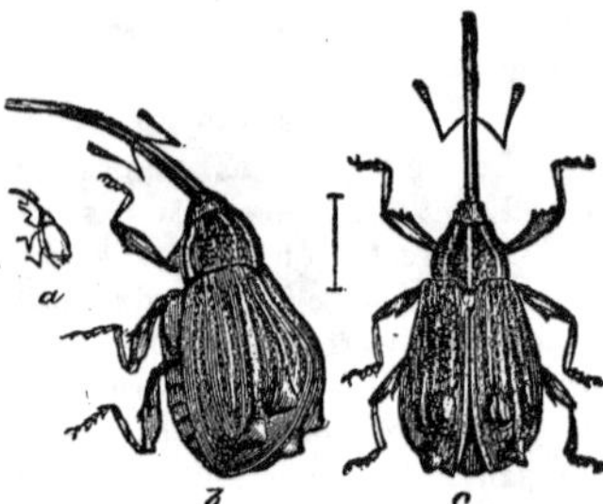

Fig. 22.—*Apple Curculio of the West.*

THE APPLE CURCULIO of the west, properly so called, is a distinct insect, and is shown magnified in fig. 22. The most striking point of distinction is its very long beak. Although the common plum curculio attacks the apple, as well as several other fruits, the insect here figured is an especial enemy to it. It has not been found in the eastern States, and may not prove generally formidable at the west.

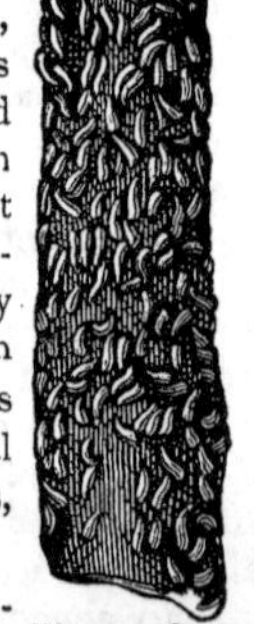
Fig. 23—*Oyster-Shell Bark-Louse.*

THE BARK LOUSE.—There are two distinct species of insects that produce scales on the bark of apple trees, known as bark lice. The imported Oyster-shell Bark-louse is shown in fig. 23. The scales, which have much resemblance in shape to minute oyster shells, are of a greenish brown color, and the many eggs beneath them are minute, oval, milk white. They hatch out in summer into minute lice, which are so small as scarcely to be seen, or would be likely to be mistaken for specks on the bark as they are nearly motionless.

The American Bark-louse is represented in fig. 24. Unlike the last described, the scales are milk white, and much more flattened; and the eggs, instead of being white, are pink or blood red.

It is the Oyster-shell louse that is so injurious to young apple trees. When the stems and branches are densely covered with them they retard growth, render the trees feeble, and in many cases entirely destroy them.

The eggs hatch by the first of summer, and the young lice may then be destroyed by a wash of potash, not strong enough to injure the bark. But as the insects are too small to attract attention at this time, the remedy is always omitted. By midsummer the scales begin to form, and from this time until the following spring, other remedies must be resorted to. One which has proved effectual is to boil leaf tobacco in strong lye until reduced to an impalpable pulp, and then mix it with soft soap. This is diluted with water till like paint, and is applied with a brush when the trees are dormant, to the stem and branches.

Fig. 24.
American Bark-Louse.

Another method is to apply a mixture of tar and linseed oil, warm, not hot, in March; it soon dries, becomes a varnish, and peels off, carrying the lice with it.

Insects which Affect Small Fruits.

Currant Worm.—There are three distinct insects which commit depredations on currant and gooseberry leaves, namely, the Currant Span worm, which comes out in the form of a miller or moth, the Imported Currant worm, and the Native Currant worm, both of the latter forming four-winged flies in the perfect state. The Span worm was first observed, in several places, as a depredator, about twelve years ago, and was seen at Union Springs, N. Y., the same year that the imported worm began to attract attention at Rochester, N. Y., where it was believed to have been imported in nursery packages of gooseberry and currant bushes from Europe.

The Currant Span worm, (*Ellopia ribearia*,) is represented in the following figure, (fig.25,) the natural size and appearance. It is about an inch long, bright yellow, with numerous black spots. The head is white, with eye-like spots. It devours the early leaves of the gooseberry and currant, and when about to change, hides under rubbish, clods, or descends into the ground, and changes to the chrysalis, No. 3. In two weeks it comes out in the form of a moth or miller, of a dull yellowish white, with dark colored spots towards the ends of the wings. The spread wings measure about an inch and a quarter. The figure, (fig. 26,) represents its appearance, but is too dark. Where the larvæ has been numerous, and have stripped the currant row, this miller may be often seen in considerable numbers, flying over the bushes and laying its eggs on the twigs. Here the eggs remain till the following season, and hatch out about the time the gooseberry and currant leaves expand, ready for devouring them.

As the eggs remain on the bushes during the time that nurserymen dig and pack them for distant conveyance, care should be taken that the insects are not thus conveyed to places where they are previously unknown.

Fig. 25.—*Currant Span Worm.*

Fig. 26.—*Moth of Currant Span Worm.*

THE IMPORTED CURRANT WORM, (*Nematus ventricosus*,) is represented in figs. 27 and 28; *a a*, the larvæ in the act of devouring gooseberry leaves; *b*, an enlarged view of one of the abdominal joints, to show the position of the black spots.

In fig. 28 are magnified representations of the male, *a*, and female, *b*, the cross lines showing the natural size. The perfect insect makes its appearance as soon as the leaves of the gooseberry and currant are fairly expanded, and lays its eggs on the under side of the leaves, along the principal veins, and not, like the Span worm, on the young twigs. If the latter deposited eggs on the leaves, they would fall to the ground, as they remain unhatched till the following season, as already stated.

The eggs of the Imported worm soon hatch into 20-legged worms, of a green color, having at first black heads and numerous black dots over the body, but after the last moulting they are entirely green, except the large eye-dots and the three yellowish joints, one next the head, and the others at the rear. They are about three-fourths of an inch long when full grown. When, as usually happens, they are in large numbers, they rapidly consume the leaves, and whole rows of bushes have been entirely stripped in forty-eight hours. Hence the importance of close watching and prompt attention in applying the remedies to destroy them. A single defoliation, while it does not kill the bushes, retards growth, and commonly greatly injures or prevents the ripening of a crop; and if often repeated, so that the bushes remain bare for a long time, or for successive seasons, the bushes necessarily perish.

Fig. 27.—*Imported Currant Worm—Larvæ.*

When the larvæ attain full size, they burrow under ground, or hide under scattered leaves, and spin an oval brown cocoon. After some weeks the perfect insect comes out, lays eggs as before, produces larvæ, which pass to the pupa state, and remain so till the following season.

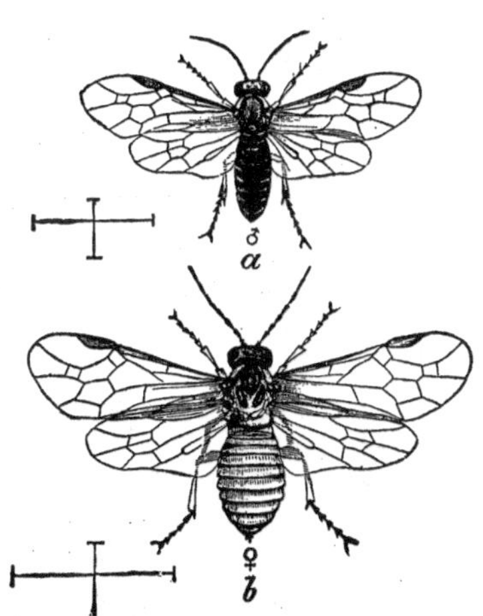

Fig. 28.—*Perfect State of Imported Currant Worm—Upper, Male—Lower, Female.*

THE NATIVE CURRANT WORM, (*Pristiphora grossulariæ*,) is smaller than the preceding, or about two-thirds in size, and otherwise resembles it somewhat in general appearance.* Unlike that, the male and female differ but slightly. The larvæ are of a uniform pale green color, *a*, (fig. 29,) without any black dots, which readily distinguishes it from the two others already described, the head becoming black. It spins its cocoons among the twigs and leaves. It appears later than the Imported Currant worm, or near mid-summer, and the second brood early in autumn. Unlike the last named, the second brood also passes to the state of winged insects the same autumn, and lays its eggs on the twigs of the bushes, where they remain till the next season.

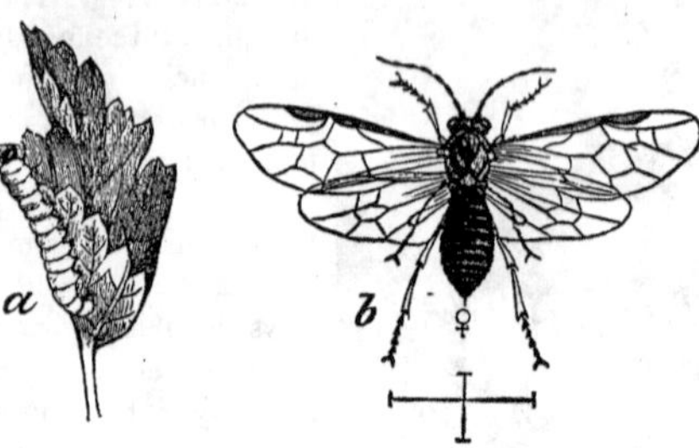

Fig. 29.—*Native Currant Worm.*

The remedy for the three species of Currant worms is the same for each—namely, killing by poison. Unlike many other insects, this remedy is comparatively cheap, easily applied, and entirely successful if used promptly. It consists in dusting powdered White Hellebore from a finely perforated dredging box, or from a box covered with fine muslin, so as to give the leaves a thin dusting of this poison. It may be had at drug stores. Do it in the morning when the dew is on, but do not wait for dew if the fruit worms have made their appearance. To prevent inhaling the dust, fasten the box to a short stick, apply it when there is only a faint breeze, and stand on the windward side. As soon as the insects devour it with the leaves, they curl up and die. It is desirable to give the leaves a very thin coating, and not to apply it in masses.

STRAWBERRY WORM.—For the account of this insect we are wholly indebted to the American Entomologist. It appears that for some years it has infested strawberry fields in certain parts of Northern Illinois and Indiana, and has also occurred in Canada. The above figure, (fig. 30,) represents, at *a*, the larva, which is about a third of an inch long; *b*, magnified forward end of larva, and *d*, hinder end; *c*, moth, magnified about twice the natural diameter. There are two broods a year, the first in June and the second

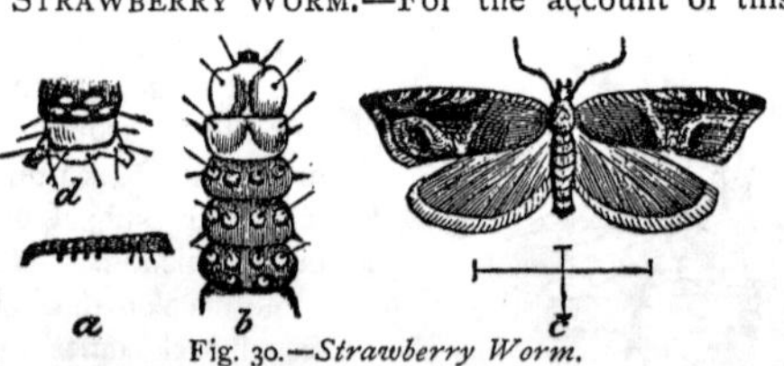

Fig. 30.—*Strawberry Worm.*

* A marked distinction will be found in the wings. There is a front section between the veins of the wings, near the tips of the forward wings in both; but while this forward section sits on *four* other sections behind it, in the imported fly, it sits on only *three* in the native.

in September. The pupa of the second brood remain in this state through the winter. The worms devour the leaves, and roll them up when they change to the pupa. In several instances they have ruined whole acres of strawberry plants. It is supposed that in some cases where the death of plantations has been charged to the hot sun, this minute insect has been the real cause. It is proposed to plow up plantations badly infested with it, and to avoid procuring plants from regions where it prevails, so as not to introduce it into new places. The American Entomologist describes it as a new species under the name *Anchylopera fragaria.*

GRAPE INSECTS.

GRAPE-LEAF PROCRIS.—The larva of this insect is found on grape leaves in June and July, and a second brood often appears in autumn. The larvæ arrange themselves in rows, from ten to thirty, side by side, and moving slowly backwards, devour the pulp of the leaf as they go, (fig. 31.) When young they leave untouched the fine net-work, as shown on the right side of the leaf; as they become older they devour all but the larger ribs, as seen on the left. When full grown they measure five-eighths of an inch long, and sometimes three-fourths of an inch. They are yellow, and are marked by transverse rows of velvety black spots on each segment, and with short, stiff hairs, mostly at each end, (*a*, fig. 32.) When disturbed, they curl to one side, and either fall or sus-

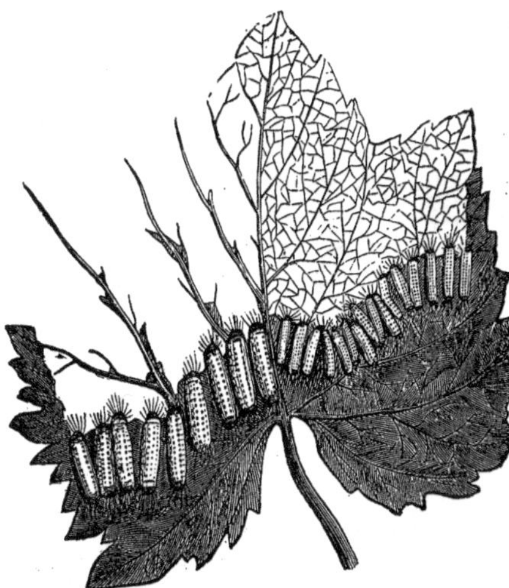

Fig. 31.—*Grape-Leaf Procris.*

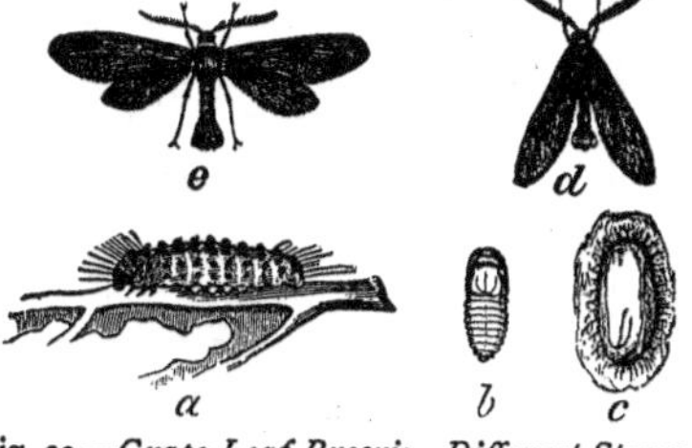

Fig. 32.—*Grape-Leaf Procris—Different Stages.*

pend themselves by a fine thread. When about to change to the pupa state they disperse, seeking a retired place, and spin a small whitish cocoon, (*c*, fig. 32,) and change to the chrysalis, (*b*.) In ten days or more they issue a small black moth, with narrow wings, expanding nearly an inch, (*d* and *e*, fig. 32.) These moths have a fan-like, forked tuft at the end of the body, and an orange ring around the neck or forward margin of the thorax. They deposit their eggs in small clusters on the under side of the leaf, from which the larvæ hatch and form the second brood. Most of the pupæ of this second brood remain till the following spring. This insect may be destroyed by drenching the leaves with whale oil soap, or with carbolic or chrysilic acid, properly diluted with soap suds.

SPOTTED GRAPE BEETLE.—The spotted Pelidnota, (*Pelidnota punctata*,) is a large brown beetle, an inch long, marked by eight dark spots, *c*, (fig. 33.) The larva (*a*) is a large white grub, about two inches long, with brownish head and feet, and a peculiar heart-shaped swelling at the hinder end, (*d*;) and it is generally found bent up in a crescent form. A magnified antenna is shown at *e*, and a magnified leg at *f*. The pupa is shown at *b*.

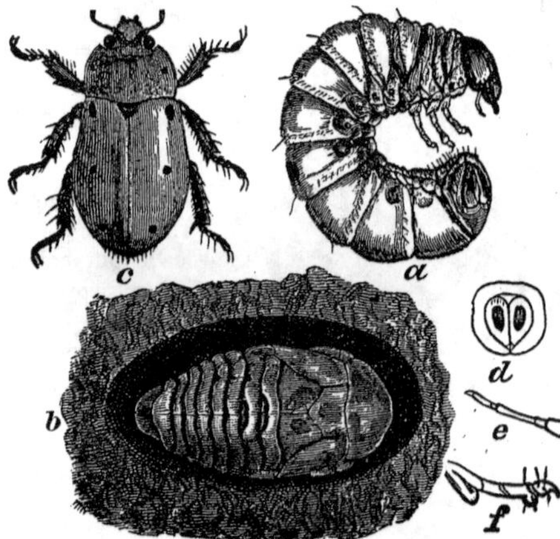

Fig. 33.—*Spotted Grape Beetle.*

The larva mostly inhabits rotten stumps, and other rotten wood, and it requires about three years to pass through the larval periods. It remains only eight or ten days in the pupa state. The perfect insect or beetle attacks the foliage of the grape in the day time from the middle to the end of summer, and when numerous has been known to strip the leaves from the main vine to the ends of the branches. It is more particularly destructive to the varieties of the Fox grape, but it also devours the leaves of the Virginia Creeper, (*Ampelopsis quinquefolia*,) which is closely allied to the grape. It is not usually a formidable insect, and being large and conspicuous to the eye, is easily caught during the day and destroyed. The grubs are preyed upon by a parasite in the larva state of a two winged fly belonging to the genus *Midas*. The Pelidnota is found throughout the United States.

THE GRAPE-LEAF GALL.—These excrescences or galls are caused by a small aphis or species of plant lice, about the twentieth of an inch long, and about half that length when first hatched. It has not been much known until within a comparatively few years; but has now become somewhat formidable both at the east and west. It was briefly described by Dr. Fitch, in his third Report on New-York insects, under the name of *Phylloxera vitifoliæ*. It appears that it has found its way into France.

Fig. 34.—*Grape-Leaf Gall.*

C. V. Riley gives an account of its habits in the American Entomologist, which we condense as follows:

The few individuals which start the race early in the year, commence on the upper side of the leaves, and by suction and irritation, cause swellings on the opposite side, and a gall about the insect. Here the female deposits her eggs, yellow in color, and numbering in each gall from fifty to four or five hundred. When they hatch, the young insects leave the gall, spread over the leaf and make new excrescences. There are several generations in a season, and the process continues as long as the vine gives fresh leaves. The leaves badly infested, turn brown and die, and the vine of course suffers from the loss of foliage. The lice then attack tendrils, leaf-stalks and tender branches, and finally work down to the roots, where they cause little knots, which eventually become rotten. It is supposed they remain during winter on the roots, and commence multiplying again the following spring. It is believed that they are often conveyed from one part of the country to another on the roots, in nursery packages, and are thus distributed.

The Clinton, Taylor's Bullit, and other varieties of the Frost grape, (*Vitis cordifolia*,) are especially liable to the attacks of this insect, but not the Isabella, and other clear varieties of the *Vitis labrusca*. It has been found sparingly on the Concord and Delaware. S. S. Rathvon of Lancaster, Penn., stated before the Pennsylvania Fruit-Growers' Society last year, that he thinks he saw it twenty years ago on the native Frost grape.

It has been recommended to destroy all vines of the Clinton and other varieties of that species, to prevent its spread. Picking off the infested leaves and burning or scalding them, is the only known remedy after the insects have obtained a foothold. The cannibal insects which prey upon them appear, however, to perform a most important service in limiting their ravages.

NOTE.—For a large number of cuts illustrating this article we are indebted to R. P. Studley & Co. of St. Louis, publishers of the American Entomologist; and also to the same valuable journal for important facts in relation to several of the insects figured and described.

PLOWING WITHOUT DEAD FURROWS.

A GREAT INCONVENIENCE to those who desire smooth fields, is caused by the frequent dead furrows resulting from the plowing of narrow lands. These furrows interfere with the working of the mower and reaper, the horse-rake and the tedder, and are troublesome in drawing in hay and grain. They are admissible only in wet fields, where ridges and furrows must be frequent for the purpose of carrying off surface water, or as a guide for sowing seed or plaster. The former is obviated by drainage, and the latter, if necessary, by a few stakes as a guide in sowing.

For these reasons the practice has been adopted by some of the best farmers, and is becoming more common, to plow around the whole field. A difficulty arises with those to whom the practice is new. If they begin at the outside, and work towards the middle, the soil is after a while banked up against the fence, by successive plowings, and dead furrows are still left at the middle and running to each corner. It is therefore much the best best, wherever practical, to begin, at least part of the time, at the middle. The question then comes up, how shall we know where and how to begin, so as to come out exactly even all around? It is our present purpose to answer this question, so that every intelligent plowman may readily understand how to go to work. A few minutes employed in comprehending the subject may save years of inconvenience with dead furrows.

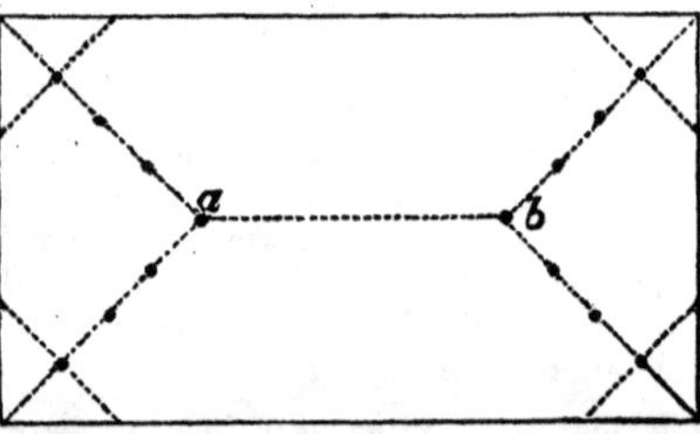

Fig. 35.—*Laying Out a Field for Plowing around it.*

Fig. 35 shows the manner of laying out the plowing for a square angled oblong field, being a quite simple operation. There are two ways of doing it. One is to measure and stick in a stake or peg an equal distance each way from the corners, say ten feet, and then stretch a cord across, as shown by the dotted lines, to each stake. Stick another stake at the middle of the line. The middle is quickly found by doubling it, sticking in a pin at the middle, and then laying it down again in place. Then stick two or three other stakes, ranging with the corner and the middle stake, and running towards the middle. Proceed in the same way with the other corner. Drive in a good stake at the point *a*, (fig. 35,) where these two lines cross each other. Performing the same operation at the other two corners, find the point to drive the stake *b*. Then beginning at *a* and *b*, lay the first furrow between them. Run back again, and the field is laid out for plowing. Go on with the work, plowing across the ends, and if the measuring has been accurate, and the plowing straight and even, the last furrow, all around, will be along the boundary fence, and the field will present the appearance shown by fig. 36, in which there is no dead furrow, no plowed ground trodden hard by the horses' feet, and the only furrow visible the one at the boundary.

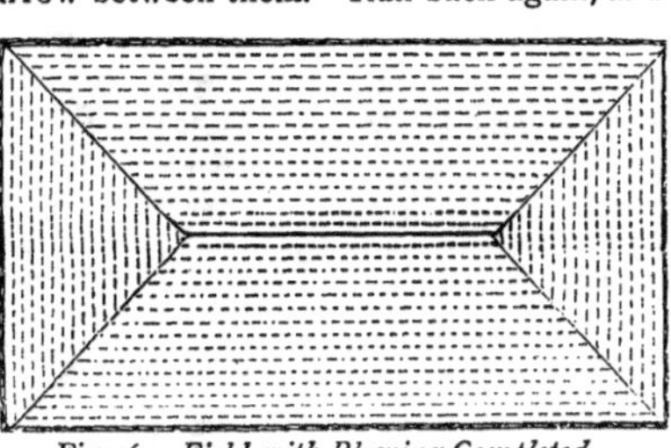

Fig. 36.—*Field with Plowing Completed.*

Fig. 37 shows the appearance of the same field when the plowing is begun at the outside. Dead furrows are left along the middle, and running out to each corner. The shaded strips show the places where the horses tread hard the plowed portions in turning about. This mode of plowing the field is admissible only when it has been too much ridged towards the middle and away from the boundary fences.

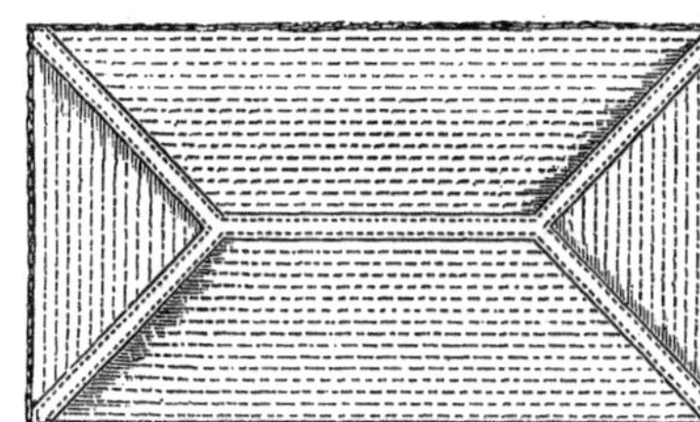

Fig. 37.—*The same Plowed outward from the middle, leaving Dead Furrows.*

Another mode of laying out the field, fig. 35, and in ordinary practice a simpler one, is to measure square across, take half this width and drive in a stake. Measure at one or two places, and drive in stakes. These will show the middle line of the field, and by ranging with them they will cut the ends into equal parts. Then from these middle points in the ends, measure exactly as far on the centre line, as half the width of the field, and drive in stakes at *a* and *b*, and proceed to plow as already described.

In order to be able to lay down a line exactly square with the sides, a common carpenter's square may be used, placing it on a box or stool, with one arm carefully in range with the side of the field. The other arm will show where to strike the line towards the middle. A larger square, or with longer sides, made on purpose, will be more accurate, is easily made, and may be used for many other purposes, (fig. 38.) Procure three strips of light pine, perfectly straight on the edges, and placing two of them lengthwise nearly at right angles, secure them by a screw or nails. Then measure off carefully from the corner, three feet on the edge of one piece, and four feet on the other. Then taking the third strip, which should be a little over five feet long, measure five feet, and mark the points with a scratch or pencil. Place these points exactly at the three and four feet marks on the two other pieces, move them till all exactly coincide, and a perfect square will be formed. Secure the points by nails or screws, and it will be in a convenient shape for use, and may be hung up against the side of a shop or shed. It need not be heavier than a ten-foot pole. It may be made quite portable, by having a screw-joint at *a*, like that of a pair of compasses, and button screws to attach the cross-piece when ready for use. The three pieces may be carried parallel in the hand together, when not in use or laid aside. A larger square may be made by substituting for the 3, 4 and 5 feet measurements, 6, 8 and 10 feet.

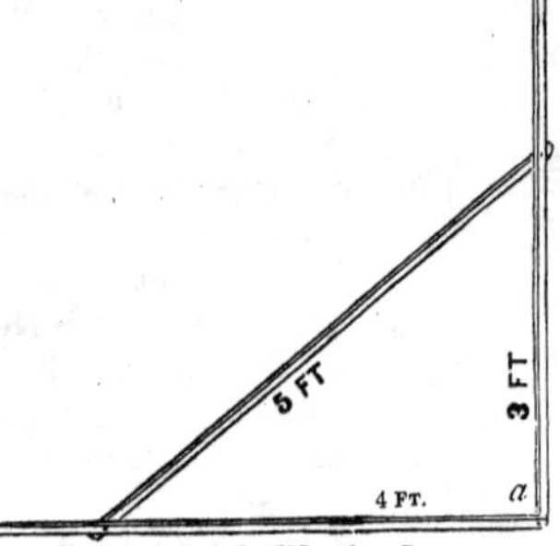

Fig. 38.—*Light Wooden Square.*

Fig. 39 represents a square field of sward plowed in the common way, by dividing it into several "lands." Each successive sod is thrown inward, the horses turning about to the right, and rendering it necessary to leave a head-land at each end, to be plowed afterwards. For planting corn on the sod, the ridges and dead furrows are not a great inconvenience, although one row of good corn is lost if planted in the furrow.

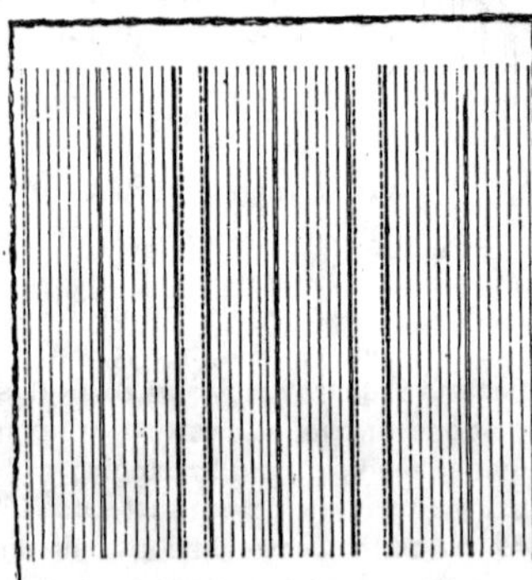
Fig. 39.—*Plowing Sward into Lands.*

This mode also possesses a positive advantage, if (as is an excellent mode,) the corn is planted by means of a horse-drill, as each day's work of plowing progresses, while the earth is moist and mellow—thus planting early, and giving the young crop the start of the weeds. This advantage, however, may be obtained by means of a good swivel plow, turning the furrows all one way, and leaving no dead furrows.

Fig. 40 shows the manner in which stubble ground is commonly plowed, the team passing around each successive "land," and throwing the furrow outwards. Each land has a dead furrow at the centre, and a branching furrow at each end, running out to the corners. In addition to this inconvenience, a portion of the plowed ground is trodden hard again by the horses' feet at four places on each land, as indicated by the shaded diagonal strips. Stubble land may be plowed as shown in fig. 39, with the exception of omitting the head-lands, and plowing across the end of each land, which will prevent the treading of the plowed ground by the horses, and leaving but single straight dead furrows, but it is somewhat inconvenient to turn the team about to the right.*

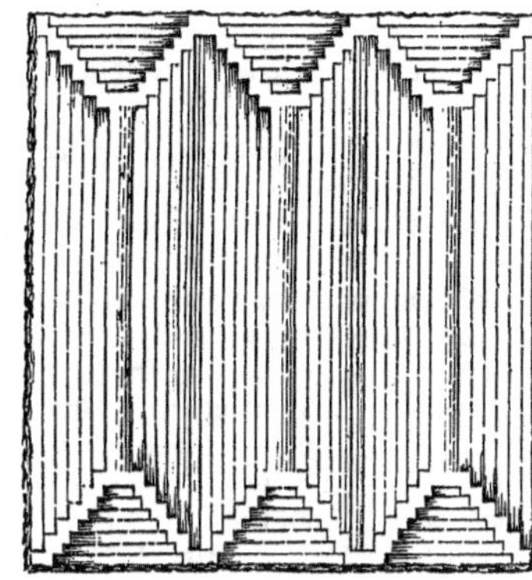

Fig. 40.—*Plowing Stubble into Lands.*

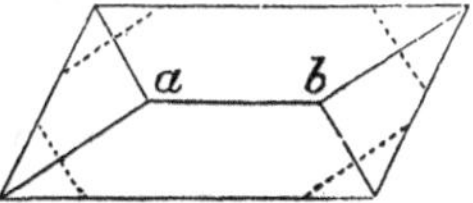

Fig. 41.—*Oblique Field.*

We give further examples of the mode of laying out fields that are irregular in form. When the sides are parallel, but the angles oblique, as in fig. 41, the two centers *a* and *b* may be found by dividing the angles, as explained under fig. 35. The line *a b*, if continued *straight on* to the two end boundaries, will be a little longer than the distance to the two longer sides, because it crosses the furrows obliquely.

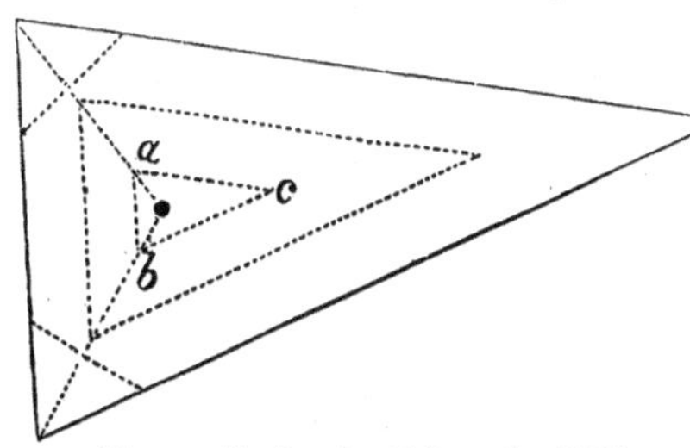

Fig. 42.—*Laying Out Triangular Field.*

The centre of a triangular field, fig. 42, is determined in a similar manner. Two angles are bisected, as shown by the dotted lines, and the

* In order to plow head-lands in a satisfactory manner, leave a strip of unplowed land on each side, as wide as the head-lands at the ends; then plow *inwards* around the whole field, beginning on the inner side, and so going out towards the fence. By leaving an ample width for the horses to turn about upon at the ends, there can be no loss or disadvantage, as this border is as easily plowed by passing around the whole field as in any other way.

place where they cross is the centre. Measure a few feet outward, equally distant from this centre, on the lines, and draw lines or set stakes parallel with the boundaries. This will form the triangle *a b c*, which is exactly similar in shape to the field, but much smaller. Begin at the middle and plow this triangle, so that the furrows shall come out parallel to its three sides, and then you will have a fair start; and all you have then to do is to plow around it till the field is finished.

It is well to measure the three unplowed sides occasionally, to see that all preserve the same width, and if they are found to vary by inaccurate plowing, the error can be easily rectified by varying the furrows.

In order to draw the centre triangle easily and accurately, the following course may be pursued. After bisecting the two angles, as already shown, put up a line of stakes on the two intersecting lines, measure from the centre stake an equal distance along these lines, and set up two stakes at *b* and *a*, and stretch a cord between them. Then, by means of the square, laid carefully against this line, sight towards the nearest boundary, set a stake there, and then measuring the distance from the square to the last mentioned stake, it will give the exact distance from the triangle to the outside of the field. Then setting the square on each of the other two sides successively, measure this distance in, and set stakes at the end. These stakes will be in the lines which form the central triangle, and measuring twice from each side, will give the exact position of the triangle.

Fig. 43 shows an irregular four-sided field. First find the two centres *a* and *b* by bisecting the angles, as shown by the dotted lines at the corners, and, as already described for the square angled field, fig. 35. Then measuring perpendicularly from *b* to the nearest side by the assistance of the square placed on that side, (which is moved along backward or forward till in the right place,) measure the same distance from the other sides at *c*, *d* and *e*, making these measurements perpendicular to the sides by means of the square. Stakes driven in at the ends will form a triangle, around which the plow is run till the field is finished. Or, if this triangle is too large, as will be apt to be the case, begin at *a* and run the furrows parallel to the three sides of this triangle, and the work will come out right.

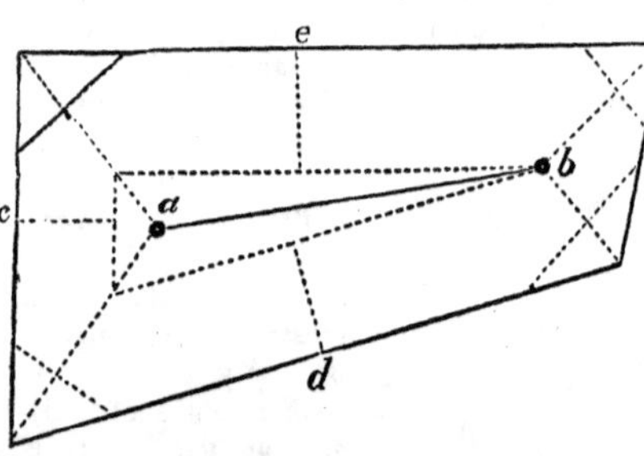

Fig. 43.—*Laying Out Trapezoidal Field.*

Fig. 44 represents an irregular five-sided field. The same course is to be adopted as before, the places for the two central stakes, *a* and *b*, being first found, and the sides of the central figure, parallel respectively to the other sides, found as described under fig. 43.*

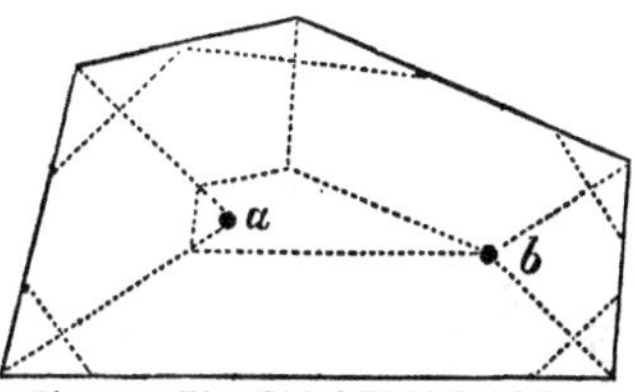

Fig. 44.—*Five Sided Field Laid Out.*

It will make the plowing easier, plainer, and more accurate, to plow light furrows from the corners in towards the centres; and also to plow light furrows to form the centre triangle. The corner furrows will show exactly where the team turns, and the work will be kept in more accurate shape.

To some all this work at measuring may seem troublesome and needless; but it will be found a great saving of labor in the end. Any one can understand the rules given by a few minutes attention; and after some practice a large field may be laid out for plowing, in an hour's time. Without such measuring the plowman may finish on one side of a fifty acre field, when he has left an unplowed strip on the other two or three rods wide, which will cost him an additional day's labor to plow, unless he finishes up in the irregular manner, with a dead furrow.

WEED-HOOKS AND CHAINS IN PLOWING.

THE USEFUL AND INCREASING PRACTICE of turning under heavy crops of clover and other green growth as manure, renders it essential to perform the work in a perfect manner, so as to leave no stems and leaves uncovered. In plowing under tall stubble or weeds, all should be completely laid beneath the inverted furrow slice. Different modes are adopted to effect this purpose. The practice of running the harrow over the crop to be plowed under, in the direction in which the plow is to pass, to assist in prostrating the crop, has given way to other and better modes. The most common means now used is to attach a chain to the plow in such a manner that its weight, as it is dragged by the plow, shall bend over the plants and sweep them into each successive furrow. One mode of attaching it to the plow, is to fasten one end to the right hand portion of the main whiffletree, and the other to the right handle. Or it may be done as represented in fig. 45, the chain forming a loop. A little trial will show how long to make this loop; if too long, the sod will cover it; if too short, it will not hold the weeds down. A short chain extending from the rear end of the beam to the left side of the loop will keep it better in place. When plowing with oxen the chain at its forward

* For the simple and easy mode here given, of determining the centres, and for placing the sides of the central figures, we are indebted to Prof. Evans and Dr. Potter of Cornell University.

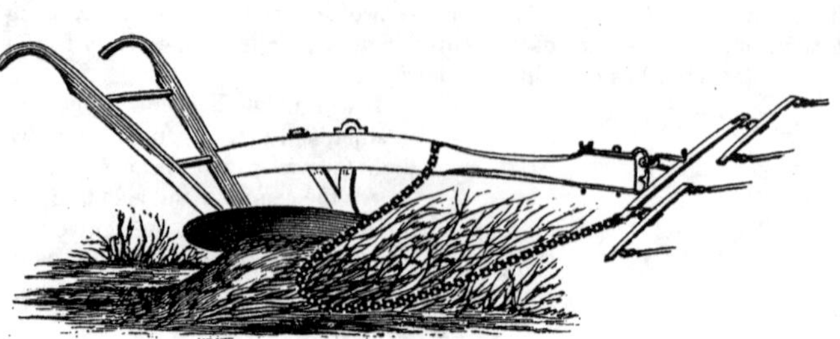

Fig. 45.—*Weed-Loop, made with a Chain.*

end is attached to a piece of wood about two feet long, screwed on the beam, as in fig. 46. This is the best mode in any case. It should be wide next the plow, so as to brace firmly.

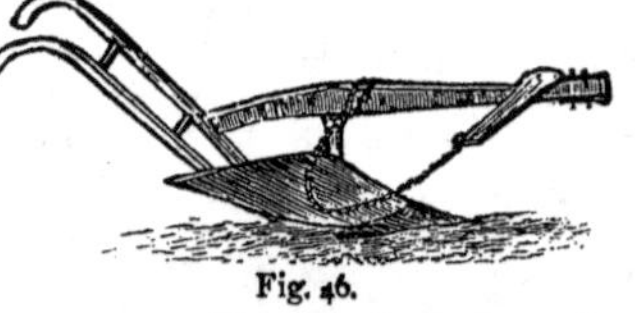

Fig. 46.

Where there is much of this work to be done, it is better to provide a weed-hook. This, like the different modes of attaching the chain, varies in form. Fig. 47 shows one which slants backwards, and *wipes* the growth from the sod into the furrow. Fig. 48 represents one in the form of a bow or hook,

Fig. 47.—*Weed Burier, or straight Rod.* Fig. 48. *Weed Hook.*

which sweeps the growth compactly together, ready for turning under. The first named method works more easily, and is not liable to be caught by obstructions. The latter is more effectual, and does best on smooth, even ground.

The weed hook may be made of rod-iron, stiff enough to retain its place, and possessing some spring when it meets with obstructions. It may be bent as circumstances may require for its best performance, after trying it a short time. It is attached to the plow beam by wedging under a band —a small groove being cut into the beam to hold it securely to its place. The steel rods which are employed in the manufacture of rake teeth will make better weed hooks than iron, and possessing considerable elasticity, will bend easily in passing obstructions, and spring again into position. Sometimes the weed hook is made to project at right angles

Fig. 49.—*Plowed without Weed Hook or Chain.*

Fig. 50.—*Weeds, &c., Buried by Chain or Weed Hook.*

from the beam near the mould-board, and bending downward in a slanting direction. In this shape it should be made of bar iron, so as to possess greater strength; but we prefer the first described forms.

In the accompanying figures, fig. 49 represents a field of tall grass or weeds partly plowed under without any assistance of hook or chain; fig. 50, the same when they are well buried by means of these appliances.

LADDERS AND LADDER STANDS.

FOR PICKING FRUIT, and for various other purposes, light and portable ladders are a great convenience. Much depends on their their being neatly made, and of the best materials, but a well devised form is also important. For moderate heights, one of the most convenient, easily made and easily carried, is shown by fig. 51. It is merely a three-legged stool, about two and a half feet high, with stout, spreading legs, a piece of tough plank for the top, and the rounds on one side placed so that one can step up easily. The legs should be about the size of a common chair post, or a little larger. This stand is always ready for use, can be carried in one hand, sits

Fig. 53. *Pointed Hooked Ladder.* Fig. 52. *Long Hooked Ladder.* Fig. 51. *Short Standing Ladder.*

firmly, and for reaching up to branches of fruit a short distance overhead is just the thing. When fruit is higher on the tree, a common ladder may be employed, the length varying from eight to fifteen feet—say three ladders, eight, eleven and fifteen. Placed in the common way against a limb, there is danger of its sliding one way or the other, at the peril of the workman and bruising and scraping the tree. This difficulty is obviated by placing a hook on the upper end of one of the bars of the ladder, to be placed over a limb, the best point being at a fork, (fig. 52.) The lower end resting on the ground, it is held firmly to its place. The bar which holds the hook should be rather longer than the other, as the cut exhibits. To prevent bruising the bark, the hook should be of wood, made broad, and padded on the lower side. The best, firmest and easiest way to make it, is to cut a thick piece of wood, as represented in the cut, and secure it to the ladder by screw bolts. It is easily padded by placing a few thicknesses of woolen cloth on the lower side, and then securing these by passing a strong cord a few times around, or better, by driving a few carpet tacks at the edges.

Another form of this ladder is shown at fig. 53, the two bars coming together at the top, where the hook, wide enough to reach across both, is screwed to them. This form has two advantages—it stands firmer, and the wedge form above allows the operator to thrust it up anywhere into the tree.

Standing Ladders.

For a height of from six to ten feet, a good, simple, self-supporting ladder is shown in fig. 54. It is made spreading rather wide at bottom, so as to stand securely. To prevent the rounds from being weakened by their length, a few of the lower ones pass through a stiffening bar, (represented in the figure,) so that all thus connected support each other. The legs of this ladder may be connected to the upper end by means of holes bored through them to receive the upper

Fig. 54.—*Standing Ladder.*

Fig. 55.—*Common Standing Ladder.*

round, or by iron straps passing around, and screwed or riveted to the upper end of the legs. This ladder is quite portable, the legs folding against the ladder when not in use.

Fig. 55 represents a common step-ladder, each leg being attached at the top by means of a universal joint, so that they may be spread out for standing firmly, or folded against the steps when not in use.

A common ladder, when not over twelve feet long, may be easily made into a standing one, by means of the contrivance exhibited by fig. 56. Two supporting legs are attached to the outside bars at the top by means of screws, the form of which is shown at *a*. The legs have an opening or slot (*b*) to receive these screws. The ladder is raised, and the legs are at once placed under, against the screws, where they remain securely till the ladder is moved.

Fig. 56.—*Long Standing Ladder.*

Fig. 57 shows the upper end of the ladder more distinctly at the place where the legs are attached. The screws should be set a little obliquely, so that the legs may spread. A blacksmith will make them at a small cost.

All ladders like this should be shod with iron or steel at the bottom, to prevent slipping, as figured and described on p. 177, vol. 5, of RURAL AFFAIRS.

A support for the fruit basket, at the side of the ladder, is represented on p. 180 of ILLUSTRATED ANNUAL REGISTER for 1871, but usually it is more convenient to attach a hook to the basket handle, so that it may be hung on a branch, or on the round of the ladder, (fig. 58.)

Fig. 57.—*Common Ladder Changed to a Standing one.*

Fig. 58.—*Hook for Basket.*

The hook may be made by bending iron rod, or from the forked branch of a tree. (In the cut it is represented much too long.)

Long ladders often become dangerous by spreading—allowing the rounds to become loose, to slip out, and to bread under the weight of the person upon it. To prevent any possibility of such danger, two or three tie-rods, shown in fig. 59, should be placed just below the round, at two or three different places. Take a piece of half-inch or five-eighths iron rod; weld on a small shoulder, just as far apart as the inside of the ladder; cut a screw and place a nut on each end, and when the ladder is put together, insert the ties.

Fig. 59.—*Tie Rods.*

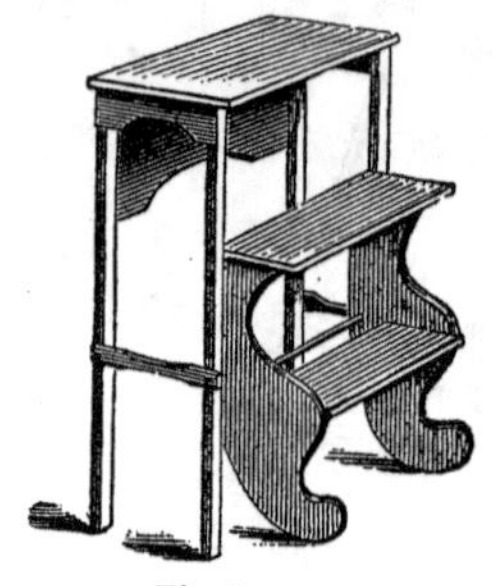

Fig. 60. *Library Ladder.*

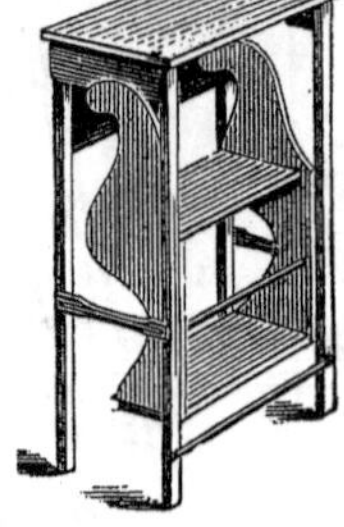

Fig. 61. *Library Ladder folded up.*

A library ladder is shown in fig. 60. The side steps turn on an iron rod, seen just above the lower step, so that when not in use these side steps may be turned up out of the way, as shown in fig. 61, when it will take up less room than a common chair. A common carpenter will make a plain one, such as the figure represents, in a day at a cost of not more than $3 or $4, and cabinet-makers sell more finished ones, with carpeted steps, for $8 or $10.

The dimensions which we have employed in having them made, are as follows: Whole height, 34 inches; size of the board, 12 by 18 inches; size of steps, 7 by 14 inches; height of steps, nearly one foot.

Fig. 62.

When at the seed store of Jas. Vick of Rochester, N. Y., we saw a new library chair, which by a single turn of the hand, was converted into a step-ladder as high as the top of the back. A number had recently been imported for sale, and were better in appearance than the accompanying cut represents—being of oak color and neatly constructed. Fig. 62 shows the appearance in the ordinary position as a chair, and fig. 63 the same, with the back

Fig. 63.

turned over and down, so as to form a support to the step-ladder, to which the chair is thus converted. It would prove a convenient piece of furniture in a library, or in a goods shop, where one wishes to reach a higher shelf.

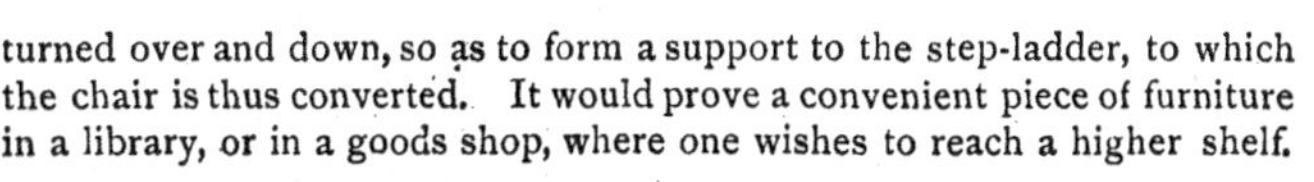

ORNAMENTAL PLANTING.

DWELLINGS ON OBLIQUE ROADS.—A correspondent has a trapezoidal piece of land, which he wishes to lay out as a building lot, of three or four acres, the road passing it obliquely, as in fig. 64. He is in a quandary whether to have the house face the road at right angles, and stand crooked with his neighbors where the road is straight, or else set his house "skewing" with the road, and he asks for information.

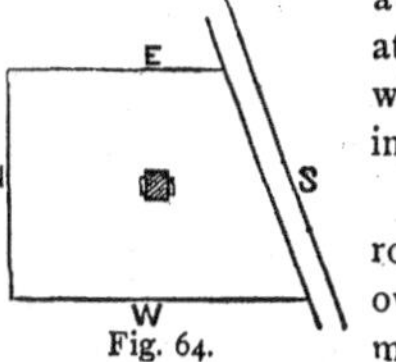

Fig. 64.

This is a common dilemma wherever diagonal roads exist, and we have many such inquiries from owners of small places. The course to be pursued must accord with circumstances. If the general course of the road is in the direction indicated, for a long distance, and the house is to be quite near it, then place the house directly facing the road, and flank it well with trees planted rather closely, so that its skewing position will be obscured by the foliage from those points where it would not appear well, as we have indicated in fig. 65. If the house is to be placed at a greater distance from the road, then let it face the finest views, nearly irrespective of the course of the road. Plant the grounds in the modern style, and with properly curved walks or roads, and the irregularity will not be at all out of place. Fig. 66 represents something of the character proposed.

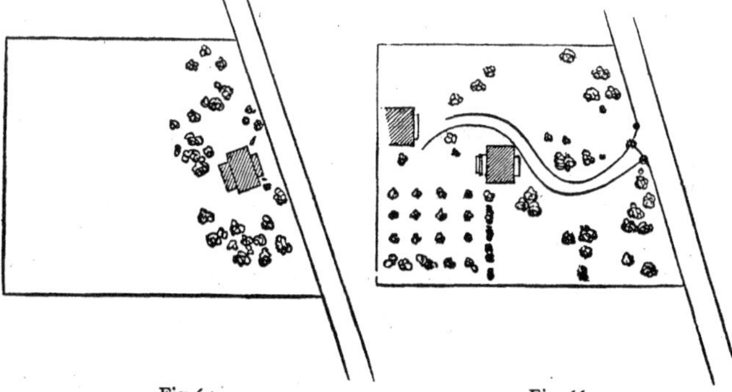
Fig. 65. Fig. 66.

In all designs of the kind it must be borne in mind that the dwelling is to face the most desirable or beautiful views, or objects of importance. Too much respect is commonly paid to the public highway, which is often a place for rubbish, instead of being made a handsome avenue. In many places cattle and swine are allowed to run in the roads, which they disfigure by their rooting or droppings. Brush and stone are often thrown into them in clearing adjoining fields. Some farmers place their farm buildings on the line of the road, and make it a receptacle of all kinds of implements and scattered lumber. Roads kept in such condition are not entitled to the respect which is shown them, by foregoing fine distant views or lake prospects in order that the dwellings may stand square and respectfully toward them. When country residents shall all unite in keeping the roads smooth and even, clear of all rubbish, neatly mowed, and handsomely planted with trees, so that they may be as agreeable a place for promenade or riding as a landscape garden, then we would advise that they be regarded as one of the best objects for a dwelling to face. The labor of keeping them in this condition would not be greater than that required to deface them, as it obviously requires as much work to carry brush, stones, lumber, tools, &c., into the road, as to carry them out of it.

We may suppose that a house is about to be built on an elevated spot, where there are four desirable points visible. One is a beautiful lake, another a distant village, a third a broad rich valley, and the fourth a street with rubbish and straggling animals. Shall we turn away from the first three in order to enjoy the last?

A little attention to these considerations, with a share of common sense, will enable any one to decide the conflicting questions of locality and position in fixing on a spot to build, and in determining the position of the dwelling.

Transplanting with Balls.—The frozen-ball method of removing ornamental trees is preferred by many to all others for some purposes. It is well adapted to evergreens growing wild, if they are of much size. In order that it may be easily and expeditiously performed, preparation should be made in autumn, or before the ground freezes hard, by digging a trench in the shape of a circle about every tree a foot deep, or as far down as the frost penetrates; and then filling these trenches with dead leaves, which are always abundant at this time of year in the borders of woods or wherever these trees are sought. The leaves will prevent the trenches from freezing in winter, and the earth within them being frozen hard, the trees are easily loosened and tipped over, and may then be readily transferred to sleds and conveyed to their place of destination, where holes, dug at the same time that the trenches were made, and similarly filled with leaves if convenient, or left open and frozen, may receive them. If holes and balls are both frozen hard, and are nearly equal in size, the first thaw will soften the ball and give it a close fit. But it is rather better to keep the hole unfrozen, so that the ball may be snugly imbedded in the mellow earth when placed there. For well rooted nursery trees this mode is not applicable; but we have found it well adapted to the removal of evergreens from the borders of woods in winter, when the work could be more deliberately attended to than during the busy period of spring.

LAWNS.—P. Barry gives the following as the requisites for a good lawn: 1. A dry ground, or one free from stagnant water. 2. A deepened or trenched soil, from 18 to 24 inches—trenched by hand or trench-plowed—which will keep the grass green during the drouths of summer, and greatly promote the growth of the trees and shrubs planted in it. 3. Evenness of surface—not level merely, for an undulating surface is quite as good as a level one—but smooth, and free from even the smallest stones. The best grass is Red Top—pure, unmixed Red Top—which Mr. B. prefers to a mixture of a fourth part or so of White clover, commonly recommended. The Red Top should be sown at the rate of four or five bushels, or fifty or sixty pounds, per acre. It should be sown very early in spring, at the first moment the ground will bear working. All preparatory work should be performed in the fall, so that the ground may settle, and defects be corrected before sowing. In the spring, at a fitting moment, plow lightly, harrow well, pick off stones, sow, and give a good rolling, which finishes the work. If the work is well done, there will be a respectable lawn by midsummer. Mow once a week, and a little oftener early in the season. One of Swift's mowers, drawn by a horse, will keep the lawn in perfect order. A hand-mower, for smaller places, will occasionally require a roller besides. If well prepared, the lawn will not need manure for a long time. A rank growth is not wanted. When it becomes feeble, top-dress with a compost of rotten turf and stable manure, decomposed to a fine mould and screened. The best way to guard against the effects of drouth is to deepen the soil.

EVERGREEN BELTS.—The Western Rural publishes an account of the timber belts on Horace Greeley's farm, at Chappaqua, N. Y., which consist of four parallel rows of Norway Spruce, White Pine, Cedar and Hemlock, respectively. The kind of Cedar is not mentioned. The trees are ten feet apart in the row, and the rows are twelve feet apart. The outer rows are covered with branches and verdure to the ground; the inner ones have the branches pruned off, so that one can look through from end to end, as through a long arcade. The whole forms a magnificent screen or barrier, through which a bird cannot fly, and it protects the garden and small fruit trees from the storms of winter.

MANURING EVERGREENS.—The Horticulturist says that "a good coat of manure, applied every fall, as far out as the branches extend, will ensure next season a deep glossy green to the foliage; the effect is sometimes so peculiarly ornamental it seems as if the shrubs and trees were of a new variety. The Norway Spruce we have often observed in some grounds, of a light, sickly green, while in other yards it is of a fine deep color; the difference comes only from treatment—one is in poor soil, the other is in rich. Those who wish their evergreens and shrubs to thrive and grow handsome every year, will not fail to remember this hint. Do not apply fresh manure; it should always be well rotted."

PRUNING EVERGREENS.—The spring of the year is the best time, just before the buds begin to swell, if considerable portions of the trees are to be removed. If done while the trees are growing rapidly, it would tend to check their growth. It is, however, an excellent time to pinch in long shoots, by simply nipping off the point. This causes the side buds to become developed, and induces a thicker growth. With the Scotch and other pines, these developed buds will push next spring; with the spruces, they will often start at once. Towards the close of summer, hardy evergreens may be moderately pruned, or when growth is approaching its termination.

HEMLOCK HEDGES.—The Gardeners' Monthly says: "Some think that as the Hemlock is a large forest timber tree, it cannot be kept down as a hedge plant; but summer pruning will keep the strongest tree in a dwarf condition for a great number of years. The pruning has to be done just after the young growth pushes out, which generally is about the end of May. It is very important the hedge should be cut with sloping sides, so that every part of the surface should have the full benefit of the light. No hedge with upright sides or a square top will keep thick at the bottom long."

HALF-TENDER EVERGREENS, such as Cedar of Lebanon, Deodar and English Holly, may, according to the Gardeners' Monthly, be grown in open air, if under the protection of evergreen belts, planted either with Norway Spruce, White pine or Scotch pine.

IMPROVING THE FORM OF FLOWERING PLANTS.—Some annuals grow in a handsome, symmetrical form; others are stragglers in a greater or less degree. These may be improved in appearance by pinching in the longer shoots in time—not cutting them back, which would be too late to obviate the mischief, and would tend to check their growth. Such plants as the Aster do not often need this shaping, but Balsams, and many others, may be much improved by a little timely attention.

LEAVES FROM LAWN TREES should be raked up in autumn on the score of neatness; and they may be applied to various useful purposes. Gardeners who employ cold frames to protect tender plants will find leaves not only a good additional covering inside, but placed around the frames contribute to the same result. They are useful for covering all beds of half-hardy plants, recently planted bulbs, &c., and may be kept from blowing away by a little brush or a few evergreen boughs. They also form a warm and comfortable litter for horse stables, more so than straw, as the leaves lie in smooth and even layers, and the thin strata of air which they enclose, render them efficient non-conductors. The manure, mixed with this mass of leaves, is not fibrous, like fresh straw manure, and is excellent for garden uses.

WINDOW PLANTS.—The Gardeners' Monthly says that a temperature of 55° will give more flowers to the common window plant than a higher

temperature, and names such old-fashioned sorts as Mignonette, Sweet Alyssum, Zonale Geraniums, Cupheas, Fuchsias, Violets, Roses, Chinese Primrose, &c., as among the best for this purpose.

PANSIES IN MASSES.—A correspondent of the Gardeners' Chronicle says that no one who has not seen the effect of pansies in large masses, can have an idea of their beauty. He planted a border, 400 yards long and 24 feet wide, with Pansies and Cerastiums, with a single row of Pyramidal Zonale Geraniums, in pots, at intervals of ten feet, and it was the admiration of all who saw it.

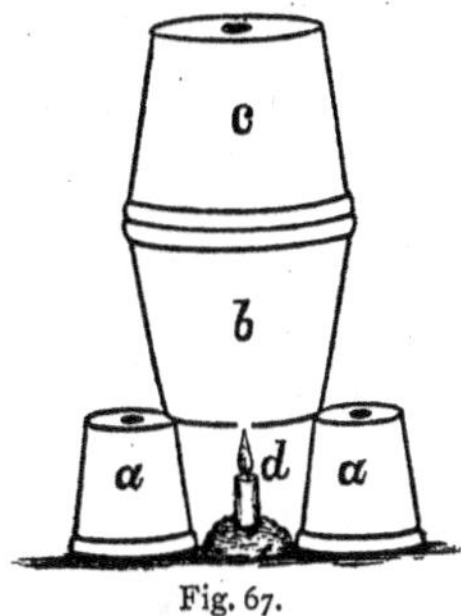

Fig. 67.

HOW TO FUMIGATE A GREEN-HOUSE.—A correspondent of the COUNTRY GENTLEMAN gives the mode represented in the annexed figure, (fig. 67.) Get two small and two large flower pots. Set the small pots bottom upward, a small distance apart. Set a larger one, *b*, on these, and put in some dry tobacco; turn the other large pot bottom upwards on this, and place a candle under, so that the tobacco will not blaze.

RUSTIC WORK.—Finish the structures and let them become well seasoned, and then brush them over thoroughly with crude petroleum, which will penetrate the pores of the wood, and render it durable, like red cedar. A common whitewash brush will answer, and any laborer can apply it. Let him be careful to anoint every part thoroughly, more particularly the joints. The light petroleum will penetrate the pores more freely, and the heavy will give the whole a more rich brown color. It is a good way to put the light on first, and after some time to wash over with the heavy. We have used a mixture of the two, which has proved quite successful.

THE BOSTON HOT-BED.

THE FOLLOWING ACCOUNT of this comparatively cheap contrivance for winter market gardens, is given in the COUNTRY GENTLEMAN, by W. D. Philbrick: The construction of our hot-beds is very simple. A situation is chosen with good drainage, and sheltered from the north and west by woods, or by a high board fence; a pit is dug parallel with the fence, and three feet from it, seven feet wide and two feet deep; this pit faces south or southeast, and has a cart path in front for hauling in manure and loam. A row of chestnut posts is set on each side of the pit, and 2 by 12 spruce plank spiked to them, so that the plank will be level, or nearly so, endwise the bed; but the front plank should be two or three inches lower than the back one, to admit of good drainage of the sashes, which are placed directly on the plank. When complete the pit will have a six inch space dug outside the plank; this space should be eight or ten inches wide, if it is intended to run the bed in severe weather; being filled with horse dung, it prevents the bed from freezing through the plank.

The chief difference between this bed and the ones described in the books, consists in the small amount of manure used—eight or ten inches in depth being a great plenty; and we generally put this in as hot and fresh as possible, covering it immediately with eight or ten inches of loam, and planting upon it as soon as the heat begins to rise. Very seldom does the heat rise high enough to injure the plants, if a little care is used to air them well for the first week, though if a greater depth of manure were used, there would be trouble from this source; 12 inches of manure we consider as much as can be safely used, and after the middle of March we never put in more than six or eight inches.

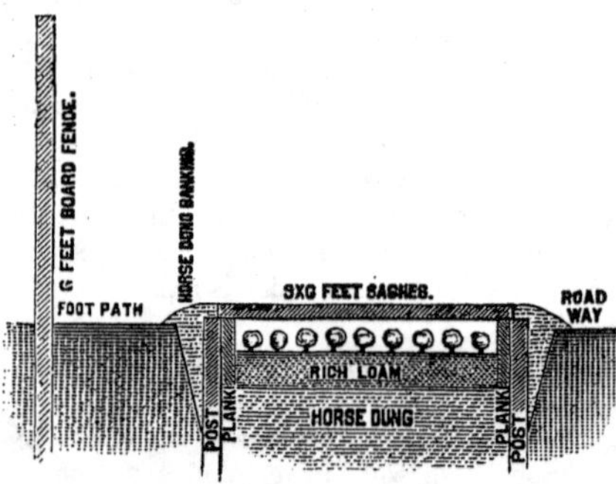

Fig. 68.—*Section of Hot-Bed.*

Beds thus constructed are run the entire winter, yielding two crops of lettuce and a crop of cucumbers or tomato plants; and although the labor of tending them is very considerable, having to be covered and uncovered every day with heavy mats and shutters, there is reason to believe that their use is profitable in skillful hands, inasmuch as the demand for horse manure for this use is in excess of the supply, and the price paid is $5 to $6 per cord in the stables in Boston—costing the farmer, delivered at his farm, fully $10 per cord for rough, strawy, long manure. A cord of manure is enough for about eight or ten sashes, and after serving this use it is taken out and used for planting. The lettuce crop sells at 50c. to $1.25 per dozen through the winter, varying greatly with the supply and quality; forty to fifty heads are planted under each sash, yielding $2 to $5 per sash; and as many farmers hereabouts run 1,000 sashes, and obtain two or three crops in the season, it will be seen that the business has reached considerable magnitude.

DRYING RASPBERRIES.

THE FOLLOWING MODE is adopted by A. M. Purdy of Palmyra, N. Y.: When there is a surplus of Raspberries, they are dried at the rate of twenty bushels a day, in a small drying house, seven feet by ten, heated by two small fires, and the whole costing about fifty dollars. The accompanying figures represent its construction. Fig. 69 shows the plan of the heating furnaces—the outer lines being the exterior of the furnace doors, through which access is had to the furnaces, F F, which are made of sheet-iron, half round, and are each about 10 feet long and 15 inches in diameter. The smoke and hot air passes through them, and through the horizontal pipes, P P, which are about five inches in diameter, into the brick chimney, C, standing against the end of the building. There should be a register in the pipe next the chimney, to control the heat.

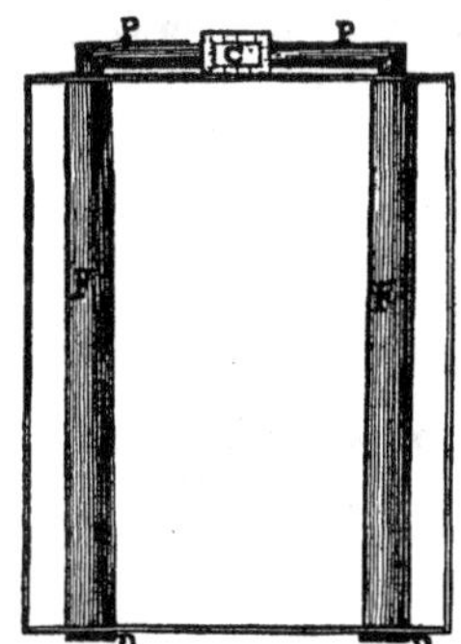

Fig. 69.—*Plan of Drying House.*

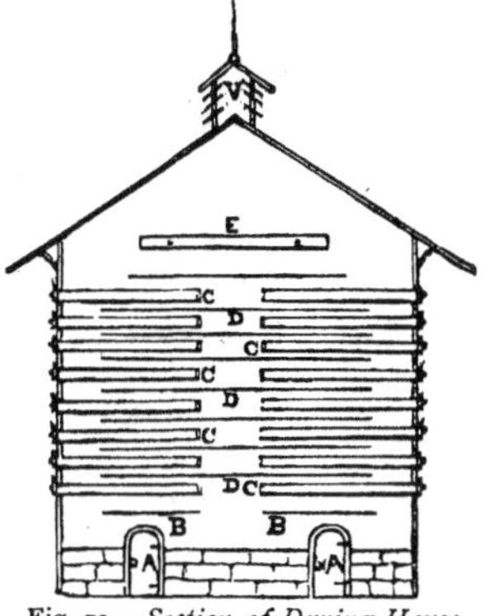

Fig. 70.—*Section of Drying House.*

Fig. 70 is a cross section of the house. It has no door, but the shelves pass into it from the outside. A A are the furnace doors; B B, pieces of sheet iron to prevent burning the bottoms of the lower drawers; C C C C, the drawers, sixteen in number; D D D, shelves between the drawers to distribute the heat; E, an extra drawer, for occasionally finishing the drying process; V, ventilator to carry off the vapor from the drying fruit.

The drawers are three feet wide and eight feet long—eight drawers on a side. Their position outside is shown in the perspective view, fig. 71. They are made of inch and a half pine for the sides and ends, and the bottoms of coarse cotton sheeting, tacked on with small nails, and supported by cross-bars two feet apart. The front of the drawers are of inch board, four inches high. The shelves or distributors, D D, are tight, seven inches apart, and they come within ten inches of the sides of the house. The drawers being all of equal size, will fit anywhere, so as to be changed from top to bottom, or otherwise. Between the two fire doors is placed a wooden box or square tube, (not shown in the cut,) running lengthwise horizontally through the house, with sliding boards at the ends to regulate the current of air, and with holes along the top. This supplies fresh air as it is heated and passes upwards. It regulates the temperature, and prevents the fruit from cooking. If at any time the house becomes too hot, pull out the lower drawers a few inches, to let in air. When nearly dry, finish up by placing the fruit from three or four drawers together, near the bottom, where they will have more warmth and fresh dry air. It is well to have the eaves extend some distance over, so as to protect the drawers when they are pulled out in rainy weather. Some would prefer to have the cleats on which the drawers slide, to extend outside to a short distance, as an additional support, but this is not essential. The house may be larger or smaller than the dimensions given, according to the amount of fruit likely to require drying.

Fig. 71.—*Purdy's Drying House.*

This mode of drying fruit of all kinds has important advantages. The drying being done rapidly, or in twenty-four hours, its freshness, flavor and color are preserved. All flies, wasps, millers, and other insects are kept away, and the fruit is clean and free from the eggs of flies. The drying is not retarded or interfered with by rain. Fine flavored apples, dried in such a house as this, ought to command a higher price in market than the brown and poor article, dried in open air, too often seen offered for sale.

HORTICULTURE IN COMMON SCHOOLS.

MANY YEARS AGO a successful teacher in a district school in the western part of the State, interested himself and his scholars in the culture of ornamental plants. The school yard, instead of the bald and repulsive appearance too often seen, was brilliant with tasteful flower-beds. The pupils understood that the flowers were objects for special protection, and they would as soon think of breaking the glass of the windows as to injure the plants. This early lesson in neatness and taste made a distinct and permanent imprint on the character of some of the young people, perhaps as useful and valuable as a knowledge of algebra and declamation. In those days there were many district schools where no refining influences of the kind where brought to bear, but where the school house in which the children are to receive so many moulding impressions, had nothing to render it attractive, or to soften the semi-barbarous habits created by the rough surroundings of the house of learning.

A few years since a fine Union School house was built in one of our villages, at a cost of some twelve thousand dollars. A thousand dollars more was paid for an acre yard. There were no trees, and there was nothing to break the severity of wind and snow storms in winter, or to afford refreshing shade from the hot sun in summer. A handsome building had been erected, but its beauty was defaced by the bleakness around it; a costly play ground had been purchased, but it possessed no more attraction than an unplanted common. One hundredth part of the cost of this building and yard, used for planting trees, would have rendered both beautiful. The writer proposed to the trustees to furnish gratis the necessary shade trees, foreign and native, to plant the grounds. The offer was declined—"the boys will tear up or break down all the trees in a few weeks." "But do you not think taste and civilization are as important as many things which they spend years in learning?" "Oh, yes, this is well enough, but we cannot control those wild boys." "The principal controls them within doors, and maintains perfect order; is it to be civilization within, and vandalism without?" "Nobody will take the trouble to protect the trees, and there is no use in planting them." "Gentlemen, I think you are quite mistaken in your views; allow me to state briefly why I urge this matter. You all admit the difference between a country where there is no neatness or planting about the dwellings, and one where comfort and taste prevail. Any one would select the latter as a place for living, even at a considerable higher price. We wish to impress this taste on our young people, and the earlier we begin, the deeper and more permanent will be the impression. Look across the street yonder at Frank Gardener's neat house, half hid in trees and shrubbery, and with plenty of bearing fruit trees in the rear, and tell me if you do not think this a more desirable place to live than Sam Slipslop's below, where not a tree is to be seen, but old boards and barrels supply their place? Frank's boys spend their spare time at home, in study, or in brushing up their pleasant home. Sam's boys are idling in the streets, or hanging around grog-shops. Which do you prefer?" "Oh, that is all well enough," said the trustees, "but it is just as they happen to take a notion—schooling has nothing to do with it." "I think you are quite mistaken—there is nothing like early impressions. See it in Frank's and Sam's boys; the character of both distinctly marked by the impressions to which they were respectfully subjected. It would be so at the school.

"If you show that you are interested in ornamental planting, the boys will soon catch the spirit. It is because you evince no interest in it, that they care little about it. I can tell you how to keep the grounds in handsome condition, and the trees uninjured in the least degree. Form a TREE PLANTING ASSOCIATION among the scholars, or if you prefer, call it a Horticultural Society. Let it be understood that certain gentlemen of admitted taste are going to visit the grounds occasionally, and that if successful, these efforts will be noticed in the village paper. They would be stimulated to excel; the officers would prize their honors; and all would be interested when they knew that the PUPILS were to have the credit of the beautiful grounds at ——— Union School. We should not only gain the advantage of a place of learning creditable to our village, but we should confer a lasting benefit on our young people, by inspiring them with a taste for rural improvement. Would not this be as useful to them as to spend years in learning Latin and Logic?"

The culture of flowers, the planting of ornamental trees, brief lessons on the requisites for successful growth, might with propriety be introduced as a recreative study into every school where the teachers have the taste, knowledge and ability to conduct it. It would civilize, humanize and smoothen—lead to useful and exalted pursuits, when without it, the tendency would be towards idleness and barbarism.

The stream cannot rise higher than the fountain. This axiom will apply in all ordinary cases to the influence of the older on the younger. Every man therefore who has children, every school teacher, and every one who feels any interest in the great nation that is now growing up from childhood, should impress these matters upon his own mind, in order that his influence may be imparted to the young.

THRESHING CLOVER.—In the absence of a clover huller, says a Maryland correspondent of the COUNTRY GENTLEMAN, my practice has been, first to mow or gather the clover heads when dead ripe, or when the heads wear a dark brown color; thresh with a threshing machine, the concave elevated or the cylinder depressed, leaving barely room for the ends of the cylinder spikes to pass clear of the concave; then attach a board in front, on the left side of the cylinder, and half the width of the cylinder. Back of the cylinder, and opposite where the clover enters, a similar board. The clover is passed through the opening in front, strikes the back board and rebounds back over the cylinder, striking the front board and passes out; thus each feed is struck or threshed twice and (if it has undergone wet and dry curing, threshed when dry and during frosty weather) thoroughly. When winnowing, if the screen is too coarse, cover it with wrapping paper, secured to the sides of the screen with tacks. In the fan shoe attach an oats and a four or six mesh riddle. The seed will pass down, the heads among the tailings, and loose chaff fly off. If not satisfactorily threshed, pass the heads through the thresher a second time.

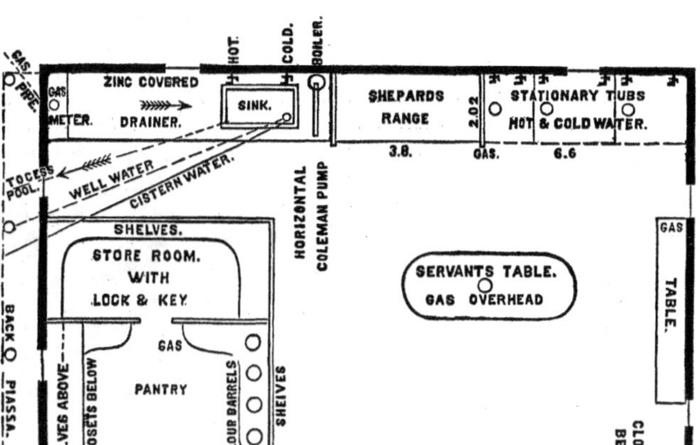

PLAN OF A KITCHEN.

J. L. DOUGLASS of Belleville, N. J., has furnished us the above plan of his kitchen, at his country residence at that place. The plan gives its own explanation.

KITCHEN TABLE AND APPENDAGES.

WE GIVE ENGRAVINGS of a useful kitchen table, with a set of shelves over it—an improvement on that described in Miss Beecher's book. Fig. 73 is a perspective view of the table, which may be about six feet long, the length being varied to suit the wants of the owner. For a small kitchen or small family it may be 4 or $4\frac{1}{2}$ feet long, with two drawers; but usually it will be found none too large if it rather exceeds six feet. It may be

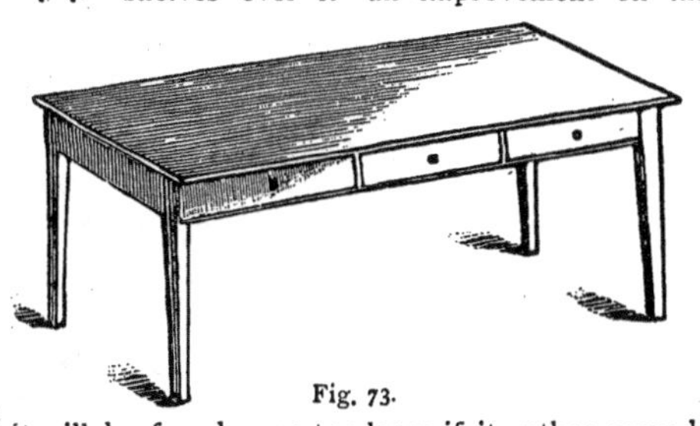

Fig. 73.

about $2\frac{1}{4}$ feet wide, and $2\frac{1}{2}$ feet high. It stands against the wall. If two of the drawers are partitioned off, as shown in fig. 74, so as to have three or more compartments, the contents may be kept in neater order, and be more easily accessible. Two of the drawers may be thus divided, and the third remain undivided, for larger articles. In order that these drawers may be sufficiently spacious, and the table not extend too far into the room, the top should not project more than two inches over the frame or drawers.

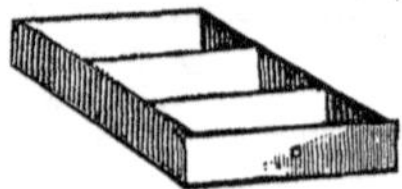

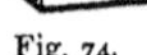

Fig. 74.

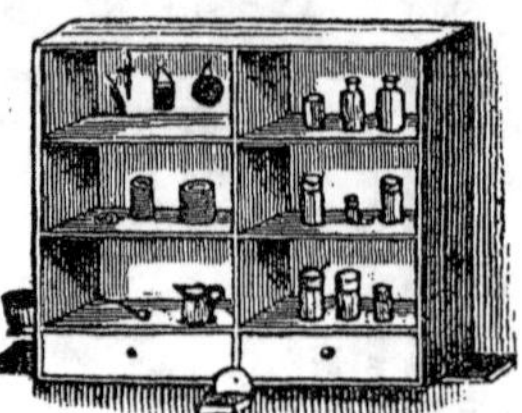

Fig. 75.

Over the table, and fastened to the wall, are the shelves, shown in fig. 75. These shelves, taken together, may be about three feet high and four feet long, varying with the wants of the owner. They may be 10 or 12 inches deep. The two drawers beneath will be found convenient for smaller articles, napkins, &c. The shelves may hold jars, dishes, bottles, &c. On one end is a coffee mill, and on the other a soap-dish, with a salt box at the middle.

Miss Beecher mentions the following articles, which may be placed in the table drawers. In the undivided one—rolling-pin, griddle, spad, coffee stick, meat fork, gridiron scraper, saw knife, skewers, apple corer, meat hammer, whetstone, &c. In the middle or divided drawer place kitchen knives and forks, and iron and other spoons, for the front division. Middle division—kitchen table cloths. Back division—bags of all kinds, strainers, pudding bags, &c. In the third drawer, clean dish-cloths and towels in one division, and clean lamp towls, holders and dust cloths in the others. Remember to have all these kept in their places, and insist that the cook shall not change them. In this way all may be quickly found, and there will be no confusion.

On the shelves above the table, one division is occupied with tin boxes with close fitting caps or covers, varying from 8 inches high and 3 inches in diameter, to 4 inches high and 2 inches in diameter. The larger boxes hold sugar, starch, &c.; the medium, tea, coffee, salt, ginger, &c., and the smaller, spices, mustard, &c. Junk bottles keep vinegar and molasses. Wide mouthed jars, soda and saleratus. All these to be largely and distinctly labelled, and each kept invariably in its place. On another shelf may be graters, dredging box, pepper box, sieves, bottle brush, quart, pint and gill measures, scales and weights, corkscrew, &c. On another, teacups and saucers, bowls, pitchers, and funnels. A full assortment of all these vessels and tools, always within reach of the hand, and everything always in its place, will save a vast amount of labor, and innumerable steps, which must result from an indiscriminate scattering of things around the kitchen.

PLOWING WITH THREE HORSES.

THE PRACTICE of using three horses for plowing, possesses such advantages that it is rapidly extending among farmers, and we have many inquiries relative to the best mode of attaching them. Two horses alone are hardly strong enough for such deep and thorough work as the best farming commonly requires; and a single plowman can cut a wider and deeper furrow with three horses, and consequently do more work in a day. When four are employed, an additional hand for driving is commonly necessary; and another disadvantage is, that the two forward horses, being at a distance from the plow, draw on a nearly horizontal line, and with much of the waste of power resulting from a line of draught in so unfavorable a direction. A brief explanation of the principle on which a horizontal or disadvantageous line of draft operates, on the one hand, and an inclined or rising line, on the other, may not be out of place here: If there were no friction, the draught traces might be on a level; but as there is always some friction, the draught line should rise at an inclination, thus tending to lessen the pressure of friction between the plow and the soil. This upward inclination should always be increased, so far as may be practicable, as the friction or resistance increases. Hence the great reason that short traces result in a great saving of strength. An experiment was tried for the purpose of testing the correctness of this theory; first with traces of such length that the horses' shoulders were about ten feet from the point of the plow; and again with the distance increased to fifteen feet. The short traces required a force measured by the dynamometer equal to 225 pounds; but with the long traces it amounted to 350 pounds, or 125 pounds more. The draught traces may, however, be made too short for the size of the animals. In this case the plow will be thrown too much upon its point in the effort to keep it in the ground. To prevent its flying out, the plowman is compelled to press down constantly upon the handles, thus increasing the friction which it is desired to avoid. Let the line of draught be so adjusted that the plow may pass equally all along upon its sole or bottom causing it to run with an even, steady motion. The traces should therefore be of just such a length that the share of the plow, (or more properly, the center of resistance,) the clevis, and the point of draught at the horses' shoulders, (or the ring of the ox-yoke,) shall all form a straight line. In the accompanying figure (fig. 76) A represents the place or point of the forward end of the traces at the horses' shoulders, (or the ox-yoke ring,) and is in right position.

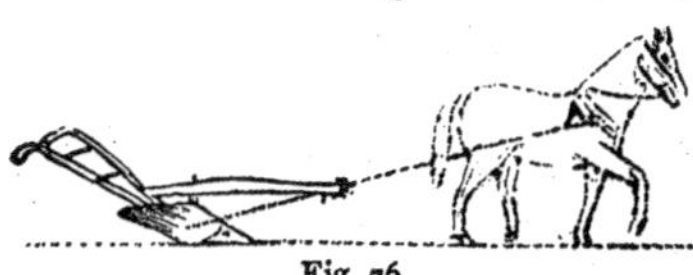

Fig. 76.

Now, in using four horses, it is impossible to give the whole line of draught this continuous ascent in a right line, but it will be broken at

A, (fig. 77,) and of course the force of the forward horses tends to draw the rear ones down towards the ground, on their fore-legs, thus causing not only unnecessary fatigue, but occasioning a disadvantageous line of draught.

Fig. 77.—*Plowing with Four Horses.*

All these objections are obviated by using the three-horse team, which works so well that three horses attached to the plow are preferred to a four-horse team, without taking into the account the additional cost of the horse and of the driver.

The mode of attaching the three horses to the plow is not universally understood, and we have many inquiries in relation to the subject. The simplest and most common form of the whiffletrees is shown in fig. 78, where the two horses are hitched to the shorter end of the long evener or main bar, and the one horse to the longer end, so that all three have an equal share of draught, 1 representing the tree for the single horse, and 2 and 3 those for the two. (The single horse should be to the left.) They are all to be made as short as practicable, and of such a length that the centre of the middle one may be exactly in front of the clevis on the plow-beam, in order that there may be an even draught. For the purpose, also, of having all work evenly side by side, the chain attaching the tree 1 to the main evener, should be long enough to reach forward to a straight line with the others.

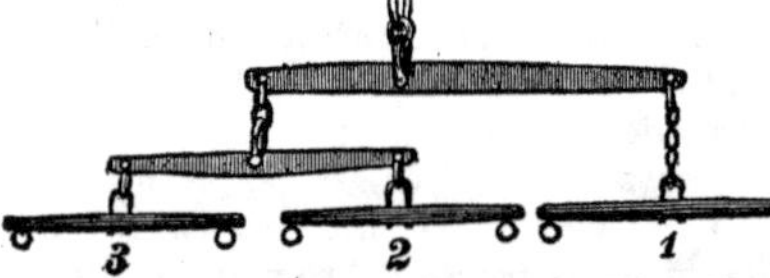

Fig. 78.—*Whiffletrees and Eveners for Three Horses.*

There are several other contrivances for equalizing the draught of three horses, possessing greater or less merit, but being more or less complex, require a full trial to test their comparative value. Among these is Potter's Three-Horse Clevis, represented by fig. 79. It consists of two wheels in one, the larger circumference being twice the diameter of the smaller, and each having a groove in which a chain, fastened to the wheel, runs. The single horse draws on the larger

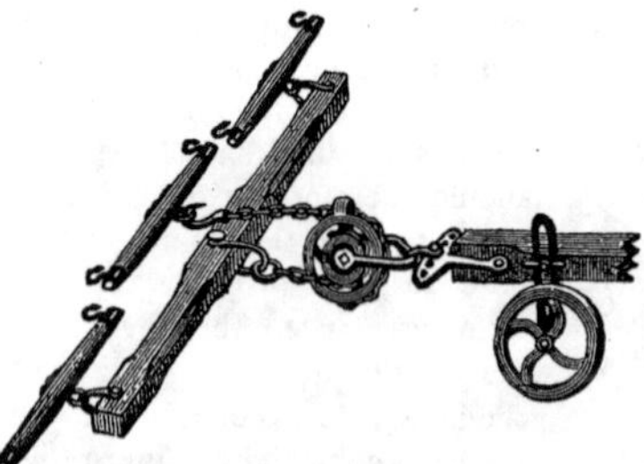

Fig. 79.—*Potter's Three-Horse Clevis.*

wheel, against the two horses with the shorter purchase on the smaller. A special advantage of this contrivance is that the draught of the horses is not varied if they do not draw evenly. A simpler, but less perfect, contrivance substitutes a short lever, placed in a vertical position, for the wheel, the single horse being attached to the longer end of the lever, and the two to the shorter.

Another mode of constructing eveners for three horses, which has been used to some extent in Western New-York, apparently with good success, is shown in fig. 80. The double-tree is similar to those commonly employed for three horses, but rather longer; the single-trees are about two feet on the outer arm and one foot on the inner. Small, semi-circular single-trees, made of iron, are attached to the inner ends of the common single-trees, and the traces as shown in the cut. It will be seen that each horse draws half of each curved single tree; but the common single-tree being half as long at the inner ends, the horses all have the same draught, and an equal amount on each trace. The curved single-trees may be about a foot and a half long before bending, half an inch thick, an inch wide at the middle and five-eighths wide towards the ends; they are attached to the ends of the wood single-trees by a bolt, so as to play freely. Care must be taken to adjust the traces of proper length for all the horses to work evenly.

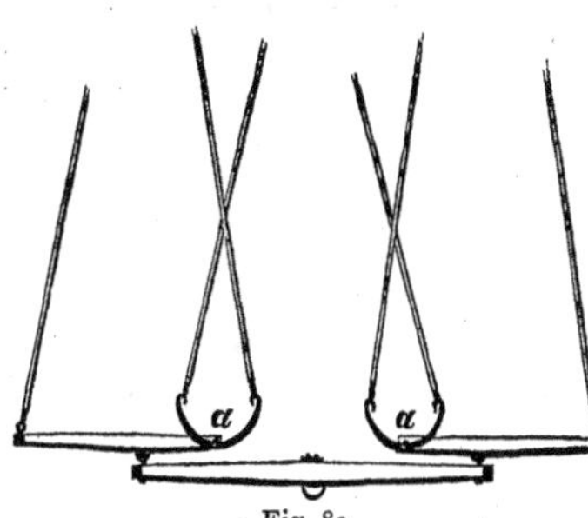

Fig. 80.

Different modes are adopted for attaching the lines to the three horses, so that all may be guided and controlled equally. One is shown in fig. 81, exhibiting only the right-hand line. The main or long rein is fastened to the bit of the right horse. A branch from this connects with the right-hand bit of the middle horse. About 2 feet farther back another branch rein is attached, connecting with the right-hand bit of the

Fig. 81.

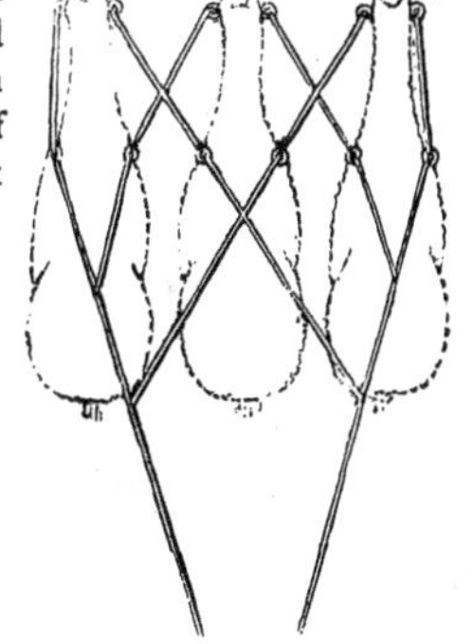

Fig. 82.—*Reins for three Horses.*

left horse. The left-hand main line is precisely similar to this, only reversed in position. The view of the whole of these reins is shown in fig. 82.

Another mode is represented by fig. 83, the long lines running straight to the middle horse 2, the side reins being branches of these on each side, connected at *c d*, and terminating at the outer horses' bits, at 1 and 3, *a b* being the bit of each horse. This is more compact than the one last described, but in operation they are not materially different from each other.

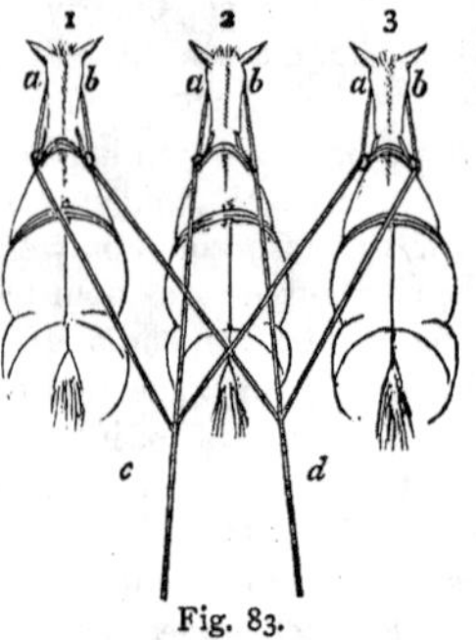

Fig. 83.

There are several other modifications, which may be adopted, according to the circumstances of owners and their several conveniences. A simple mode, for temporary arrangement, is to attach common double or branched lines, one to the left of the left horse and to the left of the middle horse ; and the other to the right of the right horse and to the right of the middle horse ; connect the heads of the three by short connecting straps, about a foot and a half long, and it is done. This, however, is not to be commended, as it is bad to tie to the bridles, which jerk and see-saw the horses' heads. An improvement consists in attaching the short straps to the hame rings of the middle horse, fig. 84 ; and then, placing the slowest horse in the middle, these straps will keep the others from out-drawing him.

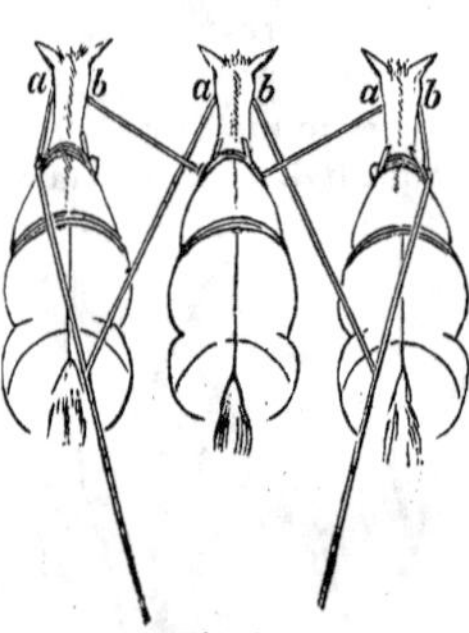

Fig. 84.

Still another mode, which may answer well in certain circumstances, is to put the common double lines in the usual way on the two right-hand horses, and secure the third or left-hand one by tying his bridle halter to a point a little behind the left hame ring of the middle horse, or to his large trace buckle, the length of the halter being so adjusted as to keep them all even in walking.

Fig. 85. *Plowing with two horses, the Plow and Furrow between them.*

The proper adjustment of the plow is a matter of vital importance. The right-hand horse walking in the furrow, and the two others on the left, on the unplowed land, a different position for the line of draught is required. In plowing with two horses, the plow follows between them, fig. 85 ; but with three it must not follow the middle horse or centre of the team, but the space between the middle and right horse, fig. 86. This position of the plow is effected by different contrivances. One is have a movable beam, the rear end of which works to the right or left on an iron arc, so as to throw the forward end to the right or left, as may be desired. When two horses are used with such a plow, the beam stands in the usual position; but when three horses are attached, the rear is moved to the right,throwing the forward end to the left, or several inches to the eft of the furrow, thus giving a side direction to the plow point towards the right. Another mode is to bolt a wooden block on the left side of the forward end of the plow beam, on which to place the draught clevis, as shown in fig. 86. An excellent and much better contrivance than either is Holbrook's patent plow clevis for three horses,* shown in fig. 87, which is arranged with a head-piece, A, and side rod, B, for draught. The double series of holes through the head-piece allows a ready adjustment with great accuracy. The wheel D may be used at the side for deep plowing, or under the beam for shallow plowing. A shorter head-piece is used for plowing with two horses abreast, without the side rod

Fig. 86—*Plowing with three horses, the right-hand horse in the Furrow, the Plow between the two right horses.*

Fig. 87.—*Holbrook's Three-Horse Clevis.*

Three Horses on a Wagon or Carriage.—The same advantages, in a greater or less degree, which result from the use of three horses to a plow, are obtained from their employment in drawing vehicles, the comparative advantages being greater as the roads are worse or more muddy, requiring a more upward draught. When thus attached, two large thills may be employed, between which the middle horse walks. Two neck-yokes are used, the middle horse being hitched to the inner ends of both, and having twice the length from the bearing that the outside horses have. The whiffletrees are arranged in the same manner, two eveners or double-trees, (fig. 88,) being used, and the single-tree of the middle horse attached to both the

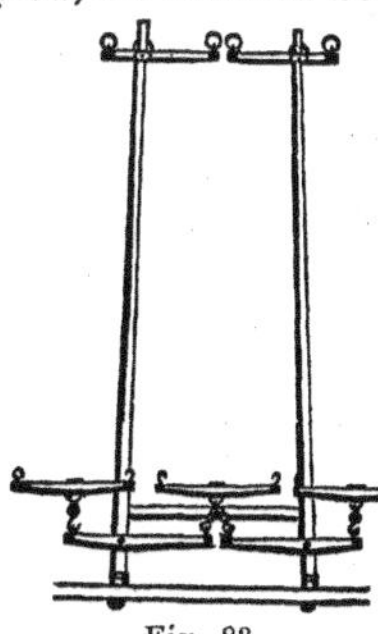

Fig. 88.

Fig. 89.

* Made by F. F. Holbrook & Co., Boston, Mass.

inner ends, which of course are twice the length of the outer ends, to which the other horses are hitched. The mode of coupling the neck-yokes is shown in fig. 89, which is much better than a single long yoke. The flexible joint between the two, effected by rings and coupling, is a great improvement.

STANCHIONS FOR CATTLE.

A CORRESPONDENT sends us a sketch and description of a mode of constructing stanchions, which resembles the one figured and described in RURAL AFFAIRS, vol. IV, page 76. The movable pieces are secured by an iron loop, *a*, (fig. 90,) resembling a clevis, thoroughly securing the animal, and making the stanchions safe and strong. The upper and lower horizontal timbers are 6 by 4 inches—both morticed for the reception of the upright pieces. The tops of the movable pieces are cut slanting, so that when pushed into position, the iron loops slip on of their own accord. They are lifted with the hand when the cattle are to be liberated. The loops are made of quarter inch round iron. The other dimensions are given in vol. IV, as already referred to.

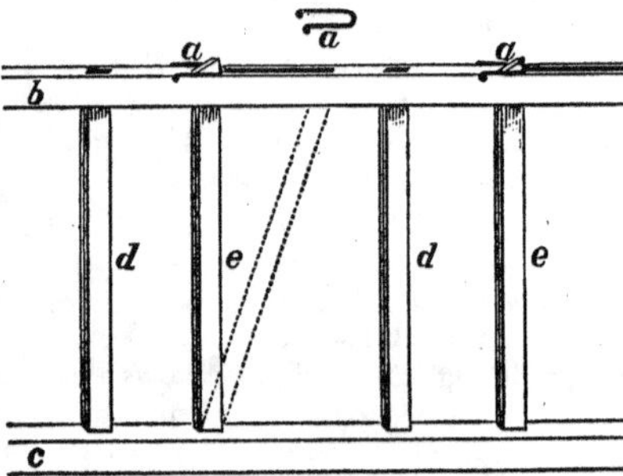

Fig. 90.

C. G. Taylor of Illinois gives the following minute details for making and managing stanchions, which may be specially useful to those having no experience with them: My bill of lumber is as follows: For the bottom part, two pieces, 2 by 8 inches, the length to be governed by the number of cows, allowing 3 feet 4 to 6 inches for each space. This is room enough for common sized cows, and affords ample room for the milker by having the cow stand a little one side while being milked. The boards for main upright, 1½ by 12 inches, and 6 feet long. The movable upright, 1½ by 6 inches, 5 feet 9 inches long. The board to fill up space and to keep the fodder from falling out of manger, 1 by 6 inches, 5 feet 6 inches long. The two top strips are 2 by 6 inches, being the same length as the two bottom pieces.

To put them together I lay flat one of the bottom pieces and one of the top, so as to be 6 feet high when raised to place, and divide in space so as to have 6 inches space at the end near the wall. Then the 12-inch wide board, leaving a space of 6 or 7 inches, according to size of cow's neck. Then the movable upright board, 1½ inches back, (or short.) The inch board, 5 feet 6 inches long, I do not use until the stanchions are raised to place. Upon the ends of these boards, and even with the ends of the 6 feet boards, I place the other bottom and top pieces. Pin, or use 6 inch bolts, which are the best. The inch and a half short at the lower end of the movable upright piece is to allow it to play on the pins or pivot.

The one-half inch at top is left for the fastener or clapper to drop into when closed. A pin or bolt is placed in front of a movable upright, to prevent it from falling forward over the cow's head. The fastener is about 15 inches long, and a little thinner than the place it is to occupy, so as to play easily. It never breaks, is easily handled, and always ready for use. After raising and fastening the stanchions, the edges of the boards each side of cows' necks are rounded off a little, and the inch board put in place, as represented in the accompanying cut, (fig. 91.)

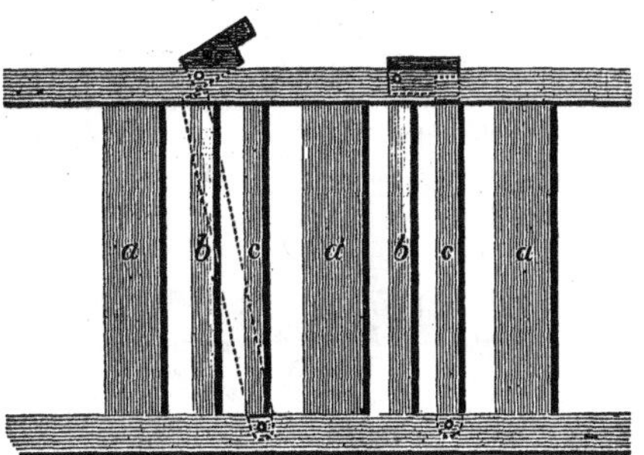

Fig. 91.—*a a a, 12 inches wide; b b c c, 6 inches wide; space between a and b, and b and c, 5 inches wide; space between c and a, for cows' necks, 7 in. wide.*

When any one wants to shut up young cattle—and they should always be shut up at night—prepare stanchions as above and nail on a thin, narrow strip on one or both sides of the neck space. As their necks grow, widen the space. I use pine lumber.

The distance from stanchions to partition should be 2½ feet for feed manger. If farther than that, the cows cannot reach the meal or bran when fed. Between the cows' heads on the floor is a board 6 or 8 inches high, reaching across the feed manger, making a box, so that each cow can quietly get her own mess, and not be disturbed by her neighbor. The floor slopes one-fourth of an inch per foot to the gutter, which is 2 inches deep and one foot wide, and placed 4 feet 6 inches clear from the stanchions. This gutter admits the shovel, and with the help of the wheelbarrow the manure is easily taken to the compost heap. The gutter receives all the droppings and slops of the floor, and leaves a dry place for the cows to lie down. With a little litter the bags and teats are kept clean.

With the lumber and floor ready, I can put up stanchions enough for a dozen or more cows in a day, and not work as hard as I have in making the diagram here presented. When stanchions are made in this way, no fodder can possibly be wasted, and each cow gets her own share of feed. There are no wrong places in space to thrust her head through. When the fastener or latch is up, there is but one place to admit the head, and that is about 18 inches wide at top.

SECURING CORN FODDER.

IN THE LAST number of the REGISTER we have described the different modes commonly adopted for cutting and securing corn-fodder sown thickly in drills. Having recently given a thorough trial to another mode, which for hand-cutting appears to be superior to any other, we give the following description: The corn is cut by an instrument represented in the accompanying figure, fig. 92, and the mode of using it is shown in the cut, (fig. 93.) The operator, taking the knife in his right hand, bends a mass of the standing corn with his left hand against his right leg, at the same time, with a sweep of the knife, cutting all off, accompanied with a quick step to the left. Two or three such strokes fill his left arm, the contents of which are placed in a small shock. When

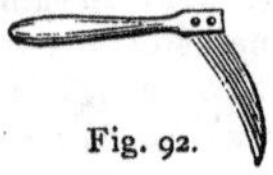

Fig. 92.

Fig. 93.—*Cutting Corn-Fodder by Hand.*

completed, the shock is firmly bound as shown in the left hand of the cut, where it will remain safely for many weeks, and become well dried. It may then be pitched on the load and drawn in, and either deposited in small stacks, as already described, or allowed to remain until needed for winter feeding, if deep snows are not likely to cover it.

The great advantage of this mode consists in the fact that the stalks have to be touched or *handled but once.* When cut with a scythe or reaper, it is necessary to gather up the stalks after they are laid on the ground. By the mode here described, they are never laid on the ground. They dry in a more perfect manner than if exposed some days on the earth. The rapidity with which an active man will thus cut and set up from half an acre to an acre in a day, seems at first almost incredible; and is only exceeded by the reaping machine and horse-rake, which do the work in a more imperfect manner.

FRUITS AND FRUIT CULTURE.

NOTES ON THE PEAR.

RESTORING MICE-GNAWED TREES.—One of the best pear orchards in the country, consisting of many thousand trees, stands on the grounds of W. R. Grinnell, on the east bank of Cayuga Lake. Five years ago last spring, seventeen hundred of the standards were set out. They had grown thriftily two years, without a vacancy or failure in the plantation, when we happened to visit Mr. G., early in the spring of 1868. He remarked: "I have met with a heavy loss; over one thousand of my standards have been killed—hopelessly killed—by the mice." "What is the amount of the loss?" we inquired. "Not less than three thousand dollars; each of those handsome young standards were worth more than three dollars." A deep snow had fallen late in March, and the whole mischief was done during the two or three days that it remained—in an almost incredible manner, no trouble of the kind having before occurred. We expressed the opinion that these trees might be all saved, and recommended the remedy figured (fig. 94) and described on page 38 of the the American Fruit Culturist. With a mixture of hope and doubt, the work was undertaken. Many of the trees had been stripped of the bark by the mice for a distance of six inches up the stem, and others nearly a foot. Each operator could finish sixty to eighty trees in a day. All were thus treated, and nearly all survived

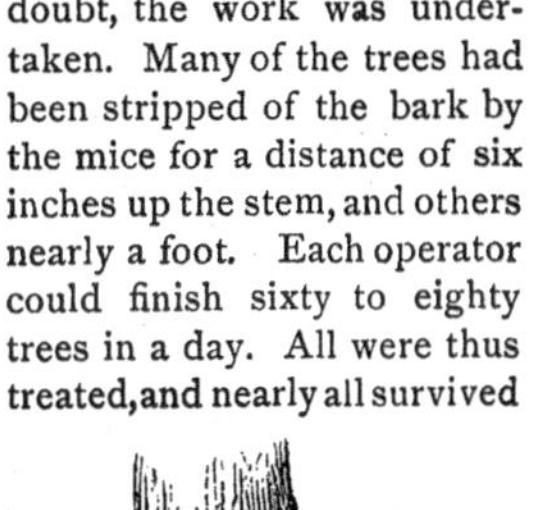

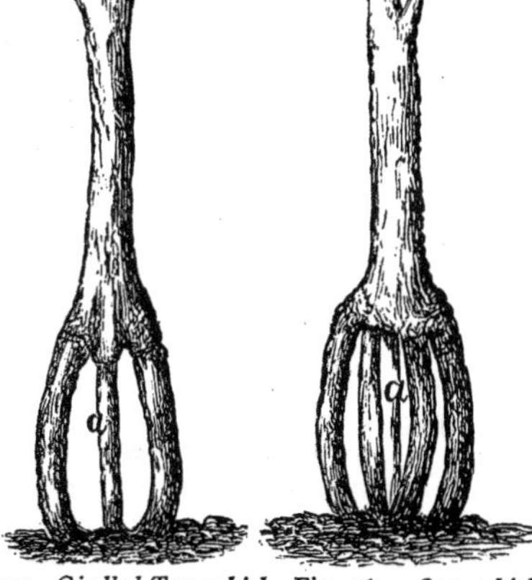

Fig. 95—*Girdled Tree which has had two connecting shoots inserted two years before.*

Fig. 96.—*One which has had four inserted; a a, denuded stems.*

Fig. 94.

and are now growing well. All, where the work was perfectly performed, lived; but through the poor work of a bungler, some sixty in all, out of about twelve hundred, have perished. The annexed figures, (figs. 95 and 96,) show nearly the present appearance of these trees. The stems are now about an inch and a half in diameter; the inserted twigs had grown to the diameter of about an inch. They were small shoots when inserted—about the fourth of an inch in diameter. The few which were half an inch, succeeded best. These shoots were in all cases cut from the tree above, and where they failed to form a connection with the upper bark, they mostly grew like a graft below, and thus new trees were formed of the same kind.

In most instances but two connecting shoots were inserted into each tree, instead of more, in order to save labor. Mr. G. remarked to us, "There is one thing in which your advice was defective. You recommended but two or three connections; there should have been four to every tree, in order to brace it firmly. But I have allowed the suckers to spring up, and have thinned all out but two on opposite sides. Next spring I shall insert them, and then I shall have four strong connections, which will make each tree firm and perfect."

In inserting these connecting shoots, the earth is drawn away below, where necessary, so as to allow the chisel to point upwards; the shoot is sharpened at both ends, bent like a bow, and the ends crowded in by bending the shoot back again. A little wax is needed at each insertion, but not on the middle portion of the connecting shoots, which should be firmly bound with bass to keep them in their places.

SUCCESSFUL PEAR ORCHARD.—M. B. Bateham gives an account of the pear orchard planted some years ago by A. Fahnestock, six miles below Toledo, on the Maumee river. The soil is a strong clay loam, well underdrained and subsoiled. Ten acres are occupied by a thousand trees, all of which, with scarcely a failure or defective tree, are of fine size and shape. Mr. Fahnestock says: "A large portion of my trees are perfect beauties of form, as well as in health and vigor. They are branched from within 2 feet of the ground, and are 10 to 12 feet in width at the base, regular cones or pyramids in shape, from 18 to 20 feet high." Of the varieties there are 100 Seckel, 100 Anjou, 200 Flemish Beauty, 200 Sheldon, 200 Bartlett, 100 Buffum, and the rest sorts in smaller quantities on trial. The Buffum is found to grow too fast, the shoots averaging 3 or 4 feet annually, the wood soft and spongy, the trees liable to blight. It is obvious they are in too rich a soil. Trees of the age mentioned should never be allowed to grow more than 2 feet yearly—less would be better. The trees of other sorts mentioned as of 12 feet spread and 20 feet high, grown in seven years, have had rather a more rapid growth than we should regard as safe, although as yet there has been but little blight among most of the varieties. Trees in this orchard which send their roots down four feet or more into the clay subsoil, are found to be more healthy than those having roots near the surface. The Bartletts outbear, four to one, any other sort.

MARKETING PEARS.—Dr. Houghton remarks in the Gardeners' Monthly, that to be successful, pears for market must be of large size, fine appearance and excellent quality—that small ones, on account of the uncertainty of

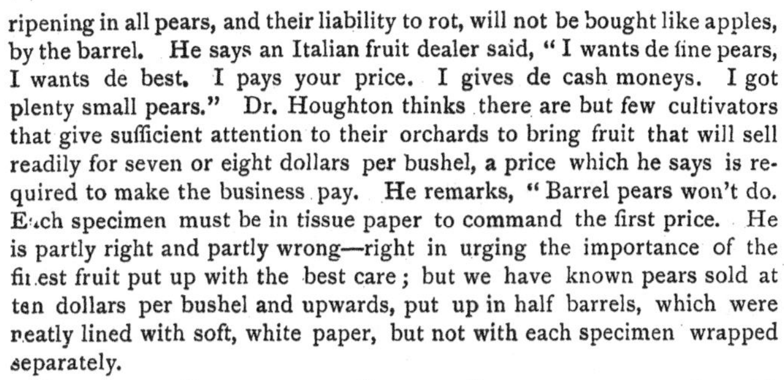

ripening in all pears, and their liability to rot, will not be bought like apples, by the barrel. He says an Italian fruit dealer said, "I wants de fine pears, I wants de best. I pays your price. I gives de cash moneys. I got plenty small pears." Dr. Houghton thinks there are but few cultivators that give sufficient attention to their orchards to bring fruit that will sell readily for seven or eight dollars per bushel, a price which he says is required to make the business pay. He remarks, "Barrel pears won't do. Each specimen must be in tissue paper to command the first price. He is partly right and partly wrong—right in urging the importance of the finest fruit put up with the best care; but we have known pears sold at ten dollars per bushel and upwards, put up in half barrels, which were neatly lined with soft, white paper, but not with each specimen wrapped separately.

CONTAGIOUS CHARACTER OF BLIGHT.—In a recent conversation with an experienced and successful grower of the pear, he stated that he had never failed to arrest the progress of the fire blight by prompt and thorough excision—taking care to cut low enough to be fully below all diseased streaks. On one occasion, finding that he was losing a new and valuable sort, he concluded to bud other trees from it before destroying it, selecting shoots for the buds that appeared to be unaffected. But he found they had already received the poison and conveyed it to the stocks. Every tree, without exception, which had buds inserted from this diseased tree, took the contagion, and either died down to the roots or was badly affected. He always made it a practice to wash his knife thoroughly after cutting a diseased tree, before using it on other trees. The failure in excision may sometimes arise from the use of a poisoned knife on the lower and healthy portions. Caution should always be used to avoid thus inoculating healthy trees or healthy portions with the poison.

PEARS FOR MARKET.—The editor of the Horticulturist, who gives much attention to the market profits of fruit, advises to plant no more Bartlett pear trees south of New-York, but to set out freely Beurre d'Anjou, Beurre Bosc, Beurre Clairgeau and Lawrence—the first and last especially. They ripen when there is a healthy demand for pears. The Bartlett, quite at the north, ripens late enough to escape the great throng of early autumn fruits. In another place the same journal states that at Mr. Quinn's, at Newark, N. J., who has one of the finest pear orchards in the country, bearing profuse crops of excellent fruit on dwarf and standard trees, his great success is attributable to three principal points: 1. Constant cultivation—no grass, no weeds, no crops between the rows. 2. Yearly pruning, giving handsome, symmetrical trees, and healthy shoots. 3. Especial pains in selecting and packing—which gives him $3 to $5 per barrel more than other pears as good, but carelessly put up.

DWARFS CHANGED TO STANDARDS.—The objection which formerly existed to changing dwarf pear trees to standards, by banking up the earth was the fact that but few roots were emitted from the base of the pear stem, just above the place of union, and these, not forming an even support to the tree, were apt to render it inclined or lop-sided. Dr. Hull of Alton strongly recommends, in the Prairie Farmer, the practice of *lipping* a small portion of the bark and wood where the roots are wanted, causing their free emission around the stem, and obviating the difficulty which we have referred to. He thus obtains the early bearing quality of dwarfs, and renders them permanent and long lived by conversion to standards.

Vicar of Winkfield Pear.—The Fruit Committee of the Horticultural Society held at Dayton, Ohio, highly commended the specimens of this pear exhibited. They were stated to be "in excellent condition, and the flesh firm and luscious." The care and skill in keeping them is mentioned, and they were doubtless allowed room on the tree for the full growth and perfection of each specimen. When crowded, neither size nor perfection can be developed.

Grafting Pear on Apple.—It occasionally succeeds with some varieties. We have seen the Seckel doubled in size by working on the apple, at the same time that its quality was lessened. But the union is imperfect, and the graft generally breaks off in a few years. Some varieties do tolerably well for a time, but we cannot recommend the practice, except to such as merely wish to amuse themselves with unsuccessful experiments.

Pears for Western Michigan.—At the late Convention of Fruit Growers for Western Michigan, the following pears were recommended as best adapted to that region: Bartlett, Bloodgood, Seckel, Flemish Beauty, Vicar of Winkfield, Sheldon, Howell, Lawrence, Clapp's Favorite.

Apples, Peaches and Plums.

Cultivation of Orchards.—An inquirer asks: "What would you advise me to do with my young orchard for the first summer, in the way of cultivation?" Answer—keep the surface frequently and constantly stirred, clean and mellow. If a crop on the ground is of secondary importance, keep the whole surface bare—or leave wide bare strips where the rows of trees stand. In a small archard, or in a new fruit garden, this mellowing of the earth may be kept up by means of a one horse cultivator; in a larger one, a two horse cultivator, Shares' harrow, common harrow or Smoothing harrow may be employed. Where crops are planted, let them be such as require frequent hoeing or cultivating; but never sowed grain. Such low crops as beets, carrots, beans or potatoes are generally preferred; but it is most important that the ground should be frequently stirred. For this reason corn, although growing tall and shading the trees, is much better, if hoed several times, than beans with only a single dressing. But all crops are more or less like weeds, and a clean, bare surface is best.

Care of Young Trees.—Newly set fruit trees, even when they have been carefully dug up and as carefully set out again, often suffer much from subsequent neglect. A little additional labor, not costing a tenth of the expense and work of procuring and transplanting them, will do much toward their subsequent success. Trees set out in the autumn need particular attention the next spring. The soil has become settled and hard about them, and as soon as dry weather comes, a hard crust will form unless the surface is kept loosened and mellowed. Keep the crust constantly broken; let the soil be entirely free from all weeds and grass, and perfectly mellow throughout the season, and the trees will not only be more likely to live, but they will grow with far greater vigor; and the nurserymen will not be so likely to be blamed for sending "bad trees," when the only fault was the neglect of the planter.

PRUNING.—Young trees should never be pruned in spring after the buds begin to open. Nothing checks their growth more than pruning too late. If the proper heading-back has not been done before growth commenced, do not do it afterward. Much of the objection to shortening back the shoots of young, newly transplanted trees, is owing to too late a performance of the work. But if done in good time, it is eminently useful.

LIST OF ONE HUNDRED APPLE TREES.—The following list was made out for a planter in Western New-York, for family supply, as well as for market. Different cultivators will vary this list, but all will approach it more or less:

FOR SUMMER.

3 Early Harvest,
2 Early Astrachan,
1 Summer Rose,
1 Early Joe,
1 Early Strawberry,
1 American Summer Pearmain,
1 Benoni,
2 Primate,
2 Sweet Bough,
1 Golden Sweet.
—
15 Summer.

AUTUMN.

3 Autumn Strawberry,
2 Duchess of Oldenburgh,
1 Porter,
1 Gravenstein,
1 Dyer,
1 Maiden's Blush,
2 Fall Orange,
3 Twenty Ounce,
2 Fameuse,
2 Munson Sweet,
1 Haskell Sweet,
1 Bailey Sweet.
—
20 Autumn.

WINTER.

10 Baldwin,
10 Rhode Island Greening.
5 Roxbury Russet,
3 Golden Russet,
3 Tompkins County King,
3 Fall Pippin,
2 Swaar,
2 Peck's Pleasant,
2 Westfield Seeknofurther,
2 Yellow Bellflower,
2 Wagener,
6 Northern Spy,
5 Hubbardston Nonsuch,
10 Tallman Sweet.
—
65 Winter.

FRUIT CULTURE—OLD ERRORS CORRECTED.—1. Instead of "trimming up" trees, according to the old fashion, to make them long-legged and long-armed, trim them *down*, so as to make them even, snug and symmetrical. 2. Instead of manuring heavily in a small circle at the foot of the tree, spread manure, if needed at all, broadcast over the whole surface. 3. Instead of spading a small circle about the stem, cultivate the whole surface broadcast. 4. Prefer a well pulverized, clean surface in an orchard, with a moderately rich soil, to heavy manuring, and a surface covered with a hard crust and weeds or grass. 5. Remember that it is better to set out ten trees with all the necessary care to make them live and flourish, than to set out a hundred trees and have them all die from carelessness. 6. Remember that tobacco is a poison, and will kill insects rapidly if properly applied to them, and is one of the best drugs for freeing fruit trees rapidly of small vermin—and is better used in this way than to make men repulsive and diseased.

HARDY APPLES.—A Wisconsin correspondent of the Gardeners' Monthly says the Fall Orange, Sops of Wine, Red Astrachan, Oldenburg, Fameuse and Autumn Strawberry, do well in Wisconsin, both as to tree and fruit; and that the Fall Pippin, tender when young, succeeds well after the tree attains age. The Yellow Bellflower is the most popular cooking apple in Central Wisconsin. Northern Spy does well in some places.

APPLES FOR MINNESOTA.—The Western Farmer, through a correspondent, gives the following list of sorts for an apple orchard of 500 trees, for that cold region for fruits: 125 Tetofski, 75 Duchess of Oldenburgh, 50 Haas, 50 Saxton, 25 of the new Minnesota crabs, 10 Red Astrachan, 10 Fameuse, 3 Tallman Sweet, 2 Fall Orange, 50 Perry Russett, 100 Ben Davis.

PROFITABLE ORCHARD.—The Boston Cultivator gives an account of the orchard of Capt. Pierce of Arlington, Mass., consisting of 86 trees, 38 being of the Williams Red. These trees have averaged over $600 per annum. The orchard is cultivated in the best manner, the spaces between the trees being occupied with potatoes and squashes. He has no faith in growing trees in grass.

WATERING TREES.—The best watering you can give young trees is to promote the moisture of the soil by keeping the surface clean and mellow. Never water the *roots* after setting out, before the leaves expand. Trees are sometimes killed by overdrenching them before there is a chance for the water to be carried off by the leaves. If the bark is shrivelled, wet the *stems* frequently, or encase them slightly in straw, and wet the straw once a day. This will often restore shrivelled trees.

SHEEP IN ORCHARDS.—The Western Rural mentions two experiments of pasturing sheep in orchards with excellent success. The short grazing and the top-dressing of sheep manure increased the growth of one orchard from so feeble a state that no grafts could be cut from it, to a thrifty growth of a foot to eighteen inches in the yearly shoots. A great improvement in the fruit was reported.

SUCCESSION OF PEACHES.—Edmund Morris gives, in Tilton's Journal, the following list of peaches for market, to yield a succession for more than two months: Hale's Early, Troth's Early, Early York, Crawford's

Early, Reeve's Favorite, Oldmixon, Ward's Late, Fox's Seedling, Late Crawford, Delaware White, Freeman's White and Smock's Yellow.

PEACH BORER.—M. B. Bateham, (Secretary of the Ohio Horticultural Society,) says that he finds carbolic soap an efficient remedy for the peach borer, having used it on 3,000 trees with entire success. Five pounds of soap is dissolved in eight gallons of hot water, and then a barrel of water added—enough for a thousand trees, at a cost of half a cent a tree—applied to the stems early in July.

FELT AS MICE PROTECTOR.—There are four principal modes for protecting fruit trees from the depredations of mice under snow—namely: Clean ground; mounds of smooth earth; treading the snow hard about the tree, and hollow cylinders of tin, sheet-iron or felt. The tin cylinders are durable; the felt cheap. The dimensions of the pieces of felt must depend on the size of the tree. If the tree is two inches in diameter, a breadth of seven inches will go round it; if three inches through, the felt must be ten inches. One foot high will answer, but a foot and a half is safer. A sheet of felt will make enough for quite an orchard, and may be had at a low price of any slate roofer.

SUCCESS FROM GOOD CULTURE.—A correspondent of Colman's Rural World mentions the case of some neighbors who plant peach orchards and get about one crop, after which weeds, insects, &c., prevent their ever getting another. Another neighbor planted 125 Hale's Early peach, and in twenty-eight months shipped from them 640 boxes, of a third bushel each. The next year the amount was nearly doubled; the third year his net proceeds were nearly $1,200. Last year the frost killed his crop. Weeds and grass are never seen in his orchard.

DESTROYING THE CURCULIO.—At the winter meeting of the Western New-York Horticultural Society, J. J. Thomas, being called upon for a statement of his experiments with the curculio, said that he began to make thorough work with this insect in 1866. His plum orchard of 80 trees had previously borne but a few quarts yearly. By a thorough destruction of this insect he had a profuse crop—the number killed was over 1,600 that year. The following winter killed all the fruit buds, a circumstance never before known to the plum crop. There were consequently no plums, and no curculios visible. They appeared to have been thus much reduced in number, for the following season, 1868, only 400 were destroyed, and a heavy crop of plums, as usual, saved. In 1869 about 1,200 were killed, and in 1870 nearly 5,000, and fine crops the result every time. Perhaps the work was rather too thorough, as some of the trees overbore. The actual cost was six or seven cents per tree, counting all the labor, each year. The mode of killing was jarring down on sheets, which were stiffened with light rods, so that one operator carried them in one hand, and a heavy hammer in the other. Expedition and thorough work was greatly assisted by placing an iron spike in each tree or large limb, on which a sharp blow might be struck.

Small Fruits.

Distances of Small Fruits.—The large growing grapes, as the Isabella and Concord, should have a space of at least 12 feet in the row, and the rows should never be less than 8 feet apart. When they become old, they should have more distance; the first few years they may be nearer. Smaller sorts, as the Delaware, do not require quite so much room. Raspberries may be about 4 feet apart, or rather better, 3 by 5 feet. Blackberries, being larger, should have nearly double this space, but it they are kept well pinched in while growing, they may be brought down nearly to the space for raspberries, and bear more besides. Currants and Gooseberries will do at about the same distance as Raspberries, or a little less. Strawberries in rows 2 or 3 feet apart, and a foot or so in the row. For horse culture in large grounds the rows must be nearly 3 feet asunder; but in garden beds plant about 1½ by 2 feet. Quinces may be about 8 or 10 feet apart.

Cultivation of Strawberries.—The treatment may be varied with circumstances, provided the great leading requisite is constantly kept in view, namely, *to allow no weeds to get above the surface.* This is the great cardinal essential, which must not be departed from. After the plantation is set out in clean, well prepared soil, stir the ground often enough to destroy the sprouting weeds before they get to the light. The work may be then done with less than a tenth of the labor required after the weeds are several inches high; and all the labor of this frequent stirring is more than repaid by the increased growth and vigor given to the plants, to say nothing about the weeds. If the plantation is small, the work may be done with a garden rake; if large, with a one-horse cultivator, or perhaps better, with a fine toothed one-horse harrow. If this is attended to thoroughly through autumn, the plantation may be mulched at the begining of winter with straw. It will be better, especially for heavy soils, to remove the mulching in spring and mellow the surface one or more times before the plants blossom. This may be done by raking the mulch into every alternate row, and then, after the denuded spaces are stirred, to rake it back again and do the other rows. The mulch being replaced by flowering time, the berries will be kept clean. Some cultivators, who have small plantations, do not disturb the mulch in spring, but loosen the soil through it with a pronged hoe—but whatever course is adopted, see that the weeds do not grow.

Strawberries—Comparative Productiveness.—During a recent visit to the grounds of H. E. Hooker of Rochester, who is well known as one of the most intelligent and successful cultivators of fruit at that place, he gave us the following list of Strawberries, which he preferred for family supply: Large Early Scarlet, Wilson, Triomphe de Gand and Russell's Prolific. The Early Scarlet is valuable for its earliness, good quality and reliability. Taking the Wilson as the standard of productiveness, the Scarlet bears about one-fourth as much. Triomphe de Gand varies from one-fourth to one-half the crop of the Wilson, and the Russell, if well fertilized, about one-half, but sometimes three-fourths as much. Green Prolific, although not of very high quality, and too soft for market, is valuable for its great productiveness, being nearly or quite equal in this respect to the Wilson, and many would therefore find it valuable as a berry for family supply. Jucunda is somewhat uncertain in its crop, but comes nearly up to Triomphe de Gand in productiveness.

GOOSEBERRIES.—Tilton's Journal of Horticulture recommends Downing's, Houghton's Seedling and Mountain Seedling, among the American sorts which succeed best in this country. The Mountain Seedling makes a better bush than Houghton, but the fruit is not equal in quality. Those who would try the more uncertain English gooseberries, may select the Red Champagne, (small but high flavored,) Crown Bob, Warrington, Laurel, Green Walnut, Ironmonger, Early Sulphur and Green Gage, and plant in a deep soil, north side of a fence, and mulch several inches in summer with salt hay.

STRAIGHTENING UP BLACKBERRIES.—The Kittatinny blackberry, which has the valuable advantage over some other sorts of extreme hardiness, (obviating laying down in winter,) requires care to keep it within bounds and in proper shape. Neglected, it grows in the form shown by fig. 97, and usually more spreading than the figure. It requires, as every good cultivator knows, pinching-in during its growth in summer, to keep it snug and compact and to induce abundant bearing. But even after full pinching, the stems often lean over as shown in the figure, and should be well straightened up in spring. Those which have not been pinched, should be cut back so as to appear like fig. 98, which shows the size and form of the bush after summer pinching.

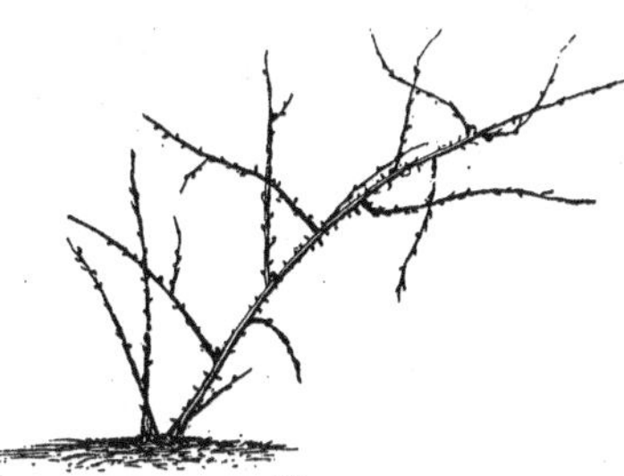
Fig. 97.

It may then be readily and firmly straightened up by taking out one spit of earth from the upper side, as represented, when a pressure of the foot on the opposite side will place it erect, as in fig. 99. The earlier in spring the better for this work, but it may be performed at any time before there is much new growth. Rows of Blackberries, instead of spreading 6 or 8 feet on

Fig. 98.

Fig. 99.

either side, catching the clothes of every one who passes, and becoming a nuisance in the opinion of all who never saw them properly kept, ought to be brought into as compact and unobtrusive a form as a currant bush.

SUCCESSION FOR RASPBERRIES.—An Illinois correspondent of the Horticulturist gives the following list of some of the newer varieties of the Raspberry, named in the order of ripening: Doolittle, Ellisdale, Mammoth Cluster, Philadelphia and Clarke. The Doolittle is represented as a moderate bearer; the Ellisdale as better and hardier than the Purple Cane; the Philadelphia productive, but of second quality; and the Clarke of good quality, but it does not set full large berries. The Ohio Ever-bearing follows the Doolittle, with a good crop, and yields a fair crop in the fall.

Dr. Hexamer stated, at the New-York Farmers' Club, that the excellence of the Mammoth Cluster consists in its holding the good size of its berries to the end; the Ellisdale he regarded as earliest, the Davison's Thornless three days later; Doolittle and Seneca ten days later still. These are among the best out of many sorts.

GRAPE SEEDLINGS.—Novices sometimes complain that the seeds of grapes do not grow when planted. It is important that they are not allowed to become much dried, and that when planted, the surface be kept moist. A correspondent of the Rural New-Yorker says that formerly not a fourth part of the seeds which he planted, germinated and grew. Accidentally covering part of the bed with leaf mould and rotten leaves, he found they grew freely and abundantly under this mulching. He then made an entire bed of rotten leaves and mould, watering the bed after the seed was planted, every other day. Nearly all grew and made vigorous plants. The varieties planted were Catawba, Isabella, Concord, Clinton, Delaware, Ives, &c.

PRUNING ISABELLA GRAPES.—The Isabella is a strong grower, and the soil should not be manured or very rich, as a rich soil promotes a more rampant and succulent growth, which is less likely to bear well, and is more liable to be winter killed. But if the soil is already rich, let the vines have a longer run, as a compensation. Old vines, which have grown thick with brush, may be treated in two ways, according to circumstances. One way is to select two of the youngest, longest and thriftiest vines, cut off all the rest as short as practicable, and stretch these two out for bearing arms; or if they are not long or strong enough for this, take four to six, and stretch them out in the form of a fan on the trellis. But if there are no good ones, cut all down so as to leave a few buds to grow, as near the ground as possible, and train the shoots or canes growing from these, either as two horizontal arms or into a fan shape. The pruning may be done in autumn or winter, or very early in spring. If much cut in autumn or winter, they are rendered tenderer, and should be therefore slightly covered for protection. To have good, early and well ripened and sweet grapes, the vines should have plenty of room, so as to form large, healthy shoots and large leaves—these will give large, fine grapes. As a general rule the shoots which bear the bunches should be a foot apart on the trellis, and if likely to be thicker, the supernumeraries should be rubbed off before the first of June, so as to let the others have a good chance to grow freely.

PROPAGATING CURRANTS.—As soon as the leaves ripen, cut off the new growth and make cuttings about 6 inches long. Set them in rows 15 inches apart and 2 inches in the rows. Just as winter sets in, cover them over with coarse litter—taking it off in spring, and keeping them well hoed, and by fall they will have large fine roots.

CURRANTS—HEAVY MANURING.—A resident in Canada says that the best currants he ever had, produced in great abundance, were obtained in a dry season, by covering the whole surface of the ground with cow manure as a mulch, 3 inches thick. On looking under, the soil was always moist. Heavy pruning has to follow the luxuriant growth thus produced.

CURRANT WORMS.—A correspondent says that he treats his currant worms occasionally in summer by sprinkling the bushes with cresylic soap suds, made quite strong and followed with a coating of freshly slacked lime. The lime alone, if applied when the foliage is wet with dew or rain, is generally effectual if thoroughly applied. He believes that gypsum or gas lime made fine, will kill the worms. The currant worm is moist and tender and soft, like the snail and pear slug, and the remedy for the latter used by nurserymen is lime well scattered over the trees.

PROPAGATING RASPBERRIES.—In answer to an inquiry relative to the increase of raspberries, the Small Fruit Recorder says that black caps should have the tips nipped by midsummer, and when these branch out and form tips that are bare of leaves from 4 to 6 inches, bury these tips in the ground at an angle of 45 degrees, and before winter they will form fine roots. This layering or burying is generally done in August and September. The sucker raspberries will furnish new plants the second year, springing up in the form of suckers, which may be taken up in autumn for setting out. For bearing *fruit* well, the suckers should be kept hoed off closely on their first appearance, treating them precisely as weeds. If the suckers grow for increase, they tend to exhaust the old plant, and it will not bear so well.

THE ROCHESTER BERRY BASKET.—A neatly constructed and handsome berry basket, manufactured by Collins & Co. of Moorestown, N.J., is represented in the accompanying cut, (fig. 100,) of three sizes—quart, pint, and half pint. It is well ventilated by openings in the sides. The raisers of fine berries are enabled to sell their fruit at a higher price, in consequence of the neater appearance presented by this basket over those of less attractive form.

Fig. 100.

COST OF MAKING HAY.

A CORRESPONDENT in Rensselaer county, who regards himself "a novice in farming" and in hay-making, wishes to know the estimated expense of curing hay and getting it into the barn, by the use of the present machinery employed for this purpose. Also an estimate of the comparative cost of making hay by machinery and wholly by hand.

Estimates of this kind can be only approximate. Much will depend on the following controlling influences: 1. The character of the meadow, its smoothness, and the weight of the crop per acre. 2. The weather—whether dry, or with frequent showers. 3. Strength of the horses employed. 4. Perfection or good order of the machinery used. 5. More than all the rest, the man who directs the movements, and his ability to keep everything in perfect order and running like clock-work. The last item alone will make a difference of at least 50 per cent.

CUTTING.—A good machine is of the first importance. This is not entirely dependent on the manufacturer, but on the owner, in keeping it in perfect order. Those who use their machines roughly, and leave them to rust in the field, will find that they cannot make a good day's work. Keep every part in good running order, and a good team may cut 12 acres a day. Many men, however, will manage to get something or other out of order, so as to go over only 6 or 7 acres. Ten ought to be a full average—requiring the horses to travel about 20 miles, with a cutting bar 4 ft. 2 in. Any of the well known mowing machines will doubtless do excellent service, as the Clipper or the Kirby machines, or the several other excellent mowers made by different manufacturers. The cost of cutting, machine, team and man, will average about 75 cents per acre—many will charge only 50 cts.—or at 2 tons per acre, 25 to 37 cts. per ton.

Fig. 101.

Fig. 102.

TEDDING.—In large meadows, it will be a matter of economy to use either the American (fig. 101,) or Bullard's (fig. 102) tedder. They work with great rapidity, three or four times as fast as the cutting; they generally prepare the hay perfectly for drawing in the day of cutting, when it is done early in the day; they make better hay, and often avoid a loss of labor and hay by eluding storms. In other words, they place the manufacture of hay more completely under the owner's control. In this way they lessen the expense by simplifying the work, and obviating cocking and opening.

RAKING.—A half grown boy will handle without difficulty any good wheel horse-rake. Taking an average of 8 feet at a passing, he will rake as many acres as the horse travels miles—say 20 or 25 in a day. A poor horse and a stupid boy will not do half this amount.

GATHERING AND DRAWING.—At the west, an implement is much used called the Hay-Gatherer. It is not unlike the old revolving horse-rake, but with much larger teeth and timber. It takes the hay from the windrow and draws it at once with great rapidity to the stack. The Hay Sweep, figured and described on p. 180 of the 3d vol. of RURAL AFFAIRS, performs a similar purpose. The former requires one man to manage it; the latter two boys. If the ground is smooth, either of these machines may be employed to draw the hay directly to the barn, where the horse fork will deposit it in the bay. For stacking in the field, either of these gatherers will work with great rapidity, the horse-fork being suspended by either of the several modes figured and described in Thomas' Farm Machinery.

Fig. 103. Fig. 104.

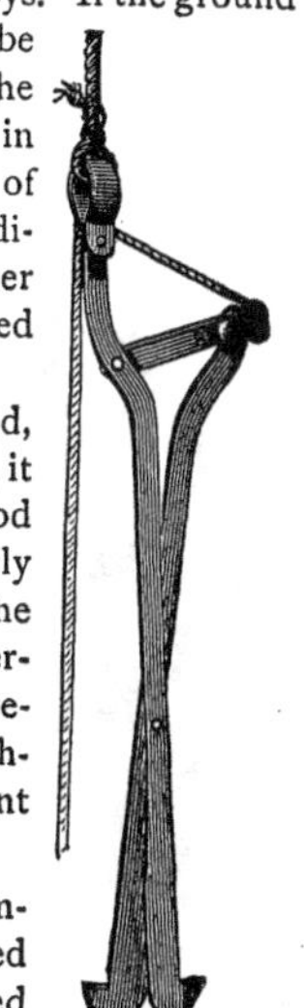

Where but one mowing machine is owned, and the amount of meadow is considerable, it may be kept running all the time in good weather. By using the tedder freely, nearly all cut in the forenoon may be got in the same day. That cut the first part of the afternoon may be put into the windrow, and the remainder cut late, left without detriment, without stirring, till next day, a small amount of dew effecting little or no injury.

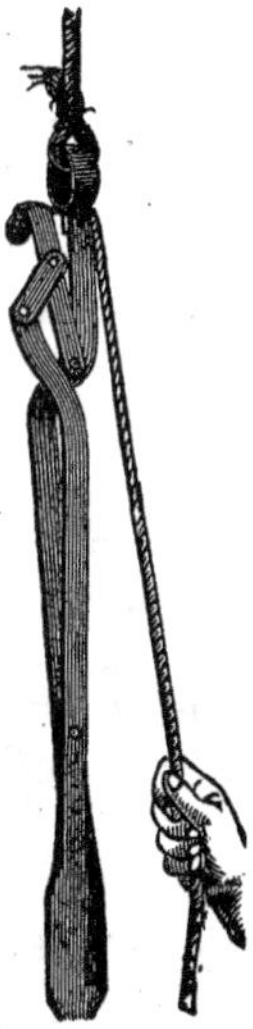

The stacking or drawing in may be commenced in the morning with the hay raked the previous day into windrows—followed by the last cutting the previous afternoon; and completed by drawing the early cut hay the same day. Six horses would be required to work to advantage, viz., two on the mowing machine,

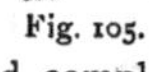
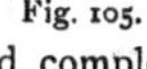

Fig. 105. Fig. 106.

one alternately on the rake or tedder, one to work the horse-fork, and two for the gatherer or sweep. The Harpoon forks work with great rapidity for pitching common timothy or other hay of a similar character. Among these, Rogers' & Nellis' fork, shown in the annexed cuts, (figs. 103 and 104,) one open for lifting hay, and the other shut for plunging in for another load, is an efficient one. Another form is Sprout's, (figs. 105 and 106,) which is both an excellent hay knife and a hay lifter, one figure representing it closed for thrusting it into the hay, and the other opened, to secure the load.

A correspondent of the COUNTRY GENTLEMAN at Muncy, Pa., sends us the accompanying representation (fig. 107) of the mode which he successfully employs in the use of the horse hay-fork, by what he calls the *double-hitch.* It avoids the heavy friction of the fork-load of hay against the beam and mow, which has induced many to throw aside horse-forks. It represents pitching into a window, by which the hay may be deposited at any point from the window to the farthest part of the barn, filling it to the very peak, and requiring but little if any mowing away. The dotted lines that appear upon the roof indicate the direction of the rope to the pulley inside.

Fig. 107.

At 1 is shown a post in the ground, with braces extending to the barn (2 and 3) to support the same. The rope passes through pulley A at the foot of the post, thence through pulley B at the top of the post, thence through pulley D that runs loose upon the rope, thence through a pulley fastened at the inside peak at a proper point near the other end of the barn, and

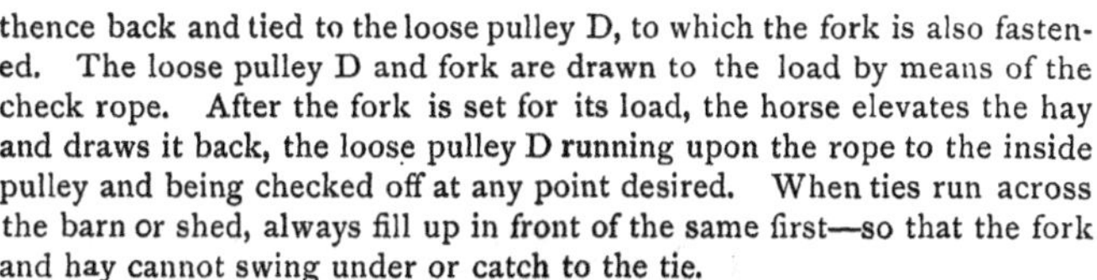

thence back and tied to the loose pulley D, to which the fork is also fastened. The loose pulley D and fork are drawn to the load by means of the check rope. After the fork is set for its load, the horse elevates the hay and draws it back, the loose pulley D running upon the rope to the inside pulley and being checked off at any point desired. When ties run across the barn or shed, always fill up in front of the same first—so that the fork and hay cannot swing under or catch to the tie.

The same correspondent gives a description of his mode of securing the pulley by means of a grapple: "There should be no swivel, but a bail sufficiently large to attach both the rope and fork, a place to oil the pin, and withal light, so as not to sag the rope when returning the fork to the load. I use a pulley as seen in the cut, (fig. 108,) which has all the above qualities, besides being durable. I also use a steel grapple, (as shown in fig. 109,) which, when attached to the rafter, places the pulley in the right position so as not to chafe the rope. I have no trouble in returning the fork to the load, as the two forks and pulley, as I use them, together only weigh 20 pounds. The cuts show the the form of the

Fig. 108.

Fig. 109.

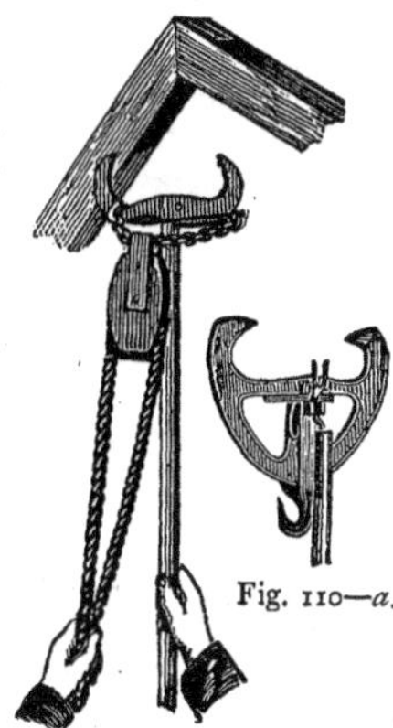

Fig. 110—*a*.

Fig. 110.

pulley and grapple which are used here—fig. 110 representing the mode of attaching the grapple to the rafter of the barn. There is no patent on either, and they can be made by any skillful mechanic."

Another form of grapple is represented in figure 110-*a*, and is made by A. J. Nellis & Co. of Pittsburgh, Pa. By means of this contrivance, pulleys may be affixed to beams or rafters in one minute, or changed again, without the use of a ladder.

The Douglas Hay Loader promises to be a valuable labor saver, and is shown in the following figure, (fig. 111.) It is attached to a common hay rigging, and the wagon is driven outside the windrow. Motion is communicated to a drum below, from pulleys attached to the hind wheels. The spurs catch the hay and carry it up in a stream between the belts and the apron on the rigging to the top, where inclined slats turn the current into the wagon, the spurs disappearing behind them. Should this machine continue to succeed after a thorough trial, it will prove especially useful on large hay farms.

Fig. 111.

By keeping everything in order, and with no interruption by rain or accidents, skillful managers have cut, made and drawn in hay at a cost of only 50 cents per ton, in extreme cases; but more commonly the cost is about 75 cents per ton. On small farms, where the same team cuts, rakes, draws in, &c., attended with frequent transfers from one kind of work to another; where a tedding machine is not employed, and where the drawing is done on a wagon, the ordinary expense will go as high as $1 or $1.25 per ton—the weather being favorable. The old mode of cutting by scythe, hand-raking, cocking and drawing in and pitching with hand-forks, would, at present prices of labor, cost about $3 a ton. This result will be varied greatly with light or heavy meadow, and by various conveniences or the reverse.

NEW OR ADDITIONAL MACHINERY.

HOADLEY'S STEAM ENGINE.—On large farms, or for itinerant threshing, the portable steam engines are rapidly finding their way into use. They have important advantages over horse-power, in the steadiness and perfect control with which they may be run; and when used for itinerant threshing machines, they obviate the necessity of the farmer to employ his horses to run the machine, and he may use them exclusively for drawing away the grain as fast as threshed. There is

Fig. 112.

no cessation of work to rest the horses, and the threshing may go on continuously without stopping as long as men can attend to it. The result is that with the same horse power a much larger quantity of grain may be separated from the straw in a day. By carelessness fires have been sometimes occasioned by these steam engines, but by the use of spark preventors, and a fair amount of care, the danger may be obviated.

Fig. 113.

Among the best engines manufactured at the present time is that made by J. C. Hoadley& Co.of Lawrence, Mass., represented by fig. 112.

GAAR'S STEAM ENGINE, made by Gaar, Scott & Co. of Richmond, Ind., is represented by fig. 113, showing the compact manner in which the smoke pipe is lowered and laid aside for travelling.

Fig. 114.

WOOD'S STEAM ENGINE, shown by fig. 114, is among the very best made in the country, the manufacturers having had long experience in the business, and being among the pioneers in the introduction of steam for farm work. The engraving shows the manner in which the smoke-pipe is folded for conveyance. It is manufactured by Wood, Taber & Morse, Eaton, N. Y.

HOLBROOK'S HORSE HOE.—This is among the best of the more recently made implements for working in corn, potatoes, and the various drilled root crops, (fig. 115.) The length of the teeth frees it from danger of clogging. It works admirably in tearing out quack or witch grass. It may be expanded to 3 feet, or contracted to 15 inches. The teeth may be set to throw the soil out, or to turn it in from the row, and the depth may be varied from 3 to 7 inches. Extra large rear plows may be inserted behind for hilling. Messrs. Holbrook & Co. write us : " The large centre or double mould-board plow of the horse-hoe, when used alone, works very well for digging potatoes. One of the best farmers in the State used it very successfully as follows : First, raked off the vines, and then ran the hoe with large double mould-board, through every other row ; then gathered up the potatoes, and went through the remaining rows in the same manner, which dug nine-tenths of the potatoes, as fast as the horse could walk. He then put on the three common or small plow teeth, and crossed the field, which threw out the remainder of the potatoes, and left the field in good condition for seeding again."

Fig. 115.

PERRY'S SCARIFIER, (fig. 116,) made by F. L. Perry, Canandaigua, N. Y. This we have found a very efficient implement for loosening the soil to a considerable depth between rows. Its sharp, chisel-pointed teeth are sc curved that they penetrate the soil in a nearly horizontal direction at first, giving the cultivator an easy draught for the depth of its work. It possesses great efficiency in adhesive loams. One horse drew it with ease in penetrating to a depth of eight or nine inches. Its depth of running may be regulated from one to ten inches.

Fig. 116.

SHARES' HARROW, (fig. 117.)—This valuable implement has been in use several years, but is still unknown to many good farmers. The forward part of the tooth cuts and slices the earth, and the rear portion turns it over. It works admirably on inverted sod, which it pulverizes to a depth twice as great as the common harrow ; while the inclination of the teeth like the form of a sled runner, renders impossible the tearing

Fig. 117.

up of the grassy portion of the sod. When the teeth are cast-iron they soon wear and become dull; but steel teeth are permanently efficient. Its great value is in preparing inverted sod for planting corn and other crops.

TURF PARING PLOW, (fig. 118.)—A. B. Allen gives the following account of this implement: The share is thin and flat, made of wrought iron, steel edged. It has a lock coulter in the centre, and a short coulter on the outside edge of each wing of the share; cutting the turf as it moves along into two strips about a foot wide, and as deep as may be required. This depth is regulated by the wooden slide or shoe under the beam, which is better for this purpose than a wheel. After the turf is pared off in strips, it is cut into any required length for sodding. I have found those from 15 to 18 inches long the most convenient to handle. This plow is much used in Great Britain for what is called paring and burning. There the sod, after being pared, is cut into pieces and thrown into heaps, which, after drying, are burned, and the ashes spread broadcast on the land. These ashes prove an excellent fertilizer, and thousands of acres of a stiff clay soil have been rendered much richer, more friable, and more easily cultivated by this simple process of paring and burning. The price of the paring plow here is $25. It is strong, effective, and does its work rapidly. It is as easily handled as the common turning plow. It is made by R. H. Allen & Co., New-York.

Fig. 118.

BICKFORD & HUFFMAN'S DRILL, (fig. 119.)—This machine has been greatly improved of late years, and the dropping is done by means of a cast-iron contrivance, which will not wear out in a lifetime; the seeds are never crushed; and the smaller seeds, as wheat and barley, may be sown, or larger, as peas and corn. The rapidity of discharge is controlled by wheel-work. The ends of the tubes are shod with steel, and they may be made to plant at any desired depth. It has an attachment for sowing guano or plaster, and another for grass seed.

Fig. 119.

EMERY'S CORN PLANTER.—One of the best corn-planters drawn by a horse is the one represented by fig. 120, and known as Emery's. Seeds which fall by their own weight are dropped by a wood cylinder, having adjustable cups to measure or count the seeds, while light seeds, as beets, carrots, broom-corn, parsneps, &c., are forced out with regularity by a revolving brush. Although in use many years, it is still unknown to many good farmers.

Fig. 120.

CAHOON'S SEED SOWER.—We have given the hand sower represented in the accompanying cut, (fig. 121,) a thorough trial, having used it for some years. Its great value is in sowing grass seed, which it does more than twice as rapidly as by the old way, and with great evenness. It also requires less skill. For sowing heavy grain, like wheat and barley, it becomes rather heavy, and requires harder labor, but this is less necessary, as the work is commonly done by the seed drill drawn by two horses. In using it for grass seed, or for any other purpose, it is of great importance to keep the rapidly running parts well oiled—as when a little dry, it runs so much harder as to be insufferably fatiguing. The quantity of seed sown to the acre is regulated with accuracy.

Fig. 121.

STEEL PLOW CUTTER STOCK.—The cut (fig. 122) represents the steel plow cutter and stock, one of the newer contrivances manufactured by Holbrook & Co of Boston. Its substantial character is obvious; it is self-clearing, of easy draft, stiff, light and strong. The stock is made of malleable iron, of such form as to combine strength and freedom from bending. It takes the chief part of the strain of the cutter, and comes above the wear.

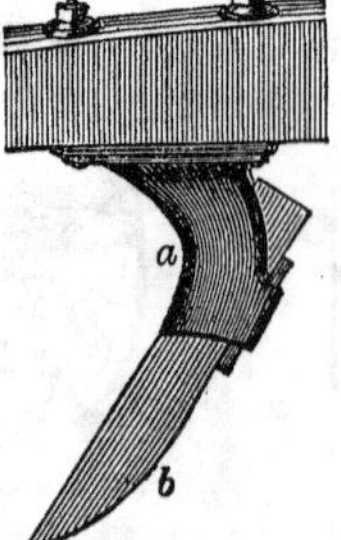

Fig. 122.

The steel blade is 5-16ths of an inch thick, by 2½ wide, about half the thickness of the common cutter, and being so thin, passes through the ground more easily, and controls the plow less than the common cutter.

Fig. 123.

HOLBROOK'S CLEVIS FOR A ONE-HORSE PLOW, to enable the horse to walk in the furrow, is shown by fig. 123. It needs but little explanation, and is similar in general principal to the three-horse clevis, described in another part of this volume.

Another new contrivance made by Holbrook & Co., is the attachment of this clevis to their *one-horse swivel plow*, a movement at turning changing the clevis or the point of draught to the right or left, as wanted.

HAPGOOD'S ONE-HORSE SWIVEL PLOW is so contrived as to throw the beam enough to one side or the other as the mould-board is changed from side to side, by what is termed an eccentric movement. This contrivance is of a very simple and ingenious character, and on using the plow on our grounds, it appears to answer the intended purpose perfectly, that is, plowing without a dead furrow. It is made by the Ames Plow Co., Boston, Mass.

THE SMOOTHING HARROW.—The most improved form of this implement is shown by fig. 124, exhibiting the mode of attaching the draught, and the form of the three sections hinged together. It takes a sweep of nine feet, and is easily drawn by two horses, harrowing 20 acres a day. The backward inclination of the small steel teeth prevents them from ever clogging, and renders the draught easier; and it also allows the harrow to pass over young wheat, until it is over a foot high, without injury. The same peculiarity permits it to run broadcast over corn in hills, in drills, or sown broadcast, clearing out the

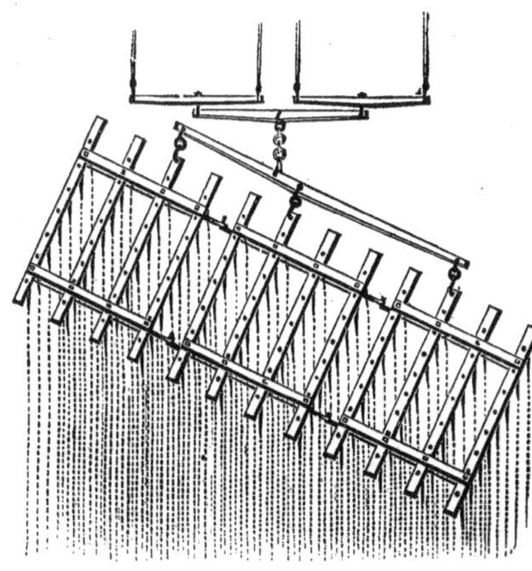

Fig. 124.

fine weeds, mellowing the surface, but causing no harm to the larger plants. It is also used for fitting the surface for seeds of any kind harrowing in grass seeds, and for finely pulverizing spread manure.

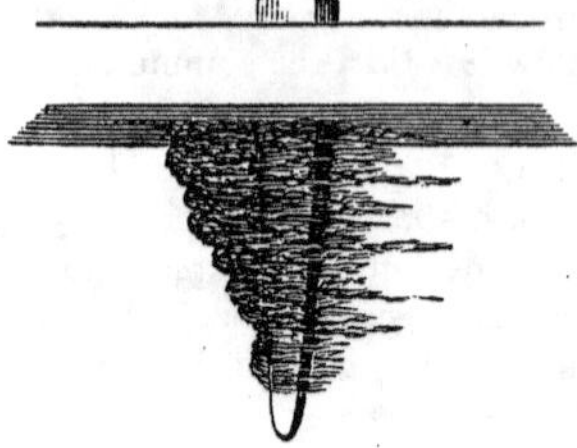
Fig. 125.

The difference between the operation of the sloping teeth of this harrow, and the common vertical teeth, is shown by fig. 125, representing the coarse, square teeth usually employed, clogged with weeds, roots and earth.

HOLBROOK'S SWIVEL PLOW, (fig. 126.)—This implement has been much improved from the old side-hill plows. We have given the two-horse plow made by F. F. Holbrook & Co., a thorough trial, and among other experiments several acres which had been in sod eight years, in many places so steep that no wagon could be driven over them, were successfully inverted to a measured average depth of 7 inches, drawn by a pair of horses. The facility with which the sod was laid down, and the complete pulverization of the surface, were entirely satisfactory, and excited the admiration of neighbors who came to witness its operation. The time required to change the mould-board from right to left and left to to right, at the ends of the furrows, was usually less than the time for the horses to turn about, and much less than with the common plow, as the whole is done at one operation, no passing across the end of the lands being required. On level ground its operation is equally successful, and it entirely obviates dead furrows. This plow combines more excellent qualities as a swivel plow than any other we have had an opportunity of testing—among which its thorough pulverization of the sod stands conspicuous.

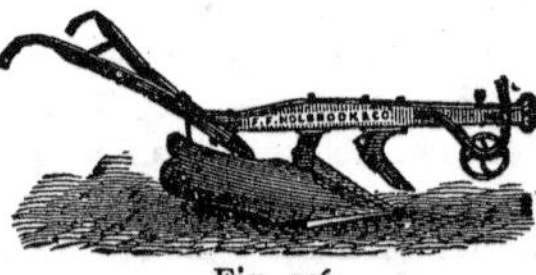

Fig. 126.

HOLBROOK'S STUBBLE PLOW is shown in fig. 127, with their new cutter stock attached. Some of our readers will remember the admirable work which it performed at the Utica trial a few years since. The manufacturers have since made some additional improvements. The remarkable power which it possesses of pulverizing the soil at the same time that it completely inverts all weeds, grass and rubbish, is among its excellent points.

Fig. 127.

A ONE-HORSE SUBSOIL PLOW, light, strong, and of easy draught, is manufactured by R. H. Allen & Co. of New-York, which is a capital implement for deep loosening the soil in gardens and limited grounds, and for working between rows.

HOLBROOK'S HAND CULTIVATOR, (fig. 128.)—In good soil, and for garden and drill crops, this implement is used instead of a hand hoe, and performs work with rapidity. It has the advantage over the horse cultivator of working in small patches, where a horse could not readily turn about, and it is more perfectly under the control of the operator, and may be pushed very near the row, without danger of cutting the plants. It expands from 8 to 14 inches, and is particularly adapted to beets, carrots, onions, turnips and other drill crops. It is manufactured by F. F. Holbrook & Co., Boston, Mass.

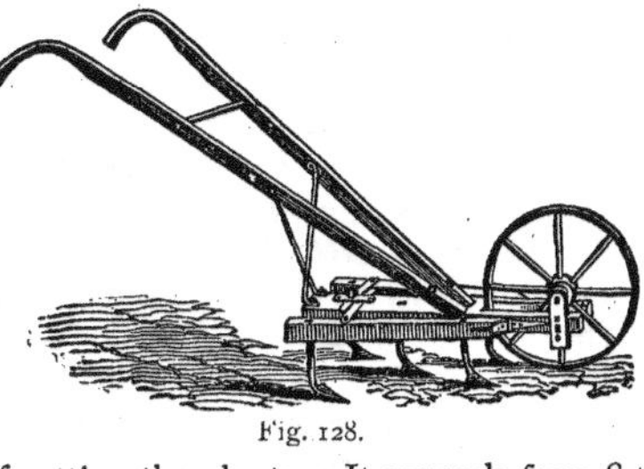

Fig. 128.

HARRINGTON'S SEED SOWER AND HAND CULTIVATOR, (fig. 129.)—This implement, made by the Ames Plow Company, Boston, Mass., combines the sower and the cultivator in one machine. As a cultivator it destroys weeds in rows, and is easily expanded and contracted. It is changed to a sower by adding the hopper and its attachments, and removing the cultivator attachment.

Fig. 129.

HOLBROOK'S DOUBLE-TREE, (fig. 130.)—This dispenses with single-trees in plowing with two horses, and renders the whole arrangement less cumbersome and more manageable. The horses cannot step over the traces in turning. The small iron contrivance at the middle moves freely and keeps the traces of both horses even. It may be so adjusted as to give either horse the advantage. Our plowmen who have tried it are greatly pleased with it. With a well broken team that works moderately even, there seems to be no objection to it whatever, but with unsteady horses, who draw quite unlike, we are told that it does not succeed so well.

Fig. 130.

Messrs. Holbrook & Co. write us under a late date—"Last week we made a series of experiments with three horses abreast on our plow, and came to the conclusion that the present rig, slightly modified, is well adapted to our swivel plows, and will save our making the three-horse clevis. Our long three-horse evener, as now made, measures 4 feet between clevises; we shall hereafter bore two holes, an inch apart, in each end, inside the present clevis bolts, and shall also bore corresponding holes for the middle clevis, to keep the proportion right. By using short whiffletrees, say 27 or 28 inches, we are thus able to adjust the team so as to give the plow more or less land, and still keep the clevis in the centre of the beam, where it must necessarily be in using a swivel plow to good advantage. We think, from our observation last week, that in using this rig for harrowing, it would be desirable to use a long evener, about 4½ feet in length. Of course the same could be used in plowing by boring the requisite holes in it. By making the evener 6 inches more than 4 feet long, it would give a plow 3 inches more land, which would be too much; but in harrowing in hot weather horses work easier spread well apart."

POTATO PLOW.—The annexed cut (fig. 131) is a representation of the potato digger made by R. H. Allen & Co. of New-York, which digs potatoes easily and cheaply, by simply throwing them out of the ground, and shaking them from the earth by the rods which pass backwards. It works well in clean and mellow ground, particularly light soils. It does not succeed so perfectly in heavy, adhesive earth. The prongs are made of iron or polished steel, as may be desired. If they ever break, they are easily replaced by others by a simple contrivance. The high curved standard readily clears itself from weeds. Its cheapness and simplicity specially recommend it.

Fig. 131.

CARHART'S TWO-HORSE CULTIVATOR.—This cultivator (fig. 132) has no wheels, the depth being regulated with accuracy by means of a pair of light runners, placed in the rear of the two outward forward teeth, and which are elevated or lowered by a touch of the hand. The omission of wheels allows it to run close to trees and boundaries. It is strongly made and managed with ease. We have found its operation to be very satisfactory, while

Fig. 132.

its simplicity and cheapness are in its favor. It pulverizes the soil quite equal to the best wheel machines. It has been also used as a potato digger, and operates well. The two outside rear teeth are taken out, and all the front ones, and the two runners or regulator teeth are set in the centre forward, by which five teeth are used, and the exact depth for the potatoes is secured. It is made by C. C. Bradley & Son of Syracuse, N. Y.

Fig. 133.

BRADLEY'S HORSE-HOE, manufactured by C. C. Bardley & Son, Syracuse, N. Y., is shown in fig. 133, and is especially useful in cultivating and hilling crops sown in hills or drills. It has been also used for digging potatoes. The wings to the large front teeth, combined with the narrow teeth, make it a good pulverizer and cultivator.

HAND SEED-DRILLS.—Among the best seed sowers to be used by hand, are Allen's Planet Drill and Holbrook's Seed Drill. The former is made by S. L. Allen & Co., Philadelphia, Pa., and is represented by fig. 134. It is neatly constructed, plants uniformly at the adjusted depth, on uneven as well as on level ground, the planter standing directly under the seed-box and running wheels. It is regulated by a graduated scale. It is probably the best hand implement for sowing corn thickly for fodder, and other seed in large quantity, and answers equally well for the concentrated fertilizers. Holbrook's Seed Drill has an iron frame, and combines neatness with lightness and strength. It is changed to seeds of different sizes by a slight turn in the dial. The seed conductor being enameled white, the operator can see the seed as it drops, and before covered.

Fig. 134.

Fig. 135.

Fig. 136.

BUCKEYE THRESHER AND CLEANER, (fig. 136.)—Since the first complete combined machines for threshing and cleaning, invented and made by Pitts, they are now widely manufactured throughout the country, and of high excellence—variously modified from the original. Among some of the best is that known as the Buckeye, made by Blymyer,

Day & Co. of Mansfield, Ohio. It is distinguished for its compact form as the cut indicates, and it works with efficiency.

WAGON RACK FOR HAY.—In the third volume of RURAL AFFAIRS, we figured and described two modes for constructing hay and grain racks, and here give another form, (fig. 137,) which has some peculiar advantages, the chief of which is the facility with which it may be placed on or taken off the wagon, in separate pieces, so that one man may do the work alone. The bed-pieces may be 2 by 8 inches—sometimes they are made 4 by 5 inches. They are usually about 14 feet long, but sometimes as long as 18 feet. If made of white oak, they need not be quite so large as the dimensions given ; but they are best of pine, as they are so much lighter. Small cleats are nailed to the outer sides of the bed-pieces, at the wagon stakes, to prevent sliding. The cross-pieces, which connect the bed-pieces, should be of the best white oak, or other equally hard and tough wood, as they receive the lower ends of the racks into oblique mortices. The racks or side frames consist of three boards each, bolted to cross-pieces, which are about 6 eet long, made of oak. When placed on the bed-pieces, they rest upon them, the ends being thrust into the mortices. A stronger modification is to make the side-pieces 7 feet long, and to brace them firmly by thrusting the ends under the bed-pieces on the opposite sides, where they are pinned or bolted.

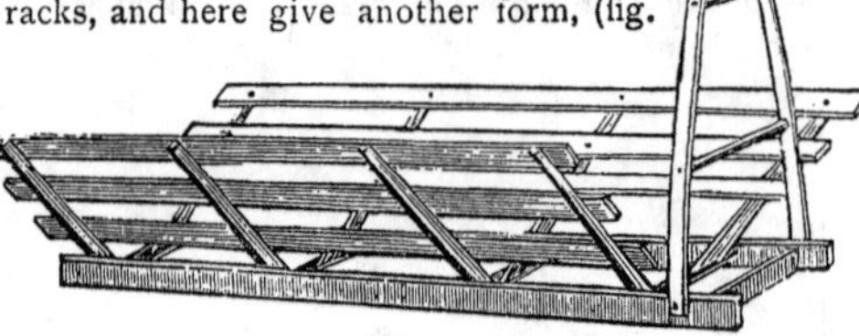
Fig. 137.

PAINE'S WHEEL JACK, (fig. 138.)—This is simple, light and always ready. Being made of strong wood, with iron lever, it does not get out of order. The notches adapt it to various heights of axles without changing. It locks itself, without the trouble of fastening. It is made by the Ames Plow Company.

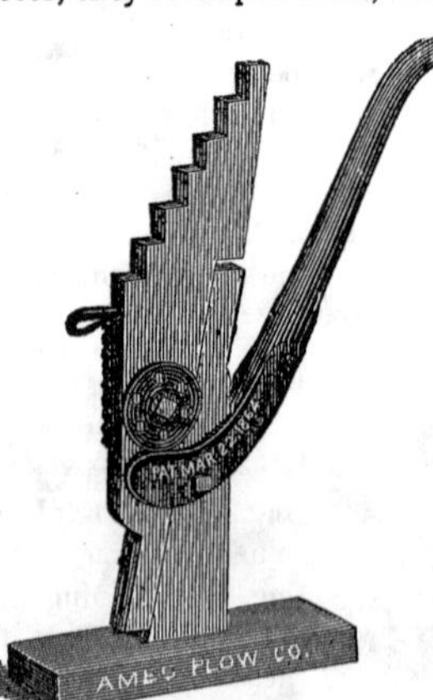

Fig. 138.

TETHER FOR HORSE OR COW.—Persons who keep single horses, and have small grounds, often desire to give them a bite of grass where it can be spared, but wish to confine them to proper limits. To tie them with a cord or halter to a stake, endangers the tangling of their feet, and the cord or strap is injured by the wet grass. A contrivance to obviate this difficulty is shown by fig. 139. It is perfectly safe, and can never catch the animal's feet, or throw him down. A. B. Allen, Tom's River, N.J., who has given it a full trial, says :—"A friend of mine recently lost a fine cow from entangling in the rope attached to her neck and to a stake in the ground, and I have had any amount of trouble with this mode heretofore. A year ago I got this. My cow was rather wild, and I was doubtful whether it would answer. You would have laughed heartily, as I did, to see her movements when first attached. She started on a furious run, but the cord on her horns gently guided her in a circle, and she soon tired of that. Then she attempted backing and pulling, but making little progress, soon gave it up, and then accommodating herself to the range went to feeding as quietly as a puppy, and so continues. One great merit in it is, it acts or rather restrains gently, and the pole moving with perfect ease at the least touch, and the cord being elastic and supple, the animal is constantly guided within its range." Sold by R. H. Allen & Co., New-York.

Fig. 139.

LAWN MOWERS.—A great improvement has been made in the appearance of lawns in some parts of the country, by the recent introduction of lawn mowing machines. On large grounds the mowing is done by horses, the mowers costing from $100 to $200; on grounds of an acre or two the best hand mowers answer well, and cost about $25. Some of these have been greatly improved of late years, so as to obviate the necessity of using

Fig. 140.

the lawn scythe; and they have another great advantage namely, little or no skill or practice is required to use them while making a perfect and even green carpet. We have tried a number of different kinds, but have been especially pleased with the one known as the Philadelphia, (fig. 140,) as manufactured by R. H. Allen & Co., New-York. It cuts a strip a foot or more wide, as fast as a man will walk, with far less effort than scythe-mowing; and one hand will readily go over two acres in a day—cutting from three to four times as fast, and a great deal better, than mowing. It needs sharpening for about every ten acres, but varying much with the condition of the grass, as to dust, or a well washed surface. One important convenience in the use of the Philadelphia mower, is the facility with which it may be run over the ground when not in use for cutting, by simply turning it over, which makes it more portable than a hand truck.

CONTINENTAL WASHING MACHINE.—Of the different washing machines of which we have made trial, we have been particularly pleased with that made by the Brinkerhoff Manufacturing Company, at Auburn, N. Y., known as the Continental, (fig. 141.) We find that it rubs or wears the clothes in a very slight degree, or almost imperceptibly; it will wash a small or large garment equally well; it works with great ease, and its operation is rapid. About one-half of the labor commonly used for a wash-board will do the washing well with this machine. On examining the principle on which it operates, we find that its leading advantage consists in a constant succession of pressures, combined with a gradual turning over of the clothes, bringing all parts under action.

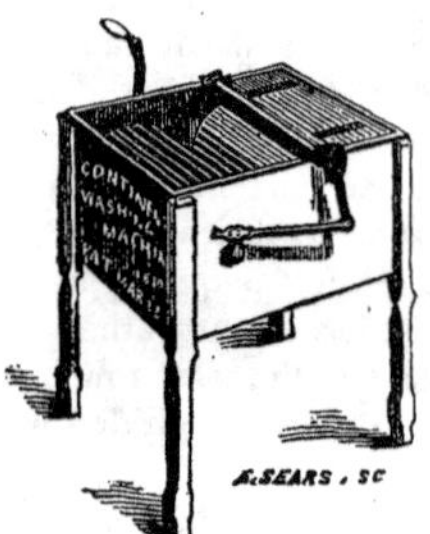

Fig. 141.

POULTRY HOUSE.—Every farmer should have a good, convenient poultry house, properly constructed, sufficiently large to contain the number of birds he desires. It should be warm and dry in the winter, well ventilated and kept scrupulously clean. The house should not be over crowded, but just large enough. Nothing is made by over-crowding the hennery; on the contrary it will prove detrimental. The fowls must be fed regularly and at stated periods. They must have plenty of pure water at hand at all times—this is of as much importance to the health of the brood as proper food. If possible, they should also be given, in addition, a plat of grass for a run. Place within the hennery a dust heap; this may consist of wood or coal ashes, sand, or dust from the streets. It should be kept under a cover, so that it will not become drenched with rain or snow, and to it the fowls should have access at all times, to dust and thereby rid themselves, in a great degree, of the numerous parasites which infest them.—*Poultry Standard.*

CHEESE MAKING.

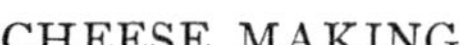

AS A CLEAR AND CONCISE historical review of the factory system of cheese making has appeared in a previous number of the REGISTER, (1864—RURAL AFFAIRS, vol. IV,) we are called upon now to give only a short exposition of the present condition of cheese-making, together with the newer ideas, and a description of improved apparatus and utensils. We will begin with a few remarks on

COWS AND THEIR MANAGEMENT.

The selection or rearing of a herd of cows is the first important consideration in dairying—always supposing that the dairyman has a farm at least tolerably well adapted to dairying purposes. As a general thing it will be found impossible to get a satisfactory herd without raising it. There are two stubborn reasons for this: 1st. Good dairy cows are scarce, and must be raised before they can be found in the country in sufficient numbers to stock our dairy farms. 2d. Those who own good dairy cows will not, as a general thing, part with them at such prices as other farmers can afford to pay. The wisest and most economical course is, therefore, for the dairyman to raise his own herd. To do this, he must purchase or get the use of a pure blood bull. We will not say what breed or family he must select from. We leave that to his own judgment. But he should keep distinctly in view the object for which he is breeding—whether butter or cheese-making. To this end he must cross the desired breed with his best native cows; then breed again from the best crosses, and follow this up until his herd of cows is all that he can desire.

In this way the dairyman can provide himself with the requisite number of milk-producing machines, of the best quality and largest capacity. The next thing is to run these machines economically and profitably. They will turn out milk corresponding very much to the food and drink which is given them. They must have enough, or the machine will run feebly, and the product be light. If given too much, the yield will be large, but the machine will clog. High feeding, forcing the machine, wears it out rapidly and engenders disease and premature death. If the quality of the milk-producing material is not right, the milk will not be right. It has been asserted, and some experiments have seemed to demonstrate, that the quality of the feed has no perceptible effect on the quality of the product. But the experience of almost every one who has owned a cow teaches to the contrary. Feed a cow on leeks, skunk-cabbage and turnips, giving her stagnant water to drink, and she will yield you a villainous compound that will almost sicken the hogs. On the contrary, give her a supply of sweet grasses, with plenty of clear running water, or give her early cut, well cured hay, with roots and a little corn meal, and she will reward you with a flow of deliciously flavored milk.

Milk—that is, the oil, the butter in the milk—is a great absorbent of odors. Hence your cows must be kept in a clean, sweet, well ventilated place. When you drive them from the clean pasture, do not not dog, worry or hurry them, and then shut them up in a foul, stinking barn-yard or horse stable. Do your milking in a clean place and a sweet atmosphere, if you would not have your milk "taste of the barn-yard." The milk will not only absorb the foul odors, but they will get into it through the circulating system o the cow, by her breathing these odors. Think, too, what you are taking into your own lungs, and how your own system is getting defiled, through and through, by your breathing this filthy atmosphere!

Be cleanly in the operation of milking. Have not only all your utensils thoroughly cleaned, scalded and aired, but properly clean the cow's udder and teats, and see that no foul stuff clinging to them gets into the milk. If necessary, wash them before the milking is begun, and give them time to get dry. If the teats crack and bleed, keep milk off from them, and limber them up well with some softening oil, as soon as the cow is driven up to milk. Use only tin pails, and those without sharp angles in which ferments can collect.

Treatment of Milk.

As fast as the milk is drawn from the cow, strain it into the can, but do not let the can stand in a foul atmosphere, nor with the sun blazing on it. Do not cover the top of the can with a cloth strainer, keeping the air *out* and the heat and animal odor *in*. Either use a common strainer pail, or a strainer that can be fastened to one side of the can. (See fig. 142.)

Fig. 142—*Jones, Faulkner & Co.'s Patent Can Milk Strainer, hooking on the side of the Can, inside.*

It seems to be essential that milk, as soon as drawn from the cow, should be either thoroughly aired or cooled, or both. The opinion that has for some time been gaining favor, is that airing is the great desideratum. Various contrivances have been gotten up for this purpose, but all have been too awkward and difficult to keep clean, as well as too expensive.

Too rapid cooling—especially with the use of ice—and too low a temperature, (much if any below 60°,) it is thought, injures the flavor of milk, and of the butter or cheese made from it. It chills the particles of milk coming in immediate contact with the cold surface, and condenses in the milk all the gases, retaining the animal odors, which become all the more active, and hasten taint or putrefaction, as soon as the temperature is raised. The writer of this has traveled considerably among the cheese factories of Central New-York during the past season, (1871,) and his observation seems to confirm this position. He finds that the factories which make the finest flavored cheese do not cool the night's milk much below 70° before leaving it, and do not allow it to get, during the night, below 60°, preferring that it should not be found in the morning below 65°. Revolving the subject in his mind his present conclusion is that airing is of more consequence than cooling, and that the only practical mode of airing is by forcing air through the milk—thus oxygenizing the milk, expelling the gases or animal odors, and gradually reducing the temperature at the same time. For this purpose he and others have invented a cheap and, it is believed, effective apparatus. (See figs. 143 and 144.) By the use of this the surrounding atmosphere, as it is, can be forced through the milk, or the air can be drawn through pounded ice and then forced through the milk. Its use this summer as a simple aerator, without ice, by some of the patrons of Dr. L. L. Wignt of Whitestown goes to confirm the correctness of the theory of aeration—as, in almost every instance milk treated in this way for five or ten minutes immediately after milking, without cooling, has kept sweet longer than milk cooled by the use of water or ice.

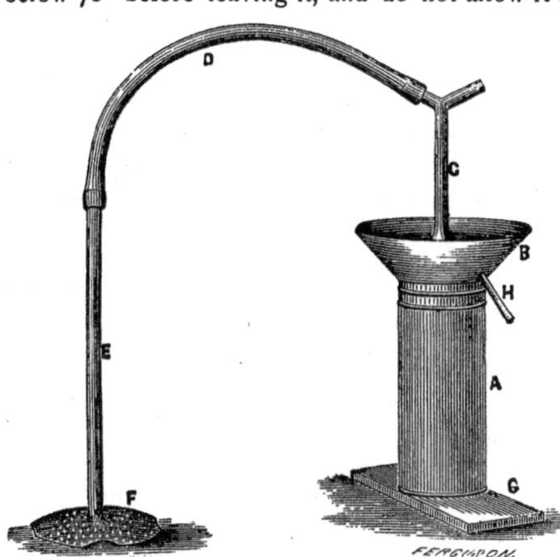

Fig. 143.—*Curtis, Miller & Wight's Force Pump Milk Aerator and Cooler*—A, *air cylinder;* B, *ice-pan;* C, *piston rod and air tube;* D, *continuation of air tube, with flexible pipe;* E, *air tube to be inserted in the milk;* F, *perforated disc for distributing the air in the bottom of the can;* G, *board supporting the pump, and on which the foot is placed in working;* H, *pipe for conducting off water from the melted ice.*

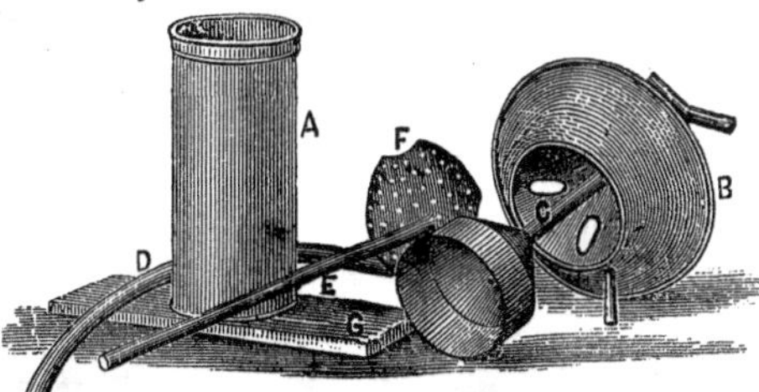

Fig. 144.—*Curtis, Miller & Wight's Force Pump Milk Aerator and Cooler in pieces*—A, *air cylinder;* B, *ice-pan;* C, *hollow piston and rod;* D, *piece of flexible tubing;* E, *tube to be inserted in the milk;* F, *air distributing disc;* G, *board on which the foot is placed in working the pumps.*

Observing all the other essentials in the production and management of milk, we think that if every dairyman would thus thoroughly air his milk as soon as drawn from the cow, and use Arnold's patent ventilator in his can-cover—taking care to protect the can from the rays of the sun on the way to the factory—we should seldom hear of tainted milk or floating curds, and the quality of American cheese would be greatly improved. One thing is certain—strictly fine, good-keeping cheese cannot be made of tainted milk. Cheese-makers have learned to manage this too common article much better than they used to, but they cannot wholly counteract the evil effects of decay when it has progressed so far as to cause taint in the milk. The first and greatest responsibility for the quality of the cheese rests with the patrons of the factory. When they *all* send their milk to the factory in perfect condition, they will have the right to demand that the cheese-maker shall turn out a strictly fine article, but not before.

Receiving and Crediting Milk.

At present there is no way of receiving and crediting milk but by weight. Of course the receiver needs to be a good judge of milk, and must use all his faculties and skill to keep out skimmed, watered, soured and tainted milk—some of which will pass him in spite of all his care and vigilance. He may exclude these, if he can detect them in time, but against poor milk—the product of poor pastures and poor cows—he has no remedy. This must go in with the rest, and count the same as so many pounds of the richest milk, and the loss is shared by those who bring better milk, the gain going to the owner of the poor stuff. So the man who brings milk rich enough in cream to make cream cheese, and the man whose milk is so poor in cream that it will make little better than skim-milk cheese, are both credited according to the number of pounds delivered.

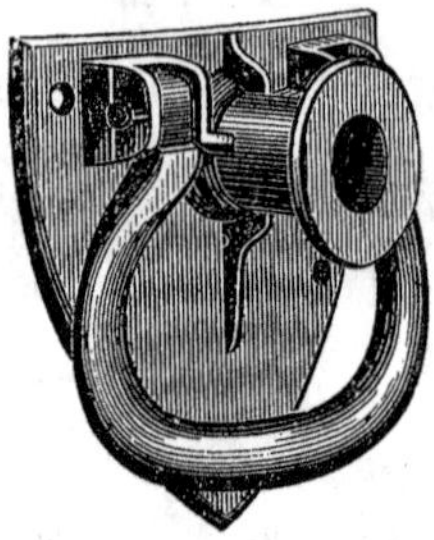

Fig. 145.—*Ralph's Patent Can Handle and Ear combined.*

There is therefore no stimulus to the patron to produce good milk, but a constant incentive to deliver the greatest number of pounds of liquid, regardless of its quality. The condition and keeping quality are all that the receiver of milk at the factory can take into consideration. In determining these he can call to his aid the glass tubes, (see fig. 146,) having one for each patron, and filling it as often as

Fig. 146.—*Jones, Faulkner & Co.'s Case for Test Glasses.*

thought advisable, and watching the milk till it sours or taints. These tubes also show, to the eye, the relative amount of cream in each man's milk; but this is of no use, except to arouse suspicion of watering or skimming, when there is little cream, and to cause watching for the legal proof. The cream-gauge (see fig. 147—the two glasses standing together, with figures towards the top only,) will show the exact percentage of

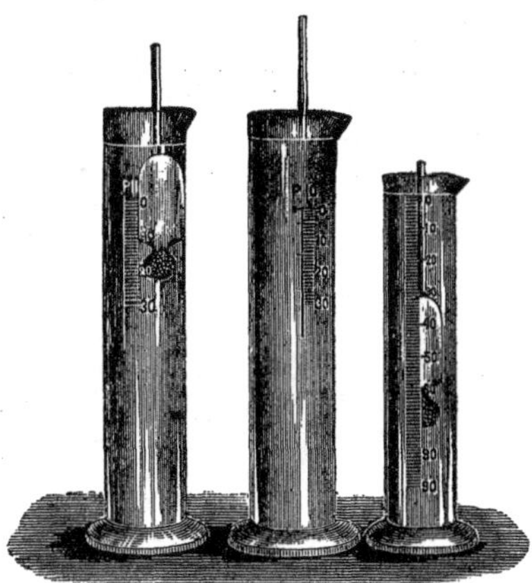

Fig. 147.—*Cream-Gauge.*

Fig 148.—*Lactometer.*

cream; and the lactometer (fig. 148) will show the density or specific gravity—nothing more. As seen in fig. 147, the per cent. glass filled with pure water, lets the lactometer sink to the water line. The middle one, containing water rendered dense by the addition of ingredients, stands at the skim-milk point; and the other cream-gauge, with a little salt added to the water, floats the lactometer at the pure milk figure. Hence the lactometer, of itself, is of little value in detecting rascality. Skimming, watering and adulterations, are best shown by the test-glasses, which are only indicators, and need to be backed up by legal proofs, which are usually ocular demonstrations. Still if the lactometer and cream-gauge both show a patron's milk to be deficient when delivered at the factory, and specimens taken at his home from the can—which is watched and known to contain no skimmed milk or water—tries all right, it would be difficult for an intelligent jury to refuse to bring in a verdict against him for either skimming or watering.

We need some simple and satisfactory test of the intrinsic value of milk for cheese-making. The cream-gauge is a satisfactory test of its value for butter-making. It has been suggested that a given amount of each patron's milk, at certain intervals, be made into cheese, and either carefully weighed or measured, and his milk be rated by this test until the next test is made. But this would only decide the percentage of cheese, without taking into consideration the quality, and is, therefore, open to the same objection, though in a less degree, that crediting by weight is. It is desirable to give credit for both caseine and cream, and no test will be satisfactory which does not do this. Perhaps the milk might be set in the cream-gauge for cream, and the skimmed milk afterwards made into cheese, to determine the amount of caseine. But to make this test satisfactory the relative value of cream and caseine must first be determined, the cream evidently being the more valuable.

A committee of the American Dairymen's Association has under consideration this subject of crediting according to the intrinsic value of milk, and we shall be gratified if it shall be able to report some simple, equitable and satisfactory system of credits.

Rennet

Is unquestionably one of the most important things in cheese-making. We can separate the caseine from the whey by the use of acids, and even by the natural process of souring, but the product is not that mellow, rich and palatable substance known as cheese. So it is said by the chemists, that the spores, or seeds, of the blue mold are identical with the active properties of good rennet; but no one has yet succeeded in making a marketable article by the use of blue mold instead of rennet. Practically, whatever theory may show, we have no substitute for the active properties of the stomach of the calf, in cheese-making. The soakings of the stomachs of the young of other animals—as the pig, the lamb, the kid, etc.—will cause coagulation, and the extract from the stomach of the pig is said to be stronger than that from the stomach of the calf; but these have never been used to any considerable extent, and we have no knowledge of experiments to satisfactorily determine the relative value of the stomachs of the young of various species of animals for the purposes of the cheese-maker.

At present, therefore, we must confine ourselves to the saving and preparation of the stomach of the calf. Of course the true or digestive stomach—sometimes called the "second stomach"—is the one to be saved. This should be healthy and active, and ought to be saved at the stage when it is just fairly emptied, and the secretions are copious, causing a keen appetite on the part of the animal. The calf should not be less than three days old, and probably ought to be five or six days old, so that all the organs may be in active and vigorous operation. It should go without eating, immediately before being killed, for twelve to eighteen hours. A good way is to feed at night, muzzle the calf or put it where it cannot lick dirt or get hold of straw, hay or other solid substance, and kill it some time during the next forenoon. The stomach should be removed from the calf as soon after killing as possible, as decomposition begins very soon, and goes on very rapidly among the warm vital organs when life has departed. The stomach should be turned inside out and carefully cleared of all

foreign substances, but not washed, and then well salted and stretched on a bow or crotched stick, and hung in a cool, dry place; or salt the ends well, tie up one end, blow up the rennet like a bladder, close the other end, and then hang up to dry. When dry, tie your rennets up in paper bags—flour sacks are as good as anything—and keep in a cool, dry place until wanted for use. Freezing does not injure them, and they are best when not less than one year old.

We are confident that the quality of American cheese would be greatly improved by the exercise of more care in saving and using rennets—never using any that are under one year old, and not perfectly healthy and sweet.

The best mode of preparing rennet is that practiced by Dr. L. L. Wight of Whitestown, Oneida Co., N. Y., who has takan the first premium at the State Fair for two years in succession, for the best five factory-made cheeses. He says: "I take pure sweet whey and steam it to boiling, and remove the scum which rises; then let it stand until it settles, and decant the clear whey, leaving the sediment at the bottom of the cask. When this whey is cold and acid, I soak the rennets in it, adding a little salt, but only just sufficient to preserve them from tainting. Of this liquid I use enough to commence coagulating the curd in fifteen minutes.' It requires about one good, strong rennet to each gallon of whey thus prepared. The mode of scalding the whey, so it is not scorched, is of no consequence.

Coloring

Is specially demanded by the London market, which takes about three-fifths of all the cheese exported from this country. The tendency, however, is toward less color, and the hope is general that the day is not far distant when no coloring will be required. It is confessed on all sides to be rather detrimental than useful, and to cause unnecessary expense and trouble, except so far as it pleases our principal foreign customers, who have educated themselves to admire an unnatural color in their cheese. But when color is used it is in a liquid state, and should be prepared from the purest and best material. Annatto is universally conceded to be the best coloring matter, and that form of it known as annattoine is at present the purest to be had in market. It is easy to reduce to a liquid state, and is put up in packages, accompanied by a recipe for its preparation.

Setting, Cutting and Heating.

The milk—having been reduced to a temperature of about 70° the night previous, and the cream prevented from rising by the use of an agitator—should be about 65° in the morning, and not below 60°. Into this is run the morning's milk, which is usually about the same in quantity as the night's, and the temperature of the whole is raised to about 82° in hot weather, and 86° in cool weather, making a mean temperature of 84° for setting in mild weather. The coloring liquid is first thoroughly incorporated with the milk, and then sufficient prepared rennet is added to cause to begin coagulation in about fifteen minutes. A slight agitation is kept up, until the milk begins to roll thick and heavy, for the purpose of preventing the cream from rising; but care must be taken not to stir too long, or a smooth, compact mass of curd will be impossible, and a broken, spongy mass will appear in its place, from which many of the fine particles will be washed off in the whey, causing great waste. As soon as the curd will break smoothly across the finger, leaving the finger clean, and clear whey settles in the broken place, the curd is fit to cut. Then, in our opinion, all the cutting should be done as quickly and as gently as possible—first with the horizontal knife, lengthwise of the vat, then with the perpendicular knife, (fig. 150,) until the pieces of curd are about the size of beachnuts. If the milk is very sweet, it may be left coarser; if it shows age, and is working rapidly, cut finer still, so as to secure even action of the heat in a shorter space of time, and be ore the acid gets too much developed. Stir the curd gently, and only enough to prevent packing, while raising the temperature, gradually and steadily, but more or less rapidly, according to the state of the milk, to 96°, 98° or 100°, as experience shall have determined to be the proper point for the milk of your factory. Hold your curd steadily at the desired point of temperature until the action of the heat is nearly or quite complete, and the acid begins to show itself slightly. Then, if you do not grind your curds, draw the whey

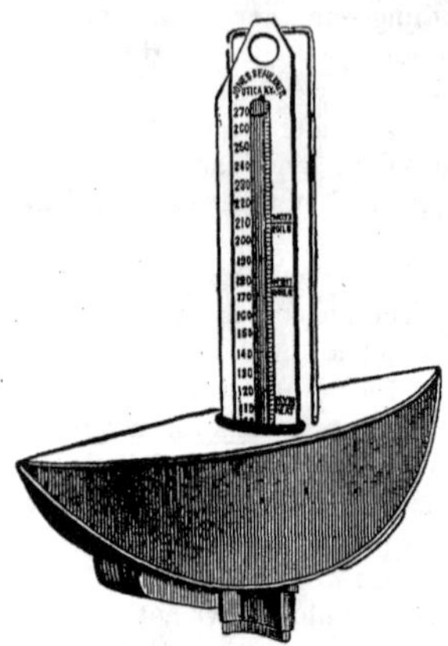

Fig. 149.—*Jones, Faulkner & Co.'s Thermometer and Float. This shows the Thermometer fixed in a Float, which keeps it always right side up in the Vat, ready to indicate the temperature.*

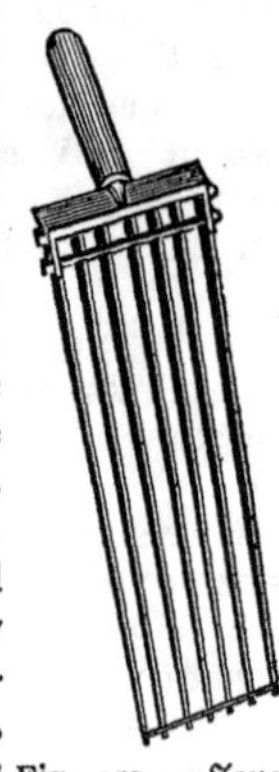

Fig. 150. — *Jones, Faulkner & Co.'s Improved Perpendicular Curd Knife, with iron head, made with any number of blades.*

Fig. 151.—*Jones, Faulkner & Co.'s Curd Pail, with slanting top and ears on the inside, so that it can be dipped against the end and into the corners of the Vat, while the bail extends beyond the edge of the pail, and saves the knuckles from bruising.*

down to the surface of the curd, and allow it to stand until ready to dip. If you grind your curds, draw off all the whey early, at least as soon as there are any tangible signs of acid; raise one end of the vat, draw the curd away from the other, pile it up along each side of the vat, and leave it to drain and take on acid, with occasionally cutting it lengthwise and across into convenient pieces to handle, and turning it so as to air the bottom and inner portions, and to give the outer portions that have cooled somewhat the advantage of more heat. When the whey that drains from it has an unmistakable sour-milk taste, or when an iron heated to a black heat and applied to the curd, will draw it out into innumerable fine threads a quarter or half an inch long, grind as soon as possible, (see fig. 152,) and apply about 2½ pounds of factory-filled salt to the curd of each 1,000 pounds of milk. It may then be put to press, or allowed to stand and air at pleasure; but if the temperature gets too low, —say below 75°—it will be found difficult to make the cheese face, especially in cool weather, when the temperature continues to run down.

Fig. 152.—*Ralph's American Curd Mill.*

If you do not grind your curds —as some of the very best factories do not—dip into the curd-sink as soon as the acid is right; allow the whey to drain off pretty well while thoroughly stirring and separating the curd with the hands, and then apply, as evenly as possible, from 2½ to 3 pounds of salt, according to the state of the weather and the amount of whey left to wash away the salt. Air the curd well, and put it to press, having some regard to both the temperature of the curd and the temperature of the atmosphere. If put to press too warm in hot weather, the curd will ferment in the centre of the cheese and cause it to go off flavor. There is no danger then of getting a curd too cool or too well aired. In cool weather the temperature must be left high enough to enable you to make a smooth face to your cheese. In truth, however, the factory should be so constructed that the temperature can be controlled at all times. Then you can do as experience shows you is best.

Temperatures, Acid, etc.

One of the most difficult things in cheese-making is to determine the exact point where the heat and acid should be arrested and the salt applied to the curd. All depends on experience and judgment; yet some acquire the skill in a few months which others can never reach. No fixed rule can be laid down, as the milk of different cows, and of different localities, as well as of different years and different seasons of the year, works differently. Each cheese-maker must determine for himself or herself the degree of heat and acid required for the time and place. Yet there is such a thing possible as a standard of excellence in the product of the dairy, which every one should have thoroughly fixed in the mind, but which every one is liable to lose if the greatest watchfulness and care are not exercised. For this reason we advise freer intercourse among cheese-makers—more frequent visits to each other's factories, and observation of the products of those that are selling best in market, that the standard in the mind may not deteriorate, but rather improve. It would be to the advantage of patrons to pay the expenses of frequent visits of their cheese-maker to other factories known to excel.

Our observation among the cheese factories of Oneida and Herkimer, during the past season, has confirmed us in the opinion that much of our cheese is made too soft, and consequently lacks in keeping qualities. There have been two reasons for this, aside from the old prejudice existing among private dairymen, that an undercooked, pasty, rank cheese is better than one suited to the London market. These reasons were—the desire to obtain a big yield, and the anxiety to have the cheese cure rapidly, that it might be sold before there was any further decline in the market.

There are several means of securing a soft and rapidly curing cheese, all of which combined not only make a cheese soon ripe, but soon rotten. These are—less heat, less acid, less salt, and more rennet. The mystery of cheese-making lies in the proper use and degree of these several agents, so as to produce a firm, flaky but not crumbly, sweet-flavored article, that will improve for at least a year, and melt in the mouth like butter, leaving a clean, sweet taste on the palate. Such a cheese will *not* be fit for market in thirty days, or less. The cheese maker should know how to produce a long keeping, or rapidly curing cheese, as may be desired, at will.

Pressing and Curing.

We do not lay as much stress on the pressing of cheese as some do. We look upon it more as a matter of convenience, in putting the product in a good shape to handle, than anything else. If the curd is all right, and the conditions for curing are all right, the cheese will be satisfactory. Still as pressing, in some form, will probably always be resorted to, it is a matter of a good deal of importance to get the best and most convenient method. Nothing, so far, has been found to answer so well as some application of screw-power. The best form of this application that we know of is the gang-press, with its improved hoop.

Fig. 153.—*A Fancy Cheese Hoop for pressing "Young America" Cheese, weighing* 8 *to* 10 *pounds.*

(See figs. 154 and 155.) It is compact, economical, and does its work satisfactorily. It has been pretty thoroughly tried during the past season, by the Rome Cheese Manufacturing Association, and by Dr. L. L. Wight of Whitestown, Oneida Co., N. Y.

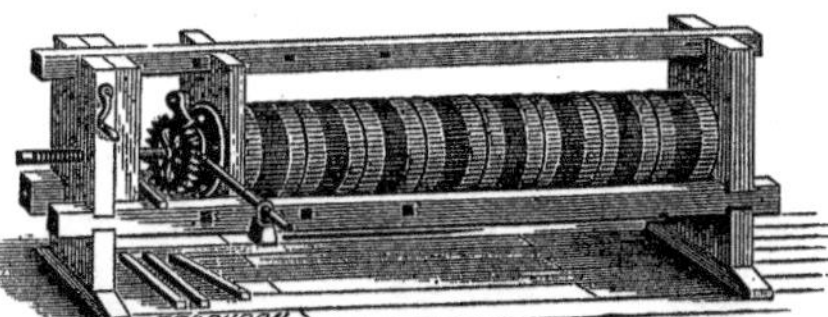

Fig. 154.—*Frazer's Gang Press in operation.*

The importance of having curing rooms so constructed that the temperature can be controlled and kept between 70° and 80°, is beginning to be more and more recognized, and more pains to secure this end are taken in building factories. With cheese not exceeding 70 lbs. weight, tables are considered preferable to rails and turners, both for convenience and to get rid of the annoyance consequent upon the pretended patent on the rails and turner. The idea of pressing a cap of bandage cloth on the ends of cheeses, and thus avoiding the use of grease, is receiving some attention. Cheeses with these caps do not look so nice, but they are cleaner to handle, and entail less trouble and expense.

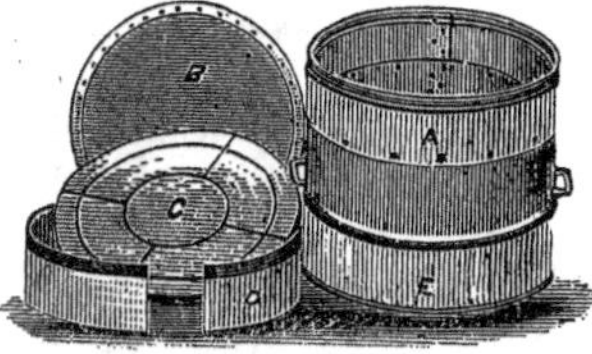

Fig. 155.—*Frazer's Improved Gang Press Hoop*—A, *hoop;* B, *closed end of hoop, the edge of which is seen below,* E; C, *follower, showing grooves, in which are holes for the passage of the whey;* D, *bandager on which the upper edge of the cheese bandage is slipped, and the bandage inserted in the hoop, the lower edge resting on the ledge seen on the inside of the hoop, just the width of the bandager from the top.*

Apparatus, Utensils, etc.

There are practically but two methods of heating. One is by surrounding the vat of milk with water which is raised in temperature until it brings the milk to the desired point; and the other is by the application of dry steam directly to the external of the vat of milk, which is surrounded by an enclosed space for this purpose. Both of these methods have their strong advocates, and our best factories use the one or the other, as they accidentally began, and seemingly with equal success, while

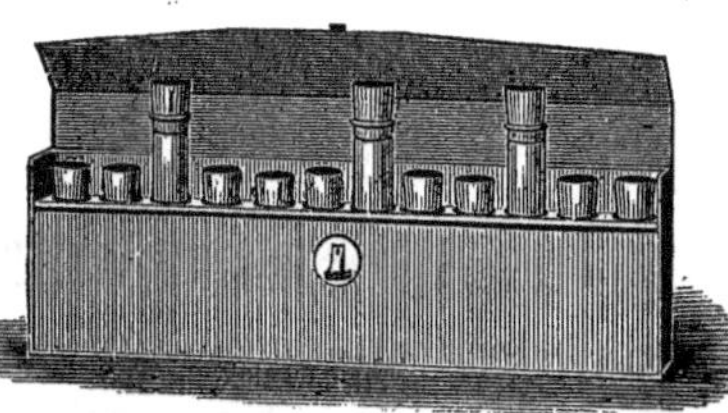

Fig. 156.—*Sample Case for carrying samples of Cheese. This is a leather pocket case with short glass tubes, furnished with corks, in which to carry sample plugs of cheese. Three of these corked tubes are shown drawn partly out of the case.*

some, like Dr. Wight's of Whitestown, use both. Theory has seemed to us rather to favor the use of water as likely to secure an evener heat and a larger yield, but we are not aware that any satisfactory test has been made to settle these points. We give an illustration of a self-heater, which has been thoroughly tried and proved efficient, (see fig. 157,) and of a dry-steam heater, which seems to us superior to anything else of the kind, because of its facilities for turning the steam under the bottom, or on the sides, at will, which no other dry-steam apparatus has, (see fig. 158.)

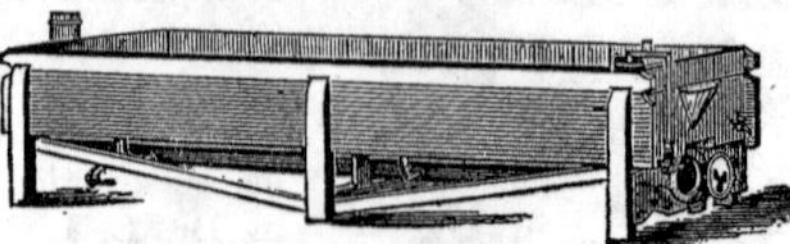

Fig. 157.—*Ralph Vat—a self-heater with fire-box running the entire length on the under side, and surrounded with water.*

Fig. 158.—*Jones, Faulkner & Co.'s **Improved** Stuart Vat—a dry-steam heater.*

The following table shows the receipts of cheese at and shipments from the port of New-York for the year of 1870, and from January 1 to October 1, 1871:

	1870.		1871.	
	Receipts.	Exports.	Receipts.	Exports.
January,	16,392	16,343	23,297	30,242
February,	16,572	16,572	16,906	47,416
March,	14,976	31,216	25,207	35,744
April,	35,681	48,399	20,936	40,841
May,	45,414	42,990	23,970	46,147
June,	117,684	115,537	141,267	118,050
July,	276,496	264,086	278,005	267,483
August,	188,258	199,917	284,695	267,451
September,	241,937	166,362	294,536	251,047
October,	232,303	115,857		
November,	185,714	78,559		
December,	180,976	101,017		
Total,	1,552,303	1,186,855 boxes.	1,108,616	1,104,520 boxes.

The average weight of the box is about 60 pounds.

The number of milch cows in the United States in in 1850, according to the census, was 6,385,094; in 1860 it was 8,728,862; in 1870 it was 11,008,925. The quantity of cheese made in 1850 was 105,535,893 pounds; in 1860 it was 105,875,135 pounds; in 1870 the aggregate is not as yet ascertained, but the following statistics of production in *Farm Dairies only* will be read with interest:

California,..	3,395,074	Massach'ts,	2,245,873	Oregon, ...	79,333	Dakota, ...	1,850
Connectic't,	2,031,194	Minnesota,.	233,977	Penna.,....	1,145,209	Idaho, ...	4,464
Delaware,..	315	Nebraska, .	46,142	R. Island, .	81,976	Montana,...	25,603
Florida,....	25	Nevada,...	None	Vermont, ..	4,830,700	New-Mexico,..	27,239
Illinois,....	1,661,703	N. Hamp..	849,118	W. Virginia,	32,429	Utah, ...	69,603
Kansas, ..	226,607	N. Jersey,.	38,229	Wisconsin,.	1,591,798	Washington,..	17,465
Maine, ...	1,152,590	New-York,	22,769,964	Arizona, ...	14,500		
Maryland,..	6,732	Ohio,......	8,169,486	Colorado,...	33,626	Total,.	50,782,824 lbs.

[T. D. C.]

THE FARMER'S REGISTER.

THE LISTS presented below, though not *complete*, are chiefly made up from the advertising columns of THE COUNTRY GENTLEMAN, during the year ending Nov. 1, 1871, and thus include the leading names in each department—those also most likely to be able to supply orders:

BREEDERS OF IMPROVED STOCK.

AYRSHIRE CATTLE.

Abbott, J J C.............Montreal, Can
Appleby, L.............Spottswood, N J
Appleton, Francis H..West Peabody, Mass
Ball, A P............. ...Derby Line, Vt
Bassett, H W.................Derby, Ct
Birnie, William,.........Springfield, Mass
Bliss, O S.............. ...Georgia, Vt
Boise, E W.............Blandford, Mass
Bradley, G C...........Watertown, N Y
Brodie, R......... ......Smithville, N Y
Brodie, James,...........Rural Hill, N Y
Brown, Henry T......... Providence, R I
Burleigh, B W ....Ticonderoga, N Y
Byrne, Patrick, ...St. Joseph, Pa
Chapman, C S.............. Malone, N Y
Chapman, Albert,. .. ....Middlebury, Vt
Clark, J K........... ....Normandy, Mo
Codman, Ogden,............Lincoln, Mass
Collins, H S..............Collinsville, Ct
Converse, J F............Woodville, N Y
Coy, E L............West Hebron, N Y
Cragin, G D........ ....Rye, N Y
Crozier, William,..... ....Northport, N Y
Curtis, F D...............Charlton, N Y
Dane, N., Jr.............Kennebunk, Me
Dixon, I......... .. Schraalenburgh, N J
Douglass, J L............ Belleville, N J
Drew, L S... .... So. Burlington, Vt
Fitch Thomas, ...New-London, Ct
Freeman, J W................Troy, N Y
French, J D W.... North Andover, Mass
Gibb, John L............Compton, Can
Hammond, G E.........New-London, Ct
Haydock, J W................ Troy, N Y
Hungerford, S D............Adams, N Y
King, W S.............Minneapolis, Minn
Landon, S... .... Eden, N Y
LeClair, Peter,..... ....Winooski, Vt
Lester, C S....... Saratoga Springs, N Y
Loring, Harrison,...........Boston, Mass
Loring, Dr. G B............ Salem, Mass
Mills, L A................ Middlefield, Ct
Morgan, J H..... ....Ogdensburgh, N Y
Myers, M E.............. Charlton, N Y
Odell, D H.......... .... Brant, N Y
Pond, C M.......................Hartford, Ct
Reed, S G.............. Portland, Oregon
Rumsey, H M............. Salem, N J
Seney, Robert,.........Mamaroneck, N Y
Sheffield, Dr. W W......New-London, Ct
Slifer, W H ....Lewisburg, Pa
Smith, E W...............New-London, Ct
Stark, W..............Manchester, N H
Stewart, H L.........Middle Haddam, Ct
Stiles, W H ....Mamaroneck, N Y
Stimson, S J.................Linden, N J
Sturtevant Brothers, S. Framingham, Mass
Thompson, S N........ Southboro, Mass
Thompson, Thos., & Son,....Dunbar, Can
Tilton, H W............. Walpole, Mass
Tuck, Andrew,........... Flackville, N Y
Turner, T T.............. Normandy, Mo
Walcott & Campbell, New-York Mills, N Y
Walling, Nelson,.......... ..Millbury, Mass
Warrington, Thos.,...... West Chester, Pa
Watson, William,.......West Farms, N Y
Wells, S M & D......... Wethersfield, Ct
Whitney, N S............. Montreal, Can

BRETON CATTLE.

Maitland, Robert L............New-York

DEVON CATTLE.

Arnold, W E.................Otego, N Y
Bowie, Oden, ...Collington, Md
Brown, S T C............. Sykesville, Md
Buckingham, J............. Zanesville, O
Cole, Walter,............. Batavia, N Y
Early, J A................Youngstown, O
Hilton, Joseph,........New-Scotland, N Y
Howard, A C................Zanesville, O
Morris, Dr. J C...... ... Westchester, Pa
Olmstead, H M......... Morristown, N J
Rockwell, J M...........Butternuts, N Y
Sessions, H M......So. Wilbraham, Mass
Terrill, M W......... ....Middlefield, Ct
Wainwright, C S.........Rhinebeck, N Y

HEREFORD CATTLE.

Corning, E., Jr..............Albany, N Y
Gibb, John L..............Compton, Can
Stone, Fred. Wm... ... ... Guelph, Can

HOLSTEIN OR DUTCH CATTLE.

Baker, Thomas, ...Barton, Vt
Baldwin, T E..............Litchfield, Ct
Ball, A P.................Derby Line, Vt
Chenery, W W............Belmont, Mass
Houghton Farm,...... .. ....Putney, Vt
Wales, T B., Jr....So. Framingham, Mass
Whiting, T E.............Concord, Mass

JERSEY OR ALDERNEY CATTLE.

Alexander, A J ......Spring Station, Ky
Anderson, W P.............Cincinnati, O
Aspinwall, J L........... Barrytown, N Y
Austin, E H........... Gaylordsville, Ct
Bagby Farm, ... Tiffin, O
Barnes, Wallace,...............Bristol, Ct

Barnum, T G .... Bethel, Ct
Barstow, J S .... Newport, R I
Bartholomew, A.............. Bristol, Ct
Bassett, H W.................. Derby, Ct
Beach, C M............ Hartford, Ct
Benedict, Chas.,.......... Waterbury, Ct
Biddle, Clement, ... Philadelphia, Pa
Bowditch, E F........ Framingham, Mass
Bradley, R................ Brattleboro, Vt
Bradley, G W .... Hamden, Ct
Bradway, J H............ Woodbury, N J
Brooks, John,............ Princeton, Mass
Brown, E B.................. Mystic, Ct
Brown, Melville,.......... Millbrook, N Y
Buck, Mrs. M E........ Poquonnock, Ct
Bull, W W..................... Plymouth, Ct
Bush, W C........... Auburndale, Mass
Bush, James P............. Boston, Mass
Chapman, A.............. Middlebury, Vt
Churchman, F M....... Indianapolis, Ind
Clift, W............... Mystic Bridge, Ct
Codman, Ogden,............ Lincoln, Mass
Coleman, G D............. Lebanon, Pa
Collamore, Davis,............ Orange, N J
Colt, Samuel C............. Hartford, Ct
Converse & Flagler,...... Arlington, Mass
Cragin, George D.............. Rye, N Y
Crozier, William,........ Northport, N Y
Curtis, F D.................. Charlton, N Y
Curwen, G F........... W Haverford, Pa
Dane, N., Jr. .... Kennebunk, Me
Darlington, R S......... West Chester, Pa
Davis, D E................ Salem, N J
Day, R L................... Boston, Mass
Delano, Charles,...... Northampton, Mass
Devries, Wm............. Baltimore, Md
Dike, Lyman,............ Stoneham, Mass
Dillon, J C................ Weston, Mass
Dinsmore, W B......... Staatsburgh, N Y
Dunlop, J S........... Indianapolis, Ind
Edgerton, James, ... Barnesville, O
Edmands, J F............ Newton, Mass
Estes, J J........... East Abington, Mass
Faile, T H., Jr......... West Farms, N Y
Farlee, G W............. Cresskill, N J
Fearing, D B............... Newport, R I
Fenner, H................ S. Orange, N J
Fitch, Thomas,.......... New-London, Ct
Frost, George,........ West Newton, Mass
Giles, John,................ Putnam, Ct
George, Thomas,......... Newburgh, N Y
Glasgow, W H............ St. Louis, Mo
Goodman, R.................. Lenox, Mass
Gould, Thomas, ... Aurora, N Y
Gridley, S R................ Bristol, Ct
Hadwen, O B............ Worcester, Mass
Halsted, J M.................... Rye, N Y
Hand, Thomas J.......... Sing Sing, N Y
Hardin, S L............... Louisville, Ky
Harwood, J A............ Littleton, Mass
Haven, John,....... Fort Washington, N Y
Haydock, J W................ Troy, N Y
Hayes, Francis B........... Boston, Mass
Hilton, S C.............. Providence, R I
Henderson, S J.......... O'Bannon's, Ky
Hoffman, J A........... Leicester, Mass
Howe, Edward,........... Princeton, N J
Howell, Dr. B P.......... Woodbury, N J
Hubbell, O S............ Philadelphia, Pa
Ide, L N................ Claremont, N H
Jenkins, J Stricker,......... Baltimore, Md
Jewett, P A............. New-Haven, Ct
Johnson, C S............. Uncasville, Ct
Juliand, Joseph,......... Bainbridge, N Y
Kelsey, H C................. Newton, N J
Kinney, J D................ Cincinnati, O
Kittredge, B.............. Peekskill, N Y
Large, S P............ West Elizabeth, Pa
Mackie, J M..... Great Barrington, Mass
Mallory, Joel,................... Troy, N Y
McCulloh, J W...................... New-York
McHenry, J Howard,...... Pikesville, Md
Mills, Lyman A........... Middlefield, Ct
Morrell, Robert,.......... Manhasset, N Y
Newell, Dr. A D... New-Brunswick, N J
Osgood, H B.......... Whitinsville, Mass
Page, Joseph F......... Philadelphia, Pa
Park, H S.................. Bayside, N Y
Parks, C C & R H.......... Waukegan, Ill
Parsons, S B.............. Flushing, N Y
Potter Bros............... Webster, Mass
Powell, James B............. Hartford, Ct
Powers, A E. .... Lansingburgh, N Y
Powers, Joseph,...... No. Haverhill, N H
Redmond, William,............ New-York
Robeson, A................ Tiverton, R I
Reeder, E................ New-Hope, Pa
Reynolds, I W H.......... Frankfort, Ky
Ridgeley, Chas.,............ Hampton, Md
Riley, F.................... Sterling, Mass
Robbins, S W............ Wethersfield, Ct
Rockwell, J T.............. W. Winsted, Ct
Root, L B............ New-Hartford, N Y
Rumsey, H M................... Salem, N J
Studder, M S.............. Boston, Mass
Seney, G I.......................... New-York
Sharpless, Charles L..... Philadelphia, Pa
Sharpless, Samuel J...... Philadelphia, Pa
Sherlock, Mrs. M J........ Elizabeth, N J
Sherwood, S S............. Paterson, N J
Stark, W............... Manchester, N H
Stephens, Romeo H......... Montreal, Can
Swain, J B.............. Bronxville, N Y
Swigert, D............ Spring Station, Ky
Tatum, George M........ Woodbury, N J
Tilden, M Y........ New-Lebanon, N Y
Torrey, J W.............. Philadelphia, Pa
Turner, T T .... Normandy, Mo
Twaddell, Dr. L H....... Philadelphia, Pa
Underhill, A A ...... Poughkeepsie, N Y
Walcott & Campbell, New-York Mills, N Y
Walker, Alex.,........... W. Elizabeth, Pa
Ware, J B................ Townshend, Vt
Waring, G E., Jr.......... Newport, R I
Welch, E B............. Cambridge, Mass
Wellington, C...... East Lexington, Mass
Wellington, H M.... Jamaica Plain, Mass
Wells, E L................ Pittsfield, Mass
Wells Phil.................... Amenia, N Y
Wheeler, A D........... Providence, R I
Whitehead, Joseph,.......... Trenton, N J
Wilmerding, G G.......... Bay Shore, N Y
Wing, John D......... Washington, N Y
Young, Richard,............. Mortons, Pa

Kerry Cattle.

Appleton, D F........... Ipswich, Mass
Green, Andrew H............. New-York
Perry, E B .... Providence, R I
Sinclair, Samuel,...... .... New-York

Short-Horn Cattle.

Alexander, A J .... Spring Station, Ky
Alvord, C T............. Wilmington, Vt
Ashworth, John,............. Ottawa, Can
Babbage, A R........... Dubuque, Iowa
Baldwin, T E... . ... Litchfield, Ct
Barbee, F J.................... Paris, Ky
Beach, C M. .... Hartford, Ct
Beattie, Simon,.............. Bangor, Can
Bedford, G M .... Paris, Ky
Bedford, Edwin G.......... ...Paris, Ky
Bellwood, John,.......... Newcastle, Can
Bidwell, B J........... Tecumseh, Mich
Blanchard, W F........... Manlius, N Y
Blanshard, William,...... Penn Yan, N Y
Brace, A W...... ...West Winfield, N Y
Bradley, A.................. Lee, Mass
Brockway, E P........... ...Ripon, Wis
Brown, Warren,..... Hampton Falls, N H
Brown, James N.'s Sons,... .... Berlin, Ill
Bussing, D S............. Minaville, N Y
Butts, George,............. Manlius, N Y
Cameron, R W........ .... New-York
Campbell, J G J.. .. Lawrenceville, N J
Carmalt, Jas. E............ Montrose, Pa
Cass, A J....... .. .... Holliston, Mass
Charles, R S............ Angelica, N Y
Christie, David,............. Paris, C W
Cleveland, H C............. Coventry, Vt
Cochrane, M H .... Compton, Can
Coffin, Charles E.......... Muirkirk, Md
Coffin, R G........ Coffin's Summit, N Y
Collier, W B............ Bridgeton, Mo
Conger, A B.......... Haverstraw, N Y
Cook, J H & Son, ... Pittsfield, Mass
Cornell, Ezra,............... Ithaca, N Y
Conrad, W C.......... Sandy Bottom, Va
Craig, J R.............. Edmonton, Can
Cutts, Hampden,......... Brattleboro, Vt
Davis, D E.................... Salem, N J
Delafield, T. .... Aurora, N Y
Dodge, Wm B............ Waukegan, Ill
Dun, R G. .. .. London, O
Duncan, W R.............. Towanda, Ill
Dunning, E J..... .... Lenox, Mass
Fitch, G N............. Logansport, Ind
Goodell, D H............... Antrim, N H
Goodman, R............. Lenox, Mass
Goe, J S.................. Brownsville, Pa
Gray, C K. ... . E. Montpelier, Vt
Greig, Major Geo .... Beachville, C W
Greene, J W .... Sayville, N Y
Griswold, A W........... Morrisville, Vt
Groom, B B..... ...... Winchester, Ky
Haight, D B......... Dover Plains, N Y
Hampton, Lewis, ... Winchester, Ky
Harison, T L............. Morley, N Y
Harwood, James A....... Littleton, Mass
Hayward, S....... ...Cummington, Mass
Hazard, Thos ...... Mendon Centre, N Y
Hills, C..................... Delaware, O
Hostetter, A..... .... Mt. Carroll, Ill
Hoyle, George V... ... Champlain, N Y
Hubbard, A C............... Danbury, Ct
Hubbard, C H... .... Springfield, Vt
Iles, Edward,......... .... Springfield, Ill
Jessup, C & Co.......... ... Bristol, Ind
Jones, T C.. .. .. Delaware, O
Judd, A T...... South Hadley Falls, Mass
Juliand, J................ Bainbridge, N Y
Kennedy, R..... .... Hamilton, O
King, William S....... Minneapolis, Minn
Kinkead, F P........... .... Midway, Ky
Kinnaird, J G.. .... Lexington, Ky
Markham, W G.......... ....Avon, N Y
Mason, V W........... Canastota, N Y
Meadows, A. .... Port Hope, Can
Miller, John, ... Brougham, Can
Miller, Robert,........... Pickering, Can
Milne, R .... Lockport, Ill
Murray, George, ... Racine, Wis
Neeley & Bro. .... Ottawa, Ill
Page, John R .. .. Sennett, N Y
Parks, C C & R H......... Waukegan, Ill
Parsons, C., Jr......... ... Conway, Mass
Perry, W N............. Rushville, N Y
Phelps, E A.................... Avon, Ct
Pickrell, J H............. Harristown, Ill
Pond, N G.......... ...... Milford, Ct
Potter Bros.............. Webster, Mass
Reed, S G............. Portland, Oregon
Robbins George L...... Worcester, Mass
Rosenberger, G W... .. New-Market, Va
Sampson, Jas... .. .... Bowdoinham, Me
Schieffelin, W H............ New-York
Scott, M T............... Lexington, Ky
Shedd & Van Sicklen,...... Burlington, Vt
Sherman, H B.................. Toledo, O
Simpson, W Jr......... West Farms, N Y
Skidmore, P A... .... Beekman, N Y
Skinner, H H.......... Silver Lake, Pa
Skinner, W E............ Hamburgh, N J
Slingerland, W H...... Normanskill, N Y
Snell, John,......... .... Edmonton, Can
Sniveley, L R............. Fairview, Md
Sparhawk, Dr................ Gaysville, Vt
Spears, J H..... .... Tallula, Ill
Sprague, G............... Oakwood, Iowa
Stone, F W......... .... Guelph, Can
Streator, S R............. E Cleveland, O
Talbutt, J H............. Lexington, Ky
Talcott Jonathan,............. Rome, N Y
Terrell, M W..... .... Middlefield, Ct
Thomson, J S............... Whitby, Can
Trabue, A E.............. Hannibal, Mo
Underhill, A A..... Clinton Corners, N Y
Van Meter, J M............ Midway, Ky
Wadsworth, J W........... Geneseo, N Y
Walcott & Campbell, New-York Mills, N Y
Ward, C K.... ....Leroy, N Y
Warfield, William,.......... Lexington, Ky
Wells, C L.......... New-Hartford, N Y
Wentworth, John,............ Summit, Ill
Whitman, A............ Fitchburg, Mass
Wilber, M J....... ..Quaker Street, N Y
Winslow, A M & Sons,... .... Putney, Vt
Wistar, R & W L......... Philadelphia, Pa
Young, W W............. Louisville, Ky

HORSES.

Alexander, A J........Spring Station, Ky
Backman, Charles,...... Stony Ford, N Y
Bagby Farm,.................... Tiffin, O
Baker, I V., Jr..Comstock's Landing, N Y
Barnes, W H.......Havre de Grace, Md
Battell, R.................. Norfolk, Ct
Beattie, Simon,............. Bangor, Can
Cameron, R W... .... New-York
Case, W H..... ....Delaware, O
Chapman, T B........... Rochester, Pa
Chenery, W W.......... Belmont, Mass
Cochrane, M H.......... Compton, Can
Conger, A B.... .. ....Haverstraw, N Y
Crozier, William,......... Northport, N Y
Fitch, Thomas,..........New-London, Ct
Gibb, John L..... ....Compton, Can
Goe, J S. .... Brownsville, Pa
Goldsmith, Alden,...Blooming Grove, N Y
Gordon, Clarence,... ....Newburgh, N Y
Haight, D B.......... Dover Plains, N Y
Harison, T L...............Morley, N Y
Hilton, S C.............. Providence, R I
Hitchcock, G C..........New Preston, Ct
Hungerford & White,........ Adams, N Y
Irwin, D B......... ...Middletown, N Y
Johnson, W Fell,......Brooklandville, Md
Kinkead, F P...............Midway, Ky
Leffingwell, W A....... Coldenham, N Y
Merritt, D H............Newburgh, N Y
Morris, Lewis G... ....Fordham, N Y
Ogden Farm,....... ....Newport, R I
Parker, J J............ West Chester, Pa
Parks, C C & R H........ Waukegan, Ill
Phillips, E T..... .... Plainfield, N J
Pickrell, J H......... ....Harristown, Ill
Reynolds, I W H.......... Frankfort, Ky
Russell, H S...... ....Boston, Mass
Sherman, H B. .... Toledo, O
Shields, H L................ Troy, N Y
Stevens, G C............ Milwaukee, Wis
Taber, George..........East Aurora, N Y
Thorne, Edwin, ...Millbrook, N Y
Van Orden, W H..........Catskill, N Y
Wadsworth, E S........... Chicago, Ill
Wood, J G........... W Millbury, Mass

SHETLAND PONIES.

Alexander, A J....... Spring Station, Ky
Anderson, W P............Cincinnati, O
Watson, William,...... West Farms, N Y

COTSWOLD SHEEP.

Albright, J.......... .... Etna, N Y
Appleton, D F........... Ipswich, Mass
Banks, Thad...........Hollidaysburg, Pa
Barbee, G L. .... Georgetown, Ky
Bedford, E G................. Paris, Ky
Burroughs, H K.......... Roxbury, N Y
Chase, L A.... ....New-York
Cochrane, M H...........Compton, Can
Coffin, H T..........Poughkeepsie, N Y
Crozier, William,......... Northport, N Y
Deuel, S T..... ....Little Rest, N Y
Hall, John,.......... ....Catharine, N Y
Harris, Jos................Rochester, N Y
Hartwell, S...............Washington, Ct
Hiester, C E...........West Chester, Pa
Hoyle, George V.........Champlain, N Y
Humphreys, John,........Robystown, Md
Ingersoll, George,....... Charleston, N Y
Jackson, George,.........Wilmington, Del
Johnson, W R.............Warwick, R I
King, W S .... Minneapolis, Minn
Loomis, Burdett,.............Hartford, Ct
Loomis, Byron,.................Suffield, Ct
McFerran, J C.... .... Louisville, Ky
Osborn, B L.........Oswego Village, N Y
Parks, C C & R H......... Waukegan, Ill
Paxton, C R............ Bloomsburgh, Pa
Perry, W N...............Rushville, N Y
Phelps, C C........ ...... Vernon, N Y
Phillips, E T....... ...... Plainfield, N J
Pratt, J M.................Goshenville, Pa
Roberts, W B.. .. ...King of Prussia, Pa
Rockwell, J M.......... Butternuts, N Y
Sayre, Cooper, ... Oaks Corners, N Y
Sherman, H B.... ....Toledo, O
Skinner, H H........... Silver Lake, Pa
Snell, John,..... .... Edmonton, Can
Ste Marie, A.............La Prairie, Can
Stone, Fred. Wm...........Guelph, Can
Tabor, A...................Aurora, N Y
Tatum, G M.............Woodbury, N J
Thorne, Edwin,.......... Millbrook, N Y
Underhill, A A.. ...Clinton Corners, N Y
Wilson, W T........... West Liberty, O

LEICESTER SHEEP.

Buckingham, J ....Zanesville, O
Curtis, F D...............Charlton, N Y
Edgerton, Jas.......... .. Barnesville, O
Hills, C................. ...Delaware, O
Hoyle, George V........ Champlain, N Y
Kirby, Joseph, ...Milton, Can
Redmond, William, ...New-York
Snell, John,.... ....Edmonton, Can
Vergon, F P.... ....Delaware, O
Walcott & Campbell, New-York Mills, N Y
Winne, Jurian,..........Bethlehem, N Y

LINCOLN SHEEP.

Chapman, J R.........Oneida Lake, N Y
Chenery, W W.. ....Belmont, Mass
Cochrane, M H............Compton, Can
Walcott & Campbell, New-York Mills, N Y

MERINO SHEEP.

Baker, I. V., Jr., Comstock's Landing, N Y
Baldwin, Theo. E...... .. .Litchfield, Ct
Bottum, N.....................Shaftsbury, Vt
Chamberlain, Wm........Red Hook, N Y
Cole, Walter,.................Batavia, N Y
Drew, L S.............So. Burlington, Vt
Godeffroy, Brancker & Co.,.....New-York
Goe, J S....... .... Brownsville, Pa
Hubbard, C H............Springfield, Vt
Pettibone, J S............ Manchester, Vt
Steele, T S.................. Shushan, N Y

HAMPSHIRE-DOWN SHEEP.

Ashworth, J............ .. Ottawa, Can
Hubbell, O S............ Philadelphia, Pa
Morrell, Robert, ...Manhasset, N Y
Newell, Dr. A D....New-Brunswick, N J

SHROPSHIRE SHEEP.

Conger, A B.......... Haverstraw, N Y

SOUTH-DOWN SHEEP.

Alexander, A J........ Spring Station, Ky
Brown, Geo. H., Washington Hollow, N Y
Buffum, Thomas B..........Newport, R I
Carson, A H............. Newport, R I
Covell, W R............Orange C. H., Va
Giles, John,.................. Putnam, Ct
Harison, T L..................Morley, N Y
Hills, C. .... Delaware, O
Hornbrook, R S & Co., New-Harmony, Ind
Houghton Farm,............ .. Putney, Vt
Hulse, Benj.............. Allentown, N J
Jenkins, J Stricker, ...Baltimore, Md
Jones, T C.. .... Delaware, O
Juliand, Joseph,......... Bainbridge, N Y
Moore, Edwin,...... .. Port Kennedy, Pa
Morris, Dr. J C ....West Chester, Pa
Parks, C C & R H......... Waukegan, Ill
Pickrell, J H.............. Harristown, Ill
Reeder, E................ New-Hope, Pa
Reybold, J F............. St. Georges, Del
Reynolds, I W H..... .... Frankfort, Ky
Sharpless, Samuel J......Philadelphia, Pa
Sinclair, S........ ... .New-York
Stewart, H L.... ....Middle Haddam, Ct
Stone, Fred. Wm...........Guelph, Can
Taylor, W J C.............Holmdel, N J
Underhill, A A..... Clinton Corners, N Y
Van Meter, J M............. Midway, Ky
Wainwright, C S.........Rhinebeck, N Y
Worth, Francis,...........Marshallton, Pa

BERKSHIRE SWINE.

Abbott, J J C..............Montreal, Can
Babbage, R A............ Dubuque, Iowa
Ball, A P..................Derby Line, Vt
Barbee, W H. . ... Frankfort, Ky
Barbee, G L............ Georgetown, Ky
Bedford, E G................. Paris, Ky
Bennett, W A.................Dover, Ky
Brown, Dr. L E ....Eminence, Ky
Brown, S H.. .... Millbrook, N Y
Cass, J F...... ....L'Original, Can
Christie, David,................Paris, Can
Cochrane, M H...........Compton, Can
Coffin, C E...................Muirkirk, Md
Colt, S C ....Hartford, Ct
Craig, J R.............Edmonton, Can
Crozier, William,.......... Northport, N Y
Crutcher, T G......... ...Shelbyville, Ky
DeForest, J J......... Duanesburg, N Y
Deuel, S T............. Little Rest, N Y
Forsyth, John,......... ....Toronto, Can
Haines, J C.............Clarksboro, N J
Homer, G W...... ..Framingham, Mass
Hubbell, O S........... Philadelphia, Pa
Jones, G W............Jones' Station, O
King, W S........... Minneapolis, Minn
Loomis, Burdett, ...Hartford, Ct
McCarty, T J., & Co....... Salem, O
Miller, John,............. Brougham, Can
Morris, Dr. J C.........West Chester, Pa
Oakley, Chas........... ... Roslyn, N Y
Parks, C C & R H.........Waukegan, Ill
Pettee, W J......... ...Lakeville, Ct
Pickrell, J H.............Harristown, Ill
Pond, N G...................Milford, Ct
Riehl, E A..... .... Alton, Ill
Scudder, M S........... Grantville, Mass
Sheffield, W R... ....Saugerties, N Y
Sherman, H B....................Toledo, O
Snyder, P....................Marysville, O
Snell, John, ... Edmonton, Can
Sprague, G...............Oakwood, Iowa
Stitt, W E....... .... Columbus, Wis
Stone, F W......... ....Guelph, Can
Ticknor, E................ St. Louis, Mo

ESSEX SWINE.

Anderson, W P............ Cincinnati, O
Bowditch, E F........ Framingham, Mass
Brown, S H..............Millbrook, N Y
Ogden Farm,................Newport, R I

CHESHIRE SWINE.

Clark & Green,............Belleville, N Y
Deforest J J..............Duanesburg, N Y
Gruver, W H.... ....Pleasant Valley, Pa
Hicks, C M............... Rushville, N Y
Hoxie, A K........Stockport Station, N Y
Perry, W N........ ......Rushville, N Y
Rockwell, J M...........Butternuts, N Y
Stiles, W H..............Mamaroneck, N Y

SUFFOLK SWINE.

Battles, A.....................Girard, Pa
Cobb, Henry, ...Amherst, Mass
Giles, John,.......... ....Putnam, Ct
Haswell H C..............Deerfield, Mass
Howard, A B..........Belchertown, Mass
Hyde, Alex........... ... .Lee, Mass
Nason, H ....Montclair, N J
Ticknor, E. ....St. Louis, Mo

YORKSHIRE SWINE.

Bush, F T............. Auburndale, Mass
Bordwell, R R C..........Penn Yan, N Y
Chenery, W W........... Belmont, Mass
Codman, Ogden,..............Lincoln, Mass
Cooper, T S. .... Coopersburg, Pa
Howe, M S.....................Penn Yan, N Y
Landon, Stephen,...............Eden, N Y
Lightfoot, T............Maiden Creek, Pa

CHESTER COUNTY SWINE.

Ashley, A B.................Burlington, Vt
Baggerley, W F ....South Butler, N Y
Bradley, John, & Co..........Chester, Pa
Battles, A ....Girard, Pa
Beal, N T...............Rogersville, Tenn
Bidwell, B J.............Tecumseh, Mich
Cox, T F........................Osborn, O
Darlington, R S........ West Chester, Pa
Early, J A. ....Youngstown, O
Edgerton, James, ...Barnesville, O
Elliott, W V.. ....Wapakoneta, O
Gordon, Clarence,........Newburgh, N Y
Hickman, G B......... West Chester, Pa
Hodgson, R H..........New-London, Pa
Hooff, Lewis...... .... Alexandria, Va
Horton, E W............ Muscatine, Iowa
Irwin, J W & M.......Penningtonville, Pa
Lehman, H F ....Hagerstown, Md
Lewis, P G............ Monroe, N Y
Mackie, J M......Great Barrington, Mass
McClintock, D................ Solon, O
McCully, Cyrus, ...Hubbard, O
Nichols, H C...........Cowlesville, N Y

Parks, C C & R H.........Waukegan, Ill
Pond, N G.................Milford, Ct
Riley, Fred..............Sterling, Mass
Roberts, J C ....West Chester, Pa
Silver, L B.................Salem, O
Sparhawk, Dr.................Gaysville, Vt
Thompson, G W.... New-Brunswick, N J
Tillinghast, J T.......New-Bedford, Mass
Tilton, H W... .... Walpole, Mass
Todd, S H...............Wakeman, O
Van Winkle, I, Jr,........ Rockaway, N J
Whitehead, M...........Middlebush, N J
Wood, Thomas,..... ... ... Doe Run, Pa
Wood, Ira B.............Iron Furnace, O
Worth, Francis,.......... Marshallton, Pa
Young, James, Jr., & Co., Marshallton, Pa

POULTRY FANCIERS.

Allen, A B........ ....Tom's River, N J
Allen, J.... .... Conneaut, O
Anderson, H S.... ....Geneva, N Y
Atkinson, W B..............Boston, Mass
Ball, H S.............. Shrewsbury, Mass
Barry, T F.............. Rochester, N Y
Bassett, G W..................Barre, Vt
Bateman, H B...............Ripon, Wis
Bates, C P......... Schuyler's Lake, N Y
Battles, C P...... .... Girard, Pa
Beattie, Simon,..............Bangor, Can
Berry, J J.. .. .. Hackensack, N J
Betts, C H............ ... Baltimore, Md
Bicknell, J Y ....Westmoreland, N Y
Blair, K... .. ..Allegheny City, Pa
Blanchard, Webster,........ Newton, N J
Bordwell, E O....... ....Penn Yan, N Y
Bradley, G W........... ... Hamden, Ct
Burgess, Edward,..... Poughkeepsie, N Y
Bush, W C & W M.....Auburndale, Mass
Butts, George,.............Manlius, N Y
Buzzell, J P...............Clinton, Mass
Cameron, R W... .... New-York
Campbell, A G ... . St. Hilaire, Can
Carpenter, F & W..............Rye, N Y
Cary, Willard,...... ....Milford, N H
Carson, A H....... ....Newport, R I
Champney, G F............Taunton, Mass
Chapman, A......... ....Middlebury, Vt
Churchman, W H........Wilmington, Del
Clark, John L............ Waterloo, N Y
Clark, E C Jr..........Ballston Spa, N Y
Clift, W.............. Mystic Bridge, Ct
Coffin, C E................ Muirkirk, Md
Cochrane T A............ Baltimore, Md
Cole, Walter,............... Batavia, N Y
Colt, S C........... ....Hartford, Ct
Comey, E C.......... ... Quincy, Mass
Cooper, J C... .... Limerick, Ireland
Coon, D D F............Marcellus, N Y
Corbett, C C... ....Norwich, Ct
Cowles, Theron,............Syracuse, N Y
Cox, J B.................. Zanesville, O
Crozier, William,......... Northport, N Y
Curwen, G F. . .....West Haverford, Pa
Darlington, R S. ...... West Chester, Pa
Deuel, S T..... ....Little Rest, N Y
Dewey, T H ..... Pomfret Landing, Conn
Dewey, G B.............. Hartford, Conn
Dibble, E B..............New-Haven, Ct
Dillon, J C . ...Amherst, Mass
Dudley Bros.......... ....Augusta, N Y
Dunbar, E B.................Bristol, Ct
Dunbar, G C.. ....East Abington, Mass
Early, J A.... ....Youngstown, O
Ellis, Robert,......... Schenectady, N Y
Engle, Hiram,. ....Marietta, Pa
Estabrook, G W........... Grafton, Mass
Estes, J J...........East Abington, Mass
Felter, G W...................Batavia, O
Felter, Jas.......... Rensselaerville, N Y
Fitz, Geo. C................Ipswich, Mass
Forsyth, John...............Toronto, Can
Frazier, E R........... Plattsburgh, N Y
Gardner, L C... ....Fayetteville, N Y
Giles, John,..................Putnam, Ct
Goodell, D H..............Antrim, N H
Gould, Thomas,. ....Aurora, N Y
Hadwen, O B... ....Worcester, Mass
Haines, J C..............Clarksboro, N J
Hall, John H..............Catharine, N Y
Hammer, E C...............Branford, Ct
Hand, T J ....Sing Sing, N Y
Hanks, W....... Middle Granville, N Y
Hatch, O L.............Worcester, Mass
Herstine, D W.......... Philadelphia, Pa
Hicks, Benj.................Roslyn, N Y
Hills, W R ....Albany, N Y
Hills, Henry N..............Delaware, O
Hitchman, D A...........Schoharie, N Y
Hodgson Bros............ ....New-York
Homer, G W.. ....Framingham, Mass
Horton, E W... ....Muscatine, Iowa
Howard, A McL..............Toronto, Can
Howlett, E P..............Syracuse, N Y
Hudson, P W......North Manchester, Ct
Hughes, J................Marshallton, Pa
Hull, E D..................Newton, Mass
Hunt, W M...............Waterloo, N Y
Ives, John S .... Salem, Mass
Judd, J W .... Box 3040, New-York
Juliand, Jos........... ....Bainbridge, N Y
Kelley, Seth,..........So. Yarmouth, Mass
King, Henry,............ Galesburg, Mich
Leavitt, G H................ Flushing, N Y
Leland, Warren, ...Rye, N Y
Lent, D B............Poughkeepsie, N Y
Lippincott, J S...........Mt. Holly, N J
Loftie, Henry,......... ...Syracuse, N Y
Long, J C Jr................ Ravenna, O
Lord, John A... ....Kennebunk, Me
Loring, C Carroll,...........Boston, Mass
Ludlow, C N..............Mt. Carmel, O
Lummis, F C.................Chaplin, Ct
Maitland, Robt. L Jr..... Red Bank, N J
McCarty, T J...... ....Salem, O
McClean, Thos............ Toronto, Can
Miles, F W.............. Plainfield, N J
Meacham, G A .....No. Cambridge, Mass
Merry, S E . ...Milan, O
Murrill, F H......... New Bedford, Mass
Morrell, Robt...... ....Manhassett, N Y
Nettleton, C P............Birmingham, Ct
Nichols, Burr H.......... Lockport, N Y
Noxon, D C.. .... Beekman, N Y
Oakley, Charles, ...Roslyn, N Y
Oat, D B............... West Chester, Pa

Ordway, W H.... Box 3138, Boston, Mass
Parker, S J........... West Chester, Pa
Parks, C C & R H......... Waukegan, Ill
Paulding, D C .... Mt. Kisco, N Y
Perkins, N B Jr. .... Salem, Mass
Perry, W N.............. Rushville, N Y
Phillips, E T........... Plainfield, N J
Pitman, Mark,............... Salem, Mass
Pogue, Joseph,............ Grafton, Mass
Polk, W H.... .... Paris, Ky
Pope & Lyman,........... Concord, Mass
Potter, F S........ No. Dartmouth, Mass
Rager, John,.. .... Waterloo, N Y
Ralph, D C........ .. .... Buffalo, N Y
Rice, J L.. .. ..Rensselaerville, N Y
Richards, J A & R R..Lansingburgh, N Y
Richardson, J D... ...Buckeyestown, Md
Roberts, J C........... West Chester, Pa
Roberts, J A..................... Paoli, Pa
Rockwell, J M.......... Butternuts, N Y
Scudder, M S.......... Grantville, Mass
Schuyler, P.............. West Troy, N Y
Sharpless, C L....... ... Philadelphia, Pa
Sharpless, Saml J........ Philadelphia, Pa
Shaw, R B........... Canandaigua, N Y
Shelton, J D...... .... Jamaica, N Y
Sherman, H B. . ... Toledo, O
Sherlock, Mrs. M J..... .. Elizabeth, N J
Sidel, J Clarence, ... Englewood, N J
Simpson, W Jr......... West Farms, N Y
Smedley, J G..... Willistown Junction, Pa
Smith, G Morgan,.... South Hadley, Mass
Snow, H H...... ...... New-Haven, Ct
Sparbeck, G M.......... Cranesville, N Y
Spalding, L A............. Lockport, N Y
Stage, D L............. Schenectady, N Y
Stephens, Sheldon,......... Montreal, Can
Stickney, E .... Burton, O
Stone, John,............. Coatesville, Pa
Stowell, H F........... Williamsport, Pa
Studley, E G........... Claverack, N Y
Stuyvesant, J R..... .. Poughkeepsie, N Y
Talcott, J ....Rome, N Y
Tannar, A...... ....Norwich, Ct
Tatnall, A R.......... .. Wilmington, Del
Tatum, G M............. Woodbury, N J
Taylor, E J............... Waterloo, N Y
Taylor, F.. .... Oakdale, Pa
Thomas, J P........ West Whiteland, Pa
Thompson, G W.... New-Brunswick, N J
Thorne, Edwin, ... Millbrook, N Y
Todd, W H........... Vermillion, O
Townsend, C H.......... .... Utica, N Y
Treadwell, C W. .... Exeter, N H
Tuttle, C E ...... Box 1156, Boston, Mass
Upham, D A...... .... Wilsonville, Ct
Valentine, C D........... Fordham, N Y
Van Keuren, C........... Rondout, N Y
Van Winkle, I........... Greenville, N J

Walcott, B D... ... New-York Mills, N Y
Warner, G H .. ... New-York Mills, N Y
Warner, John C.... Blooming Grove, N Y
Warr, Edw....... .... Utica, N Y
Warren, W R................ Albion, N Y
Warren, John,............. Flushing, N Y
Wells, C L.......... New-Hartford, N Y
Welles, J C... .... Athens, Pa
Wentworth, John,............ Summit, Ill
Wheaton, C C............. Orange, Mass
White, L E............... Taunton, Mass
Whitney, E H...... Cambridgeport, Mass
Wilbur, D L. .... Boonsboro, Iowa
Willard, F H........... Little Falls, N Y
Willetts, W .... Old Westbury, N Y
Williams, P .. .. Taunton, Mass
Williams, J P...... New-York Mills, N Y
Wolcott, R P....... Holland Patent, N Y
Wood, Thomas,......... .. Doe Run, Pa
Woodward, H...... ... Worcester, Mass
Worth Francis,...... .... Marshallton, Pa
Zollikoffer, E.... .... Uniontown, Md

American Deer.

Maitland, Robt. L............ New-York

Bees.

Bradley, A .... Lee, Mass
Hazen, Jasper,............... Albany, N Y
Langstroth, L L.................. Oxford, O
Quinby, M.... ...... St. Johnsville, N Y
Woody, T H B.......... Manchester, Mo

Angora Goats.

Chenery, W W.......... Belmont, Mass
Dinsmore, W B....... Staatsburgh, N Y
Covell, W Ross........ Orange C. H., Va
Eutychides, A.......... Spout Springs, Va
Goe, J S.... .... Brownsville, Pa
Harrison, C K......... ... Pikesville, Md
Peters, Richard,... .... Atlanta, Ga

Fish and Spawn.

Clift, W............... Mystic Bridge, Ct

Ferrets.

Osgoodby, J H............ Pittsford, N Y

Dogs.

Betts, C H .... Baltimore, Md
Bradley, G W... .... Hamden, Ct
Buckingham, J .... Zanesville, O
Cooper, T S............ Coopersburg, Pa
Estes, J J..... ..... East Abington, Mass
Hall, John,...... .... Catharine, N Y
Hallock, S P........... . Oriskany, N Y
Hulse, Benj..... .... Allentown, N J
Ives, J S. .... Salem, Mass
McGugin, S M............... Ironton, O
Stephens, S Sheldon,....... Montreal, Can
Wonson, H F........ E. Gloucester, Mass

Horticultural and Seed Register.

Nurseries.

Adair, William,........ ... Detroit, Mich
Adams, J W........... Springfield, Mass
Allen, John M.......... Hightstown, N J
Andrews, C................ Marengo, Ill

Bailey, W H. .... Plattsburg, N Y
Baker, George,.... .... Toledo, O
Battles, A.... ... . Girard, Pa
Berst & Bro......... .... Erie, Pa
Buist, R. . Darby Road, Philadelphia, Pa

Boardman, S & Co........Rochester, N Y
Bouck, D.............. .. Lockport, N Y
Bowen, P.... .... East Aurora, N Y
Chapman, J F & Co.... Fayetteville, N Y
Claggett, J B...........Brightwood, D C
Cooney, P H........... .... . Erie, Pa
Cumming, R & Co.........Pittsburgh, Pa
Draper, Jas..... ....Worcester, Mass
Duffell, S..................Yardville, N J
Ellwanger & Barry,........Rochester, N Y
Engle & Bro................Marietta, Pa
Evans, E J & Co................York, Pa
Fancher, F B.........Lansingburgh, N Y
Ferris, W L.......... Throg's Neck, N Y
Frost & Co., ...Rochester, N Y
Frost, E C. ....Watkins, N Y
Foster, Suel,.............Muscatine, Iowa
Goodale, S L....................Saco, Me
Gould Bros....... ....Rochester, N Y
Graves, Selover, Willard & Co., Geneva, N Y
Griffith, W............. ..North East, Pa
Hadwen, O B... ....Worcester, Mass
Hance & Son, A..........Red Bank, N J
Hanford, R G...... ....Columbus, O
Harrington, E W & Co......Geneva, N Y
Heikes, W F..................Dayton, O
Herendeen & Co........... Geneva, N Y
Hooker, H E & Bro...... Rochester, N Y
Hoopes, Bro & Thomas, West Chester, Pa
Hovey & Co... ... .Boston, Mass
How, H K..........New-Brunswick, N J
Howard, J R.........North Easton, Mass
Hoyt, S & Sons,.........New-Canaan, Ct
Hubbard, T S & Co........Fredonia, N Y
Kendig, R P.............. Waterloo, N Y
King & Murray,............Flushing, N Y
Kreider, John G...........Lancaster, Pa
Lacon Nursery Co..............Lacon, Ill
Lenk & Co.................. Toledo, O
Lewis, M H................Sandusky, O
Little, W S...............Rochester, N Y
Love, J W................ Geneva, N Y
Manning, J W .... Reading, Mass
Marshall, S B...............Cleveland, O
Maxwell, T C & Bros........Geneva, N Y
McCullough, Drake & Co.. Sharpsburg, O
Meehan, Thomas,........Germantown, Pa
Merrell & Coleman,......... Geneva, N Y
Mickey, S T............ .... Salem, N C
Moody, E & Sons,........ Lockport, N Y
Moon, Mahlon,............Morrisville, Pa
Murdoch, J R & A..........Pittsburg, Pa
Otto & Achelis..........West Chester, Pa
Parry, William, ...Cinnaminson, N J
Parsons & Co.......... ....Flushing, N Y
Pearson, W L.... ....Schenectady, N Y
Perry, F L.. . ...Canandaigua, N Y
Peters, C P..............Concordville, Pa
Peters, Randolph,.... ...Wilmington. Del
Phœnix, F K............ Bloomington, Ill
Pratt, DeWitt & Co .... Geneva, N Y
Puller, T J. .... Hightstown, N J
Rakestraw & Pyle .... Willowdale, Pa
Reagles & Bradley,..... Schenectady, N Y
Richardson & Nicholas, . . .Geneva, N Y
Riehl, E A..... .... Alton, Ill
Roberts, Josiah A.............. Paoli, Pa
Rollins, A S & Co.... .. Waterloo, N Y
Root, James A......... Skaneateles, N Y
Saul, John,.............Washington, D C
Smith, Clark & Powell,.... Syracuse, N Y
Southwick, T T & Co..... Dansville, N Y
Storrs, Harrison & Co,......Painesville, O
Strong, W C & Co........Brighton, Mass
Studley, E G............ Claverack, N Y
Swasey, H A & Co..........Canton, Miss
Sylvester, E Ware. .... Lyons, N Y
Teas, E Y...... .... Richmond, Ind
Van Dusen, C L......... Macedon, N Y
Wampler, John,.............Trotwood, O
Watson, B M............ Plymouth, Mass
Wickersham, Josiah,......Bendersville, Pa
Will & Clark,...........Fayetteville, N Y
Wilson, M D.............Rochester, N Y
Wilson, G W & Co.. ... Bendersville, Pa
Wood & Hall, ...Geneva, N Y

Foreign Nursery Agents.

Bruguiere & Thebaud,......... New-York
(*for Andre Leroy, Angers, France.*)
Knauth, Nachod & Kuhne,..... New-York
(*for P.& E. Transon, Orleans, France.*)
Lauer, R., *Dutch Bulbs*,....... New-York

Small Fruits and Grapes.

Allen, C M.................Beverley, N J
Allen, S L............. Cinnaminson, N J
Allis, S J................. North East, Pa
Andrews, T C.......... Moorestown, N J
Babcock, I H & Co.........Lockport, N Y
Barnett, W N...........West Haven, Ct
Bateham, M B.......... ... Painesville, O
Bohemia Vineyards,..... Town Point, Md
Boyce, W C.............. Lockport, N Y
Brehm, F C...... ....Waterloo, N Y
Briggs, I W......... West Macedon, N Y
Brown, D H.........New-Brunswick, N J
Burgess, Edw.........Poughkeepsie, N Y
Bush & Son, Isidore,.......Bushburg, Mo
Cadwallader, M...........Fallsington, Pa
Campbell, G W............. Delaware, O
Carpenter, W S... .... Rye, N Y
Chinnick, W J...............Trenton, N J
Collins, Charles,........ Moorestown, N J
Collins, John S......... Moorestown, N J
Cone, J W .... Vineland, N J
Conover & Son,...........Freehold, N J
Cooney, P H.......................Erie, Pa
Craine, J..............East Lockport, N Y
Cumming, R & Co.... ....Pittsburgh, Pa
Curwen, G F. ....West Haverford, Pa
Davis, S C.................. Medina, N Y
Dingwall, John,..............Albany, N Y
Dodge, U E............. Fredonia, N Y
Donaldson, J A.........St. Joseph, Mich
Draper, James,....... ..Worcester, Mass
Duffell, S T...............Yardville, N J
Ellwanger & Barry,.........Rochester N Y
Ferris, L M & Son,... Poughkeepsie, N Y
Goldsmith, W H......... ...Newark, N J
Griffith & Griffith,........ North East, Pa
Hall, Isaac,............ .... Frazer, Pa
Harris, T L ....Brocton, N Y
Hasbrouck & Bushnell,.... Peekskill, N Y
Hathaway, B...Little Prairie Ronde, Mich

Haynes, J H.... ....Delphi, Ind
Hendricks, H........... Kingston, N Y
Herstine, D W.......... Philadelphia, Pa
Hoag, C L & Co.......... Lockport, N Y
Hubbard, T S & Co....... Fredonia, N Y
Hunt, W M..............Waterloo, N Y
Johnson, H C.. .... Berlin Heights, O
Kinsey & Gaines,............ Dayton, O
Kline, W J.... ....Port Ewen, N Y
Lambert, George H., New-Brunswick, N J
Leeds, N.............. Cinnaminson, N J
Lindley, N H. ... .Bridgeport, Ct
Lord, E E... .... Newark, N Y
Mallory & Downs,.....South Norwalk, Ct
Martin, James F...... Mt. Washington, O
Massey & Hudson, ...Chestertown, Md
McCullough, J M & Sons,.. Cincinnati, O
McLaury, D........ New-Brunswick, N J
Merceron, F F........... Cattawissa, Pa
Merriman, A............ Granville, O
Moon, Mahlon,........ ....Morrisville, Pa
Moore, A J.............Berlin Heights, O
Oneida Community,........ Oneida, N Y
Palmer & Risley, *Hops*,.. Waterville, N Y
Parry, William,........ Cinnaminson, N J
Parsons & Co.......... ...Flushing, N Y
Patterson, J S..........Berlin Heights, O
Peck, T R. .... Waterloo, N Y
Perry, F L........... Canandaigua, N Y
Potter, E J & Co...... Knowlesville, N Y
Pullen, J Madison,...... Hightstown, N J
Purdy, A M....... ...... Palmyra, N Y
Purdy & Hance,........ South Bend, Ind
Reisig & Hexamer,.......New-Castle, N Y
Reynolds, P C............Rochester, N Y
Ringueberg, N S & Son,...Lockport, N Y
Ritz, Louis,................. Plainville, O
Robinson, E D....... Howlett Hill, N Y
Salter, H H............. .. Lima, N Y
Shaw, C W. . ...Carver, Mass
Shuler, Mrs. J D.......... Lockport, N Y
Smith, J T B.............Kingston, N Y
Strong, W C............ Brighton, Mass
Sutvan, Stokes,.........Haddonfield, N J
Sylvester, E W............ Lyons, N Y
Tatum, J C............ Woodbury, N J
Thompson, G W.... New-Brunswick, N J
Tillson, O J...............Highland, N Y
Todd, L U................Vermillion, O
Travis & Fields,.......Dobb's Ferry, N Y
Tucker, F D............. Ithaca, N Y
Underhill, S W......Croton Landing, N Y
Van Dusen, C L...... ....Macedon, N Y
Wakeman, T B & H.. ......Westport, Ct
Walton, Silas,........ Moorestown, N J
Williams, J G...........Moorestown, N J
Wilson, M N.............Macedon, N Y
Wood, L L.......... ...Vineland, N J

SEEDSMEN, FLORISTS, &c.

Acker, H E.... ...... Woodbridge, N J
Allen, R H & Co......... ....New-York
Allen, C L & Co.......... Brooklyn, N Y
Ashley, A D.......... Milton Depot, Vt
Babcock, Mrs. L D....... Clarkson, N Y
Barler, O L & Co.........Upper Alton, Ill
Barnum & Bro............ St. Louis, Mo
Barry, W C............. Rochester, N Y
Bliss, B K & Sons,.............New-York
Breck, Jos., & Sons,........ Boston, Mass
Briggs & Bro ....Rochester. N Y
Brill, Francis, ...Mattituck, N Y
Buist, R....Darby Road, Philadelphia, Pa
Buist, R Jr............ Philadelphia, Pa
Burras, O.............North Fairfield, O
Campbell, Festus,.........Pittsfield, Mass
Carpenter, C G...........Richmond, Ind
Catlin, H A.... .... Corry, Pa
Collins, Alderson & Co... Philadelphia, Pa
Curtis & Cobb,.............Boston, Mass
DeGroff Nelson & Co... Fort Wayne, Ind
Deitz, G W...........Chambersburg, Pa
Dingee & Conard,........West Grove, Pa
Douw, V P & Co. ....Albany, N Y
Dreer, Henry A...... ...Philadelphia, Pa
Fancher, F B........ Lansingburgh, N Y
Fanning, S B........ ... Jamesport, N Y
Ferre, Batchelder & Co..Springfield, Mass
Fleming, James,...............New-York
Foote, J A. ....Terre Haute, Ind
Gregory J J H.... ... Marblehead, Mass
Hacker, Wetherill & Co..Philadelphia, Pa
Hawley, R D...... ....Hartford, Ct
Hazard, J F....... Mendon Centre, N Y
Henderson & Co., Peter,.......New-York
Herendeen & Co...........Geneva, N Y
Hovey & Co...............Boston, Mass
Ives, John S .... Salem, Mass
Jones, W H............ Philadelphia, Pa
Kellogg, F S................ Chicago, Ill
Knox, W W................Pittsburgh, Pa
Landreth, David, & Son, Philadelphia, Pa
Massey & Hudson,..... Chestertown, Md
McCullough, J M & Sons,... Cincinnati, O
Mead, Alex...............Greenwich, Ct
Murdoch, J R & A..........Pittsburg, Pa
Olm Brothers,...........Springfield, Mass
Payne & DeLong,......... Lexington, Ky
Pell, O. P., Secretary,..........New-York
Peck, H S & Co...........Melrose, Mass
Philipps, Henry,........... ...Toledo, O
Reeser, C A............Pleasantville, Pa
Reeves, E A................ New-York
Reynolds, M G.... ......Rochester, N Y
Richardson & Gould, ...New-York
Rogers, C B.... ....Philadelphia, Pa
Rumsey, H M.................Salem, N J
Sanders, Edgar,............ Chicago, Ill
Saul, John,............Washington, D C
Scott, L D & Co.................Huron, O
Schwill A & Co............ Cincinnati, O
Sheppard's Seed Store,........ New-York
Snow, Dexter,.. ....Chicopee, Mass
Stevens, G M. ....Danbury, Ct
Stoms, T M & Co.......... Cincinnati, O
Strong, W C.. . ...Brighton, Mass
Such, Geo............South Amboy, N J
Thorburn, J M & Co... ....New-York
Teas, E Y.... .... Richmond, Ind
Vanderbilt, John, & Bros.......New-York
Van Dusen, C L...........Macedon, N Y
Vick, James,............. Rochester, N Y
Waring, Jr., G E......... ...Newport, R I
Washburn & Co........... Boston, Mass
Watson, B M.......... Plymouth, Mass

Wells, S M & D......... Wethersfield, Ct
Wood & Hall,...... ...... Geneva, N Y

Cranberries.

Makepiece, A D...........Hyannis, Miss
Trowbridge, F........... ....Milford, Ct
Wonson, G M........E. Gloucester, Mass

Seed Grains.

Ackert, Jacob,...........Hart's Falls, N Y
Arnold, Charles,....... .... Paris, Can
Barber, Alfred,.... ....Hancock, N H
Battles, A.... ....Girard, Pa
Boardman, H M.......... Rushville, N Y
Bryan, E T.............. Marshall, Mich
Buttles, A B................ Columbus, O
Burras, O..............North Fairfield, O
Bussing, J W.......... Amsterdam, N Y
Dickerman, J H.... .... Mt. Carmel, Ct
Fanning, S B............ Jamesport, N Y
Hazard, J F........ Mendon Centre, N Y
Hendricks, H............ Kingston, N Y
Jenison, D. .. .. Lock Berlin, N Y
Kelsey, H C .... Newton, N J
Kennedy, S....... ....Evansville, Ind
Large, S P............West Elizabeth, Pa
Masker, Aaron,........Perth Amboy, N J
Nelson, D & Co.........Fort Wayne, Ind
Newton, W..............Henrietta, N Y
Noyes, N H. .... Geneva, N Y
Pearsall, Wm.......... Moorestown, N J
Perry, W N.............. Rushville, N Y
Phelps, C C................ Vernon, N Y
Potter, E J........... Knowlesville, N Y
Small & Fisher, ... Woodstock, N B
Talcott, Jona,................Rome, N Y
Van Dusen, C L...........Macedon, N Y
Wayne, W G.. .... Seneca Falls, N Y
Wood, Caleb,............Philadelphia, Pa
Wright, Robt........... ...Sabrevois, Can

Seed Potatoes.

Baker, W L....... ....Portlandville, N Y
Bliss, B K & Sons, ...New-York
Briggs, I W....... .. West Macedon, N Y
Burgess, Edw.........Poughkeepsie, N Y
Burras, O..............North Fairfield, O
Edgerton, James,............Barnesville, O
Ellwanger & Barry,.......Rochester, N Y
Fanning, S B.............Jamesport, N Y
Gregory, J J H .... Marblehead, Mass
Griscom, W W... .... Woodbury, N J
Hicks, Isaac, & Son,..Old Westbury, N Y
Ives, J S. ....Salem, Mass
Peters, C P.......... .. Concordville, Pa
Potter, E J & Co.......Knowlesville, N Y
Pringle Bros............ .. Charlotte, Vt
Qua, Frank, ...North Granville, N Y
Reisig & Hexamer, ...New-Castle, N Y
Scott, L D & Co...... ... . Huron, O
Talcott, J.,.............. ....Rome, N Y
Thompson, G W.....New-Brunswick, N J
Uhl, Stephen,. ....Poughkeepsie, N Y
Wainwright, Geo..........Lake Como, Pa
Weld, W H....... .... Lockport, N Y

Sweet Potatoes.

Allen, J...... ... . Conneaut, O
Barrows, C H.............Willimantic, Ct
Chadwick, W W...........Mt. Healthy, O
Cummins, D.... ....Conneaut, O
Gray, S........................ Norwalk, O
Murray, M M.......Foster's Crossings, O
Riehl, E A........ .... Alton, Ill
Stoms, T M & Co...... ....Cincinnati, O
Whitall, Clement,.........Woodbury, N J
Wood, Ira R..... ....Iron Furnace, O

Implements, Machines, Fertilizers, &c.

Agricultural Warehouses.

Allen, R H & Co. ... Box 376, New-York
Ames Plow Company,...... .Boston, Mass
Barrett, W E & Co...... Providence, R I
Bartholomew, C.............. Etna, N Y
Beardslee, H W .... Syracuse, N Y
Blymyer, Norton & Co......Cincinnati, O
Blymyer, Day & Co..........Mansfield, O
Boyer, W L & Co........Philadelphia, Pa
Bradley, C C & Son,.......Syracuse, N Y
Brearley, A L & Co.........Trenton, N J
Decatur, J R & Co........... New-York
DeGroff Nelson & Co.. Fort Wayne, N Y
Douglass, John W.... ....New-York
Douw, V P & Co....... .. Albany, N Y
Gifford Bros... ....Hudson, N Y
Gill, J L & Son,..... ...... Columbus, O
Graham, Emlen & Co.... Philadelphia, Pa
Griffing & Co...................New-York
Hawley, R D... ....Hartford, Ct
Holbrook, F F & Co........ Boston, Mass
Kellogg, F S............. .. Chicago, Ill
Ladley, Geo........... West Chester, Pa
Nash & Brother ....New-York
New-York Plow Co....... ....New-York
Plant Bros., Pratt & Co.....St. Louis, Mo
Peekskill Plow Works,. ...Peekskill, N Y
Reeves, E H & Co....... .... New-York
Remington Ag. Works, ...Ilion, N Y
Rennie, Wm................Toronto, Can
Shaw & Wells,...... ....Buffalo, N Y
Titus & Bostwick, ...Ithaca, N Y
Vanderbilt, John, & Bros,...... New-York
Welch, F G & Co........... Chicago, Ill
Younglove, Massey & Co.... Cleveland, O

Horse-Powers, Threshers and Other Machines.

Albany Agricultural Works,..Albany, N Y
Brearley, A L & Co.... ... Trenton, N J
Dow & Fowler,......... Fowlerville, N Y
Geiser Threshing Machine Co., Racine, Wis
Gray, A W............. Middletown, Vt
Harder, M... .... Cobleskill, N Y
Shaw & Wells,..............Buffalo, N Y
Westinghouse, G & Co., Schenectady, N Y
Wheeler, Melick & Co.,.... Albany, N Y

Mowers and Reapers.

Adriance, Platt & Co.,......... New-York
Allen, R H & Co...............New-York
Ball, E & Co......... ...... Canton, O
Bradley, C C & Son,.......Syracuse, N Y

Clipper Mower and Reaper Co., New-York
Dodge & Stevenson M'f'g Co., Auburn, N Y
Nash & Brother, ... New-York
Osborne, D M & Co .... Auburn, N Y
Warder, Mitchell & Co..... Springfield, O
Warrior Mower Co...... Little Falls, N Y
Wilber Eureka Mower Co., Po'keepsie, N Y
Wood, Walter A...... Hoosick Falls, N Y

STEAM ENGINES FOR FARMS.

Clute Brothers,...... ... Schenectady, N Y
Gaar, Scott & Co.......... Richmond, Ind
Hoadley, J C & Co......... Lawrence, Mass
Skinner & Walrath,..... Chittenango, N Y
Wood, Taber & Morse,....... Eaton, N Y
Wood & Mann Engine Co.... Utica, N Y

DAIRY APPARATUS.

Cooper, H & E F .... Watertown, N Y
Holmes, A., *Milk Cooler*,.. Cortland, N Y
Jones, Faulkner & Co.. ...... Utica, N Y
Millar, C & Son,.. .... Utica, N Y
O'Neil, O & Co......... ... Utica, N Y
Ralph, W & Co.......... .... Utica, N Y
Weeks, G B.............. Syracuse, N Y

TILE AND TILE MACHINES.

Bender, W M...... .... Albany, N Y
Boynton, C W & Co.... Woodbridge, N J
Jackson, George, ...Albany, N J

OTHER SPECIALTIES.

Allen, S L & Co., *Seed Planter*, Philadelphia, Pa
Barnes, W., *Slicer and Corer*, Bristol, Ct
Bartholomew, C., *Ditching Plow*, Etna, N Y
Bickford & Huffman, *Drill*, Macedon, N Y
Blanchard's Sons, *Churn*,.. Concord, N H
Bowne & Schanck, *Potato Digger*, Freehold, N J
Buell, J. S., *Cider Mill*,..... Buffalo, N Y
Butterworth, R., *Cider Mill*, Trenton, N J
Church, C. A., *Stump Puller*, New-Berlin, N Y
Collins & Co., *Steel Plows*, ... New-York
Conover & Son, *Potato Digger*, Freehold, N J
Continental Wind Mill Co., *Wind Mills*, New-York
Cowing & Co., *Pumps*, &c., Seneca Falls, N Y
Dederick & Co., *Hay Press*, Albany, N Y
Dismukes, Paul, *Clover Seed Gatherer*, Gallatin, Tenn
Empire Wind Mill Manufacturing Co., Syracuse, N Y
Fairbanks & Co., *Scales*,... ... New-York
Fitch & Co., *Hay Elevator*, Lithgow, N Y
Fords & Howe, *Cultivator*, Oneonta, N Y
Gawthrop & Sons, *Water Ram*, Wilmington, Del
Goodell, D. H., *Sower*,..... Antrim, N H
Hickok, W. O., *Cider Mill*, Harrisburg, Pa
Henry, J. T., *Sheep Shears*, Hamden, Ct
Herring, S. C., *Hay Tedder*,.. New-York
Ingersoll & Dougherty, *Hay Press*, Green Point, N Y
Jones, E. F., *Hay Scales*, Binghamton, N Y
Landers, Frary & Clark, *Lawn Mower*, New-York
Lilly, J., & Co., *Water Elevator*, Binghamton, N Y
Mayne, J., *Door Rollers*, Butternuts, N Y
Osborne, Foster & Co., *Seed Sower*, Palmyra, N Y
Paddock. Dean & Co., *Bone Mill*, St. Johnsbury, Vt
Pancoast & Maule, *Steam Heating*, Philadelphia, Pa
Paris Furnace Co., *Hay Elevator*, Clayville, N Y
Perry, F. L., *Scarifier*, Canandaigua, N Y
Prindle, D. R., *Steamer*, E. Bethany, N Y
Reynolds, E. D. & O. B., *Wheel Hoe*, &c., North Bridgewater, Mass
Rumsey & Co., *Press Screws*, Seneca Falls, N Y
Schoonmaker, E. P., *Hay Rake*, Troy, N Y
Sedgwick, H., *Feed Steamer*, Cornwall Hollow, Ct
Seymour & Co., *Drill*, East Bloomfield, N Y
Shares, D. W., *Shares' Harrow*, Fairhaven, Ct
Shields, H. L., *Hay Loader*,.. Troy, N Y
Silsby, C., *Butter Pails*, Seneca Falls, N Y
Snow, B. B., & Co., *Corn Sheller*, Auburn, N Y
Sprout, S. E. & L. B., *Hay Fork*, Muncy, Pa
Swift, H. N., *Lawn Mower*, Matteawan, N Y
Thomas, J. J., & Co., *Smoothing Harrow*, Geneva, N Y
Whittemore, D. H., *Apple Parer*, Worcester, Mass

FERTILIZERS.

Allen, R H & Co..... .... New-York
Baugh & Sons,....... .... Philadelphia, Pa
Bradley, W L...... ..... Boston, Mass
Coe, Enoch,.......... Williamsburg, N Y
Cumberland Bone Co......... Portland, Me
Currie, Geo. E.............. Cincinnati, O
Davis, G W & Co........... Lisbon, N H
Decatur, J R & Co... .... New-York
Douw, V P................ Albany, N Y
Dunlop & Avery, *Plaster*, Jamesville, N Y
Foster, J T., *Poudrette*,....... New-York
Griffing & Co. .... New-York
Hand, T J... .. .. New-York
Higgins, Reybold & Co., Delaware City, Del
Hobson, Hurtado & Co., *Guano*, New-York
Lister Bros......... .... Newark, N J
Ralston, John, & Co........... New-York
Reed & Powell,........... Coxsackie, N Y
Tucker, J A .... Boston, Mass
Wattson & Clark, ... Philadelphia, Pa
Wilson & Asmus,.............. New-York
White, George E..... .. .... New-York
Wonson, G M...... East Gloucester, Mass

MISCELLANEOUS.

Babcock, H. H., & Son, *Wooden Pipe*, Watertown, N Y
Baird, Roper & Co., *Berry Basket*, Norfolk, Va
Baldwin, T. E., & Co., *Carriages*, New-York
Beadle, H. W., *Patent Agent*, Washington, D C
Beecher Basket Co.......... Westville, Ct

Books on Rural Pursuits.

[The following are especially recommended, and any of them may be had, postage paid, by enclosing the price named to this Office :]

Allen's American Cattle,....	$2.50
Allen's New American Farm Book,..	2.50
Allen's Diseases of Domestic Animals,	1.00
American Weeds and Useful Plants,.	1.75
Breck's Book of Flowers,...........	1.75
Burr's Vegetables of America,.......	5.00
Copeland's Country Life, (926 pp., 250 engravings,)...............	5.00
Cranberry Culture, (White,).........	1.25
Dadd's American Cattle Doctor, ...	1.50
Dadd's Modern Horse Doctor,......	1.50
Dana's Muck Manual,...............	1.25
Darwin's Animals and Plants, (2 vols.,)	6.00
Downing's Cottage Residences, ...	3.00
Downing's Fruits and Fruit Trees of America, (1,100 pp.,)	5.00
Downing's do. do. Old Edition,	3.00
Downing's Landscape Gardening,...	6.50
Downing's Rural Essays, ...	5.00
Drainage for Profit and Health,.....	1.50
Farm Drainage, (French,)	1.50
Farm Implements, (J. J. Thomas,)..	1.50
Farming for Boys,............. ...	1.50
Flint on Grasses,.......	2.50
Flint on Milch Cows & Dairy Farm'g,	2.50
Fuller's Forest Tree Culturist,......	1.50
Fuller's Small Fruits, (Illustrated,)..	1.50
Gardening for Profit, (Henderson,)..	1.50
Geyelin's Poultry Breeding—Commercial View,......	1.25
Grape Culturist, (Fuller,)	1.50
Gray's How Plants Grow, ...	1.25
Gray's Manual of Botany & Lessons,	4.00
Gray's School & Field Book of Botany	2.50
Harris on the Pig,.............. ...	1.50
Hints to Horse-Keepers, (Herbert,).	1.75
How Crops Feed, (S. W. Johnson,)..	2.00
How Crops Grow, do. ..	2.00
Hoopes' Book of Evergreens,.......	$3.00
Husmann's Grapes and Wine, ...	1.50
Johnston's Agricultural Chemistry,..	1.75
Kemp's Landscape Gardening,......	2.00
Langstroth on the Honey Bee,......	2.00
My Farm of Edgewood,....	1.75
Parkman's Book of Roses,..........	3.00
Parsons on the Rose,..............	1.00
Pear Culture for Profit, (Quinn,)	1.00
Practical Floriculture, (Henderson,).	1.50
Practical Poultry Keeper, (Wright,).	2.00
Practical Shepherd, (Randall,)	2.00
Quinby's Mysteries of Bee-Keeping,.	1.50
Quincy on Soiling Cattle,...........	1.25
Rand's Bulbs, ...	3.00
Rand's Garden Flowers,............	3.00
Rural Affairs, (*Six Volumes*,)	9.00
Rural Studies, ...	1.75
Strong's Cultivation of the Grape,....	3.00
Tegetmeier's Poultry Book,....	9.00
Ten Acres Enough, ...	1.50
The Horse in the Stable and the Field, (Stonehenge,).................	2.50
The Farmer's Journal and Account Book, (Perkins,)......... $1, $2,	3.50
The Percheron Horse,..........	1.00
Thomas' American Fruit Culturist, (480 Illustrations,).............	3.00
Warder's Hedges and Evergreens, ..	1.50
Woodward's Graperies and Horticultural Buildings, ...	1.50
Woodward's Country Homes,.......	1.50
Woodward's Cottages and Farm Houses, ...	1.50
Woodward's National Architect,....	12.00
Woodward's Rural Architecture,....	1.50
Woodward's Suburban and Country Houses, ...	1.50

www.ingramcontent.com/pod-product-compliance
Lightning Source LLC
LaVergne TN
LVHW020222110826
845151LV00003B/798

* 9 7 8 1 4 2 5 5 3 2 1 5 4 *